Gynecologic Cancer: Risk Factors, Interception and Prevention

Gynecologic Cancer: Risk Factors, Interception and Prevention

Guest Editors

Brandy Heckman-Stoddard
Goli Samimi

MDPI

Basel • Beijing • Wuhan • Barcelona • Belgrade • Novi Sad • Cluj • Manchester

Guest Editors

Brandy Heckman-Stoddard
Division of Cancer Prevention
National Cancer Institute
Rockville
USA

Goli Samimi
Division of Cancer Prevention
National Cancer Institute
Rockville
USA

Editorial Office
MDPI AG
Grosspeteranlage 5
4052 Basel, Switzerland

This is a reprint of the Special Issue, published open access by the journal *Cancers* (ISSN 2072-6694), freely accessible at: https://www.mdpi.com/journal/cancers/special_issues/330925EO32.

For citation purposes, cite each article independently as indicated on the article page online and as indicated below:

Lastname, A.A.; Lastname, B.B. Article Title. *Journal Name* **Year**, *Volume Number*, Page Range.

ISBN 978-3-7258-7793-5 (Hbk)
ISBN 978-3-7258-7794-2 (PDF)
https://doi.org/10.3390/books978-3-7258-7794-2

Contents

About the Editors

Brandy Heckman-Stoddard

Brandy Heckman-Stoddard serves as a senior leader within the National Cancer Institute (NCI)'s Division of Cancer Prevention (DCP), where she plays a central role in the oversight, scientific stewardship, and coordination of cancer prevention and symptom management research across multiple national networks and research bases. Dr. Heckman-Stoddard has served as a Program Officer in the Breast and Gynecologic Cancer Research, DCP, NCI, National Institutes of Health (NIH) since 2011 and as the Chief of that group since 2016. In 2023, she was named Director for the NCI Community Oncology Research Program and Acting Chief of the Community Trials and Prevention Trials Research Group. She came to NCI as a postdoctoral fellow in the Cancer Prevention Fellowship Program, completing a Master's in Public Health at the Johns Hopkins Bloomberg School of Public Health after receiving her Ph.D. from Baylor College of Medicine. Dr. Heckman-Stoddard's research focuses on drug development for breast cancer prevention and biomarker development. She is particularly interested in local delivery of agents, alternate dosing strategies and biomarkers of efficacy. She serves as Program Officer for the Early-Phase Breast Cancer Prevention Clinical Trials grants portfolio and scientific lead of early-phase breast cancer clinical trials within the NCI Division of Cancer Prevention Clinical Trial Network (CP-CTNet). She also serves as the Breast Cancer Project Scientist and Uterine Cancer Project Scientist of the Cancer Intervention and Surveillance Modeling Network (CISNET), a consortium of NCI-sponsored investigators who use simulation modeling to improve our understanding of cancer control interventions in prevention, screening and treatment and their effects on population trending in incidence and mortality.

Goli Samimi

Goli Samimi has served as a Program Director in the Breast and Gynecologic Cancer Research Group, Division of Cancer Prevention (DCP), NCI, managing the ovarian and endometrial cancer prevention grant portfolios since 2015. She also serves as Director of the NCI Cancer Prevention Clinical Trials Network (CP-CTNet), which supports early-phase clinical trials to assess the safety, tolerability and potential of cancer preventive interventions. She also serves as scientific lead of endometrial and ovarian cancer prevention trials within CP-CTNet. Dr. Samimi's research interests include molecular mechanisms of ovarian cancer etiology and progression, and the identification of women and families at high risk for ovarian cancer. Dr. Samimi leads the Traceback program, aimed to increase the identification of families at increased risk for ovarian cancer by genetic testing of previously diagnosed cases. She also serves as Project Scientist for uterine cancer for the Cancer Intervention and Surveillance Modeling Network (CISNET), a consortium of NCI-sponsored investigators who apply modeling to improve the understanding of cancer interventions in prevention, screening and treatment. In addition, Dr. Samimi leads studies investigating perceptions of cancer risk and decision-making regarding cancer prevention in patients and primary care physicians. Dr. Samimi also works with the Eunice Kennedy Shriver National Institute of Child Health and Human Development (NICHD) to support research related to the progression of benign gynecologic disease to gynecologic cancer.

Preface

Gynecologic cancers, including ovarian, endometrial and cervical cancers, account for approximately 100,000 new cases and over 30,000 deaths per year in the United States. The prevention and interception of these cancers would greatly impact outcomes and significantly reduce mortality rates. For ovarian cancer, screening in the general population is not considered effective, and therefore most cases of this cancer are diagnosed at a late stage with a poor survival rate. Endometrial cancer incidence and mortality have been markedly increasing in recent years, due in large part to increasing obesity rates and other lifestyle factors. Cervical cancer has had the benefit of the availability of both screening and prevention. Equity in the dissemination and implementation of these prevention and screening methods is essential for early diagnosis and treatment. These opportunities and challenges related to the diagnosis and high mortality of these gynecologic cancers emphasize the importance of identifying women at higher risk for these cancer types, and providing prevention/interception interventions to decrease incidence and improve outcomes. The purpose of this Reprint is to highlight original and existing research related to models, screening and early detection, preventive interventions, risk factors and risk reduction, epidemiology and health disparities in ovarian, endometrial and cervical cancers, with the aim of bringing attention to the unique challenges associated with these diseases, and bringing to the forefront exciting future directions for prevention to improve outcomes.

Brandy Heckman-Stoddard and Goli Samimi
Guest Editors

cancers

MDPI

Editorial

Gynecologic Cancer: Risk Factors, Interception and Prevention

Goli Samimi * and Brandy Heckman-Stoddard

Division of Cancer Prevention, National Cancer Institute, Rockville, MD 20850, USA
* Correspondence: goli.samimi@nih.gov

Gynecologic cancers account for approximately 100,000 new cases and over 30,000 deaths per year in the United States [1]. Effective treatment of these cancers ranges from local intervention delivery to major surgery, with varying degrees of successful outcomes. Studies have repeatedly demonstrated that early detection and screening, and preventive interventions can improve outcomes by intercepting cancer progression or, in some cases, even inhibiting or prolonging initiation [2]. This Special Issue of *Cancers* presents a series of papers including original research, reviews and meta-analyses, from leaders in the field that cover models, screening and early detection, preventive interventions, risk factors and risk reduction, epidemiology, and health disparities in ovarian, endometrial and cervical cancers. The purpose of this Special Issue is to bring attention to the unique challenges associated with these cancers, and bring to the forefront exciting opportunities for prevention to improve outcomes.

Endometrial, or uterine, cancer is particularly challenging as both incidence and mortality have increased over the past two decades, due in large part to increasing obesity rates in the population, decreasing hysterectomy rates, and an increased proportion of more aggressive subtypes [1]. The timely publication by Clontz et al. undertakes a systematic review and meta-analysis of the association between obesity-related biomarkers and endometrial cancer risk, with the goal of revealing the impact of weight loss on reducing the risk of endometrioid endometrial cancer [3]. The study examined effects of lifestyle interventions, bariatric surgery and pharmacotherapy on obesity and inflammatory biomarkers, and found that despite high heterogeneity limiting statistical significance, weight loss effectively reduces inflammatory and other obesity-related biomarkers associated with increased risk of endometrial cancer.

The Clontz study found that bariatric surgery outperformed both lifestyle and pharmacotherapy interventions; it would be interesting to see how more recent expanded use of glucagon-like peptide-1 (GLP-1) agonists or GLP-1/GIP-1 (gastric inhibitory polypeptide) dual agonists compares to bariatric surgery for sustained weight loss and prevention of endometrial cancer over time. Using disease models, Hagemann et al. explored a combined treatment strategy of semaglutide (a GLP-1 agonist) and progestins for non-surgical management of endometrial cancer [4]. Progestins are widely used hormonal interventions for endometrial cancer prevention, but their effectiveness is limited over time due to high relapse rate. In this study, the authors used endometrial cancer cell lines and patient-derived organoids (PDOs) from grade 1 endometrial cancer to test the hypothesis that combining semaglutide with the progestin levonorgestrel would improve the prevention potential of levonorgestrel alone. Focusing on mechanistic endpoints, the authors found a potential positive feedback loop between semaglutide and the progestin levonorgestrel signaling. As of the time of writing this article, a search for trials investigating GLP-1 agonists in endometrial cancer revealed seven clinical trials in this space. These trials cover GLP-1 agonists in endometrial cancer patients undergoing chemotherapy [5] or following surgical treatment [6] for weight management; GLP-1 agonists in combination with a levonorgestrel intrauterine device (LNG-IUD) in women with endometrioid intraepithelial neoplasia (EIN) or low-grade endometrial cancer for regression or to prevent progression [7–9]; and pre-surgical GLP-1 agonists in women with endometrioid intraepithelial neoplasia (EIN) or low-grade endometrial cancer to determine effects on cellular proliferation [10,11]. Mechanistic studies using representative models like the Hagermann study provide the foundation for translation into human studies, and once completed, these trials may reveal additional effects that can then be taken back to the bench for further examination using these models.

Even as these advances help to improve survival in women with endometrial cancer, the undeniable challenge of health disparities persists. Wijayabahu et al. conducted a retrospective cohort study of endometrial cancer cases using the Surveillance, Epidemiology, and End Results (SEER) database to investigate the impact of socioeconomic factors on five-year survival rates of endometrial cancer by race/ethnicity [12]. Survival rates were associated with education, poverty, unemployment and income, with non-Hispanic Black women consistently showing worse survival outcomes compared to other racial/ethnic groups, regardless of socioeconomic characteristics. These findings emphasize the necessity to further explore the persistent endometrial cancer survival disparities for non-Hispanic Black women, as well as the socioeconomic determinants influencing these survival disparities. Any interventional progress made towards improved prevention or outcomes would be dampened without a better understanding of how to apply this improvement towards health equity.

Epithelial ovarian cancer, specifically high-grade serous ovarian cancer (HGSC), faces different challenges in that early detection is rare due to lack of symptoms and ineffective screening options [13], resulting in a majority of cases being diagnosed at a late stage [14]. Clinical management of these patients involves invasive surgery and chemotherapy, with high rates of relapse and high case-fatality rates. These low survival rates in HGSC highlight the importance of early detection and preventive interventions, particularly in women at high risk, as survival rates are markedly improved for women diagnosed at earlier stages [14]. The review by Montemorano et al. presents the current landscape in ovarian cancer prevention, within the framing of the epidemiology of epithelial ovarian cancer as well as our understanding of molecular characteristics and precursor lesions [15]. The authors also describe innovative technological advances in detection and prevention, with the hope that these advances can be eventually applied to future high-risk or potentially population-based screening.

An ideal candidate for effective prevention of any cancer would be both non-invasive and non-toxic. In ovarian cancer, epidemiologic studies have highlighted the potential of oral contraceptives (OCs), which meet both of these criteria, and are strongly associated with a reduced risk [16]. While this link has been repeatedly demonstrated, the specific component(s) of the OCs and the mechanism by which they confer protection is still unknown. Zhai et al. have contributed their preclinical work investigating the various components of OCs using genetically engineered HGSC mouse models [17]. The findings demonstrate that the OC constituent medroxyprogesterone acetate, a progestin, significantly inhibits tumor development and prolongs survival in this mouse model. This publication not only reveals the potential mechanism by which OCs confer protection in HGSC, but also demonstrates the imperativeness of representative models to characterize potential chemo-preventive interventions that would provide non-invasive and non-toxic opportunities for cancer risk reduction.

The current NCCN guidelines for HGSC risk management in high-risk individuals recommend risk-reducing bilateral salpingo-oophorectomy (BSO) before age 40–45 [18].

https://doi.org/10.3390/cancers18081235

While this preventive surgery significantly reduces the risk of HGSC, the resulting early menopause increases the risk of osteoporosis, cardiovascular disease and other damaging effects [19]. Research and even some clinical practice has moved towards salpingectomy (with delayed oophorectomy), which offers a preventive option while minimizing adverse events. This approach has been undertaken as opportunistic salpingectomy in some normal-risk populations who are undergoing gynecologic surgery for benign conditions, and has shown promise for HGSC risk reduction [20]. For high-risk individuals, clinical trials are underway to determine whether delayed oophorectomy offers comparable risk reduction compared with risk-reducing BSO [21]. These trials include SOROCk (a non-randomized prospective clinical trial comparing the non-inferiority of salpingectomy to salpingo-oophorectomy to reduce the risk of ovarian cancer among brca1 carriers [22]) and TUBA-WISP-II (tubectomy with delayed oophorectomy as an alternative to risk-reducing salpingo-oophorectomy in high-risk women to assess the safety of prevention [23]). In parallel to these clinical studies, Zhai et al. utilized their HGSC mouse model to compare BSO to salpingectomy to prevent tumor development [24], and found that salpingectomy alone significantly reduced HGSC incidence, while BSO completely prevented tumor development. These preclinical studies provide the opportunity to investigate clinically relevant hypotheses in parallel to definitive clinical trials. These two studies by Zhai et al. demonstrate the value of accurately representative models that reflect not just tumor progression but also reflect precursor stages and tumor development, with adequate latency periods for prevention experiments [17,24].

With ongoing and promising research into preventive interventions for HGSC, the fact remains that identifying high-risk individuals will be imperative for implementing wide-spread prevention programs (surgical or non-surgical) that would positively impact survival outcomes. Contributions by Henrikson et al. [25] and White et al. [26] represent unique approaches towards this challenge. Specifically, they each present a version of Traceback, an approach to identify high-risk families by testing previously diagnosed cases of HGSC who never underwent genetic testing, despite testing guidelines [27]. Both groups utilized clinical data from integrated healthcare registries to identify women previously diagnosed with ovarian cancer. They used different modes of contact for outreach and offering of genetic testing, then applied standard cascade testing approaches to reach family members of those who were found to have a pathogenic or likely pathogenic variant that is associated with HGSC risk. These studies can inform sustainability and implementation of Traceback programs going forward, to increase identification of high-risk families that could be offered safe and effective prevention interventions.

Both papers on cervical cancer highlight expanding opportunities for prevention by improving early detection and enabling less invasive management of precancerous lesions. Khoja et al. emphasize that the global screening guidelines increasingly prioritize HPV-based testing as the most effective strategy for early detection [28]. However, while there is broad agreement on screening ages, inconsistent implementation can lead to a limited impact on mortality and allow disparities to persist, particularly across settings with limited access to vaccination and screening. Hamar et al. demonstrate that topical imiquimod offers a promising conservative option for treating cervical intraepithelial neoplasia (CIN), enhancing HPV clearance and reducing precancerous lesions, while potentially preserving fertility and reducing surgical risks—though conization remains more effective in some cases [29]. Taken together, the findings reinforce that combining accessible HPV vaccination, evidence-based screening, and emerging nonsurgical therapies can strengthen cervical cancer prevention.

Wang et al. contributed an interesting study of the use of tricycle antidepressants (TCAs) and selective serotonin reuptake inhibitors (SSRIs) and association with risk of gynecologic cancers across the board [30]. This epidemiologic study found that use of many TCA and SSRI was significantly associated with lowered risks of cervical, uterine and ovarian cancers, particularly among women aged 40–64, which is also the age range that corresponds with higher risk of these cancer types. Many others have explored repurposing safe interventions for gynecologic cancer prevention, such as metformin as another example [31]. It is essential for additional studies to be undertaken to better understand the potential mechanism for these associations.

Across gynecologic malignancies, the emerging evidence underscores both the urgency and the opportunity to strengthen prevention, early detection, and risk-reduction strategies. Much work remains to expand precision prevention; integrate lifestyle, metabolic, hormonal, and immunologic interventions; advance non-surgical prevention strategies; and integrate emerging technologies, as well as advance implementation work to improve care. Collectively, these directions reflect a rapidly evolving landscape that bridges molecular insights, behavioral science, clinical innovation, and public health implementation—moving toward a future where gynecologic cancer prevention is more precise and more effective across the populations most in need.

Conflicts of Interest: The authors declare no conflicts of interest.

References

1. Siegel, R.L.; Kratzer, T.B.; Wagle, N.S.; Sung, H.; Jemal, A. Cancer statistics, 2026. *CA Cancer J. Clin.* **2026**, *76*, e70043. [CrossRef]
2. Lippman, S.M.; Abate-Shen, C.; Colbert Maresso, K.L.; Colditz, G.A.; Dannenberg, A.J.; Davidson, N.E.; Disis, M.L.; DuBois, R.N.; Szabo, E.; Giuliano, A.R.; et al. AACR White Paper: Shaping the Future of Cancer Prevention-A Roadmap for Advancing Science and Public Health. *Cancer Prev. Res.* **2018**, *11*, 735–778. [CrossRef] [PubMed]
3. Clontz, A.D.; Gan, E.; Hursting, S.D.; Bae-Jump, V.L. Effects of Weight Loss on Key Obesity-Related Biomarkers Linked to the Risk of Endometrial Cancer: A Systematic Review and Meta-Analysis. *Cancers* **2024**, *16*, 2197. [CrossRef]
4. Hagemann, A.R.; Hagemann, I.S.; Mutch, D.G.; Devor, E.J.; Malmrose, P.K.; Zhang, Y.; Morrison, A.M.; Thiel, K.W.; Leslie, K.K. Enhancing Progestin Therapy with a Glucagon-Like Peptide 1 Agonist for the Conservative Management of Endometrial Cancer. *Cancers* **2025**, *17*, 598. [CrossRef]
5. A Study of a Weight Loss Intervention in People with Endometrial Cancer. Available online: https://clinicaltrials.gov/study/NCT06751589 (accessed on 13 March 2026).
6. Weight Loss Management in Endometrial Cancer Survivors (ECS Weight). Available online: https://clinicaltrials.gov/study/NCT06877572 (accessed on 13 March 2026).
7. Weight-Loss Drug for Fertility-Sparing Treatment of Atypical Hyperplasia and Low-Grade Cancer of the Endometrium (WE-FiERCE). Available online: https://clinicaltrials.gov/study/NCT06073184 (accessed on 13 March 2026).
8. A Study of Weight Loss Intervention with Tirzepatide and Progestin Intrauterine Device to Treat Endometrial Hyperplasia and Grade 1 Endometrial Cancer. Available online: https://clinicaltrials.gov/study/NCT07349641 (accessed on 13 March 2026).
9. A Study to Estimate Early Clinical Efficacy Signals of GLP-1 Agonist Administration in Conjunction with Levonorgestrel Intrauterine Device in Obese Patients with Endometrioid Intraepithelial Neoplasia. Available online: https://clinicaltrials.gov/study/NCT07107334 (accessed on 13 March 2026).
10. Tirzepatide in Obesity-Driven Endometrial Cancer. Available online: https://clinicaltrials.gov/study/NCT07065552 (accessed on 13 March 2026).
11. Tirzepatide in Women with Obesity and Endometrial Intra-epithelial Neoplasia or Grade 1 Endometrial Cancer. Available online: https://clinicaltrials.gov/study/NCT07078838 (accessed on 13 March 2026).
12. Wijayabahu, A.T.; McGee-Avila, J.K.; Shiels, M.S.; Harsono, A.A.H.; Arend, R.C.; Clarke, M.A. Five-Year Relative Survival Rates of Women Diagnosed with Uterine Cancer by County-Level Socioeconomic Status Overall and across Histology and Race/Ethnicity. *Cancers* **2024**, *16*, 2747. [CrossRef]
13. Henderson, J.T.; Webber, E.M.; Sawaya, G.F. Screening for Ovarian Cancer: Updated Evidence Report and Systematic Review for the US Preventive Services Task Force. *JAMA J. Am. Med. Assoc.* **2018**, *319*, 595–606. [CrossRef]
14. Torre, L.A.; Trabert, B.; DeSantis, C.E.; Miller, K.D.; Samimi, G.; Runowicz, C.D.; Gaudet, M.M.; Jemal, A.; Siegel, R.L. Ovarian cancer statistics, 2018. *CA Cancer J. Clin.* **2018**, *68*, 284–296. [CrossRef]
15. Montemorano, L.; Huffman, L.; Barroilhet, L. Early Detection and Prevention of Ovarian Cancer. *Cancers* **2025**, *17*, 6. [CrossRef] [PubMed]

https://doi.org/10.3390/cancers18081235

16. Havrilesky, L.J.; Moorman, P.G.; Lowery, W.J.; Gierisch, J.M.; Coeytaux, R.R.; Urrutia, R.P.; Dinan, M.; McBroom, A.J.; Hasselblad, V.; Sanders, G.D.; et al. Oral contraceptive pills as primary prevention for ovarian cancer: A systematic review and meta-analysis. *Obstet. Gynecol.* **2013**, *122*, 139–147. [CrossRef]
17. Zhai, Y.; Bedi, K.; Wu, R.; Feng, Y.; Green, M.E.; Pearce, C.L.; Pike, M.C.; Fearon, E.R.; Cho, K.R. Medroxyprogesterone Acetate Inhibits Tumorigenesis in Mouse Models of Oviductal High-Grade Serous Carcinoma. *Cancers* **2025**, *17*, 3456. [CrossRef] [PubMed]
18. Cheng, H.H.; Giri, V.N.; Goggins, M.; Yurgelun, M.B.; Karlan, B.Y.; Norquist, B.S.; Daly, M.B.; Pal, T.; AlHilli, Z.; Arun, B.; et al. NCCN Guidelines(R) Insights: Genetic/Familial High-Risk Assessment: Breast, Ovarian, Pancreatic, and Prostate, Version 2.2026. *J. Natl. Compr. Cancer Netw.* **2026**, *24*, 2–10. [CrossRef] [PubMed]
19. Dowdy, S.C.; Stefanek, M.; Hartmann, L.C. Surgical risk reduction: Prophylactic salpingo-oophorectomy and prophylactic mastectomy. *Am. J. Obstet. Gynecol.* **2004**, *191*, 1113–1123. [CrossRef]
20. Sowamber, R.; Lukey, A.; Huntsman, D.; Hanley, G. Ovarian Cancer: From Precursor Lesion Identification to Population-Based Prevention Programs. *Curr. Oncol.* **2023**, *30*, 10179–10194. [CrossRef]
21. Pennington, K.P.; Pugh, S.L.; Huh, W.; Walker, J.L.; Jewell, E.; Havrilesky, L.J.; Carter, J.; Muller, C.Y.; Drapkin, R.; Lankes, H.A.; et al. Optimization of Timing for Risk-Reducing Salpingectomy and Oophorectomy. *Obstet. Gynecol.* **2025**, *145*, 21–30. [CrossRef]
22. SOROCk. Available online: https://clinicaltrials.gov/study/NCT04251052 (accessed on 13 March 2026).
23. TUBA-WISP-II. Available online: https://clinicaltrials.gov/study/NCT04294927 (accessed on 13 March 2026).
24. Zhai, Y.; Fearon, E.R.; Cho, K.R. Bilateral Salpingo-Oophorectomy Is Superior to Salpingectomy Alone in Preventing Non-Tubal Tumor Development in a Mouse Model of High-Grade Serous Carcinoma. *Cancers* **2025**, *17*, 2759. [CrossRef]
25. Henrikson, N.B.; Jonas, M.C.; Blasi, P.R.; Buchanan, A.H.; Suwannarat, P.; Leppig, K.; Scrol, A.; Leitzel, T.; Deneal, A.N.; Canedo, D.; et al. Implementation of a Traceback Testing Program for Ovarian Cancer: Findings from the FACTS Study. *Cancers* **2025**, *17*, 1154. [CrossRef]
26. White, L.L.; Sawyer, J.K.; Zepp, J.M.; Prado, Y.K.; Reyes, A.A.; Maiyani, M.; Shuster, E.; Zucker, R.; Henrikson, N.B.; Rope, A.F.; et al. Genetic Testing Uptake among Ovarian Cancer Survivors in the Genetic Risk Analysis in Ovarian Cancer (GRACE) Study. *Cancers* **2024**, *16*, 2563. [CrossRef]
27. Samimi, G.; Bernardini, M.Q.; Brody, L.C.; Caga-Anan, C.F.; Campbell, I.G.; Chenevix-Trench, G.; Couch, F.J.; Dean, M.; de Hullu, J.A.; Domchek, S.M.; et al. Traceback: A Proposed Framework to Increase Identification and Genetic Counseling of BRCA1 and BRCA2 Mutation Carriers Through Family-Based Outreach. *J. Clin. Oncol.* **2017**, *35*, 2329–2337. [CrossRef] [PubMed]
28. Khoja, L.; Zoulfikar, M.; Hak, L.; Yousif, S.; Aljebori, M.; Stiffler, M.; Tariq, M.; Burgard, S.; Fleischer, N.L.; Wallner, L.P.; et al. Overcoming Cultural Barriers and Building Healthcare Trust: A Mixed Methods Study on Cervical Cancer Prevention Among Arab American Women Across Major U.S. Cities. *Cancers* **2024**, *17*, 5. [CrossRef] [PubMed]
29. Hamar, B.; Teutsch, B.; Hoffmann, E.; Hegyi, P.; Harnos, A.; Nyirady, P.; Hunka, Z.; Acs, N.; Banhidy, F.; Melczer, Z. Imiquimod Is Effective in Reducing Cervical Intraepithelial Neoplasia: A Systematic Review and Meta-Analysis. *Cancers* **2024**, *16*, 1610. [CrossRef]
30. Wang, C.H.; Huang, C.W.; Nguyen, N.T.H.; Lin, M.C.; Nguyen, P.A.; Islam, M.M.; Chien, S.C.; Yang, H.C. Real-World Data on the Associations of Tricyclic Antidepressants and Selective Serotonin Reuptake Inhibitors with Gynecologic Cancer Risk. *Cancers* **2025**, *17*, 1616. [CrossRef] [PubMed]
31. Febbraro, T.; Lengyel, E.; Romero, I.L. Old drug, new trick: Repurposing metformin for gynecologic cancers? *Gynecol. Oncol.* **2014**, *135*, 614–621. [CrossRef] [PubMed]

https://doi.org/10.3390/cancers18081235

cancers

MDPI

Systematic Review

Effects of Weight Loss on Key Obesity-Related Biomarkers Linked to the Risk of Endometrial Cancer: A Systematic Review and Meta-Analysis

Angela D. Clontz [1,†], Emma Gan [2,†], Stephen D. Hursting [1] and Victoria L. Bae-Jump [3,*]

1 Department of Nutrition and Nutrition Research Institute, University of North Carolina at Chapel Hill, Chapel Hill, NC 27599, USA; adclontz@email.unc.edu (A.D.C.); hursting@email.unc.edu (S.D.H.)
2 Department of Medicine, Imperial College London, London SW7 2AZ, UK; emma.gan21@imperial.ac.uk
3 Division of Gynecologic Oncology, University of North Carolina at Chapel Hill, Chapel Hill, NC 27599, USA
* Correspondence: victoria_baejump@med.unc.edu
† These authors contributed equally to this work.

Simple Summary: Obesity significantly contributes to endometrial cancer (EC) incidence and mortality. Weight loss interventions are pivotal in mitigating endometroid EC risk, showing notable improvements in obesity-related metabolic perturbations, including insulin resistance, dyslipidemia, and inflammation. This review aimed to assess the efficacy of weight loss strategies, including lifestyle modifications, surgical interventions, and pharmacological approaches, on key biological indicators associated with endometroid EC. Findings reveal the assessment of weight loss for reducing inflammatory markers linked to endometroid EC, with bariatric surgery emerging as a prominent intervention. These insights can inform clinical practice, refine therapeutic strategies, and introduce tailored interventions to promote equitable healthcare and enhance outcomes for individuals affected by endometroid EC.

Abstract: Endometrial cancer (EC) includes various histologic types, with estrogen-dependent endometrioid carcinoma being the most common. Obesity significantly increases the risk of developing this type, especially in postmenopausal women, due to elevated estrogen production by adipocytes. This review examines the impact of weight loss from different interventions on reducing obesity-related risk factors for endometrioid EC. A systematic review and meta-analysis were conducted on three weight loss interventions: bariatric surgery, pharmacotherapy, and lifestyle changes. The effects of these interventions on inflammatory biomarkers (CRP, TNF-α, IL-6) and hormones (leptin, estrogen) were analyzed. Data from controlled studies were pooled to assess the significance of weight loss in reducing these biomarkers. Despite heterogeneity, bariatric surgery resulted in an overall 25.8% weight reduction, outperforming lifestyle and pharmacotherapy interventions. Weight loss reduced CRP levels by 33.5% and IL-6 levels by 41.9%. TNF-α levels decreased by 13% with percent weight loss over 7%. Leptin levels also decreased significantly, although the exact weight loss percentage was not statistically significant. Weight loss effectively reduces proinflammatory markers and hormones associated with increased risk of endometrioid EC. The strengths of this review include a comprehensive examination of different weight-loss interventions and a large pool of participants. However, limitations include high heterogeneity among studies and only 43% of the participants being postmenopausal. Limited data on sex hormones and racial disparities underscore the need for further research.

Keywords: endometrial cancer; cancer prevention; obesity; weight loss; inflammatory markers; hormones

1. Introduction

Endometrial cancer (EC) is a significant health concern globally, ranking as the fourth most diagnosed cancer among women [1]. Alarmingly, a profound health disparity exists in EC outcomes, with Black women experiencing a twofold increase in mortality compared to their White counterparts [2–4]. EC patients diagnosed with advanced-stage disease have a 5-year survival rate of ≤20% [5]. Contributing significantly to the burden of endometroid EC (or Type I EC) is obesity, which is recognized as the leading preventable risk factor for its development [6]. Obesity not only escalates the incidence and mortality rates of endometroid EC but also heightens the risk of recurrence among cancer survivors [7,8].

Understanding the intricate relationship between obesity and the risk of endometroid EC involves exploring various underlying mechanisms. For example, during menopause, the ovaries stop producing estrogen and progesterone, making adipose tissue the primary source of estrogen. Adipocytes (fat cells) convert androgens into estrogen through an enzyme called aromatase [8–10]. Excess adipose tissue, a characteristic of obesity, leads to increased estrogen production due to the heightened aromatization of androgens. Excess estrogen can stimulate endometrial tissue growth, contributing to the development of endometrioid EC [8]. However, other conditions, such as polycystic ovarian syndrome (PCOS) and pre- vs. postmenopausal status, which can influence ovulation and hormone levels, must also be considered when examining hormone-dependent EC risk [11].

The link between obesity and endometroid EC extends beyond estrogenic pathways (Figure 1). Obesity-associated changes in insulin and leptin levels contribute to dysregulated growth factor signaling and chronic inflammation, further exacerbating the estrogen-driven carcinogenic process [12–15]. Clinical studies investigating the mechanisms driving endometrioid EC have demonstrated that pro-inflammatory signaling molecules, such as leptin, IL-6, and TNF-α, are generally elevated, with tumor size and disease stage directly correlated with BMI [5,15–17]. Reducing pro-inflammatory signaling can ameliorate tumor-promoting pathways like PI3k/Akt/mTOR and activate tumor-suppressing pathways like AMPK in endometroid EC [18].

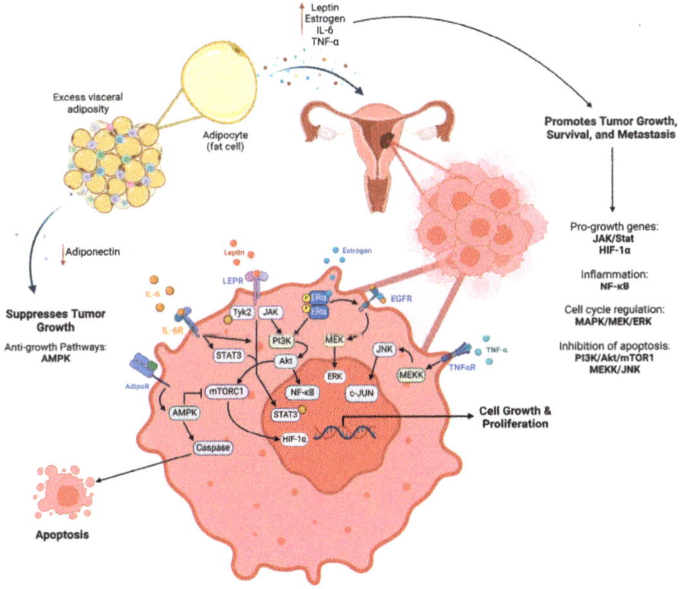

Figure 1. Obesity driven pathways in endometroid EC. Increased adipocytes from excess adipose tissue leads to increased levels of leptin, estrogen, IL-6, and TNF-α while reducing levels of adiponectin. Increased signaling factors (cytokines, adipokines) promote proliferation of epithelial cells of the endometrium contributing to tumorigenesis. Multiple cellular pathways, and gene expression, within epithelial cells become either overstimulated, mutated, or silenced enhancing tumor formation, survival, and metastasis. Created with BioRender.com.

Decreasing body weight and body fat with weight loss represents a pivotal approach to reduce the risk of obesity-driven endometroid EC. Lifestyle modifications, encompassing dietary changes, increased physical activity, and behavioral interventions, serve as foundational strategies for weight management [10,19,20]. Additionally, weight loss can ameliorate obesity-related comorbidities, such as insulin resistance and chronic inflammation, further mitigating endometroid EC risk [18]. Research examining obesity and hormone-dependent breast cancer showed that an average weight loss of 5% improved obesity-related conditions, such as insulin sensitivity; however, weight loss of 10% or more significantly improved circulating levels of estradiol, adiponectin, and leptin [10,19,20].

In contrast, bariatric surgery offers a more aggressive approach for individuals with severe obesity refractory to conventional lifestyle interventions. Bariatric procedures, including gastric bypass and sleeve gastrectomy, induce excess weight loss up to 25%, often accompanied by significant improvements in metabolic parameters [20]. Emerging evidence suggests that bariatric surgery promotion of weight reduction subsequently leads to favorable changes in estrogen levels and inflammatory profiles, potentially mitigating endometroid EC risk. Studies investigating the weight-loss effect of bariatric surgery on endometroid EC showed a significant shift from a pro-inflammatory to an anti-inflammatory phenotype with weight loss of 20% or greater. Research supports the notion that weight loss achieved through bariatric surgery can lower the risk of developing endometroid EC by addressing the key physiological changes linked to obesity [9,21–23]. However, the optimal approach for obesity management in the context of endometroid EC remains an area of active investigation, necessitating further research to elucidate the comparative efficacy and long-term outcomes of lifestyle interventions versus bariatric surgery in this population.

Bariatric surgery is effective and provides proof of the principle that weight loss, if sufficiently significant, can reverse the pro-cancer effects of obesity. However, bariatric surgery is expensive, carries risk of adverse effects, and is only available to approximately 1% of women with obesity, so is therefore not a population-wide solution [9,24]. Life-style-based weight loss interventions are more broadly available and relatively inexpensive but are challenging for most women with obesity to sustain. Thus, emerging pharmacologic strategies that achieve and sustain significant weight loss are very promising.

Weight-loss pharmacotherapies target various physiological pathways in energy balance regulation, including appetite suppression, nutrient absorption inhibition, and metabolic modulation [25]. By augmenting satiety signals and reducing cravings, weight loss drugs facilitate adherence to calorie-restricted diets and sustain weight loss efforts. Moreover, certain weight loss medications have improved insulin sensitivity and metabolic parameters, thereby mitigating obesity-associated comorbidities, such as type 2 diabetes [26,27]. While bariatric surgery induces rapid excess weight loss and metabolic improvements, weight loss drugs offer a less invasive alternative, particularly suitable for individuals with lower BMI or those unwilling or ineligible for surgery [28,29]. However, there is a notable lack of research examining the weight-loss effects of these medications on the risk of endometroid EC, highlighting an urgent need for further investigation.

Recognizing the importance of effective weight-loss strategies to reduce body weight and visceral adiposity in the prevention of endometroid EC, this systematic review aims to evaluate the impact of weight reduction on key inflammatory biomarkers by comparing weight loss results from various interventions. Specifically, it examines whether reduced body weight can affect circulating levels of C-reactive protein (CRP), interleukin (IL)-6, tumor necrosis factor-α (TNF-α), and hormones such as leptin, adiponectin, estradiol, estrone, and testosterone. These biomarkers and hormones are known contributors to the pathogenesis of endometrioid EC.

This review will provide valuable insights into the potential benefits of weight loss therapies as adjunctive components in the comprehensive management and prevention of EC. Thus, by delineating the landscape of interventions and their associated outcomes, this endeavor strives to inform clinical practice, enhance therapeutic strategies, and integrate tailored interventions that include multifaceted factors that are essential for advancing equitable healthcare and enhancing outcomes for all individuals affected by EC.

2. Materials and Methods

The protocol for this systematic review was registered with the International Prospective Register of Systematic Reviews (PROSPERO) under registration number CRD42023458858. The protocol can be accessed through the following link: https://www.crd.york.ac.uk/prospero/display_record.php?RecordID=458858 (accessed on 23 September 2023). The synthesis of reported data followed the guidelines outlined by the Preferred Reporting Items for Systematic Reviews and Meta-Analyses (PRISMA) [30].

2.1. Search Strategy

A thorough search strategy was executed utilizing the resources of the Health Sciences Library (HSL) at the University of North Carolina at Chapel Hill, encompassing EMBASE, PubMed, Scopus, MEDLINE, and the Cochrane Central Register of Controlled Trials (CENTRAL) databases. The search strategy incorporated a combination of keywords and Boolean operators (e.g., AND, OR) to compile comprehensive lists of relevant studies. Specific search terms employed included weight loss, endometrial cancer, bariatric surgery, lifestyle interventions, dietary intervention, and weight loss medications. The full list of search terms is presented in Supplementary Table S1.

2.2. Eligibility

2.2.1. Study Selection

Covidence was used to manage the study selection process to screen literature that met the review criteria [31]. Records obtained from database searchers were uploaded into Covidence for title and abstract screening with duplicates automatically removed. All eligible records were fully screened, independently, by two authors (A.D.C. and E.G.). The authors resolved any screening discrepancies through discussion, or if required, consultation with the third author (V.B.-J.).

Studies included randomized controlled trials, cohort studies, and case studies as well as retrospective and prospective studies published between 2008 and 2023. Studies must have reported pre- and post-intervention biomarkers of inflammation, adiposity, and hormones commonly elevated in patients with EC (e.g., leptin, CRP, TNF-α, IL-6, estrogen) in relation to reduced weight or BMI. Studies excluded consisted of systematic reviews, literature reviews, scoping reviews, meta-analyses, in vivo studies, in vitro studies, and publications not in English.

2.2.2. Patient Selection

The inclusion criteria focused on individuals aged between 30 and 75 years, aligning with the typical age range for EC diagnosis. Furthermore, studies targeted participants with a body mass index (BMI) of 30 kg/m^2 or higher who participated in a weight loss intervention study, encompassing lifestyle modifications, bariatric surgery, pharmacotherapy, or a combination thereof. Studies targeting participants at risk of breast cancer (BC), ovarian cancer (OC), or EC were also included. The exclusion criteria included studies involving other types of cancer, individuals with a BMI below 30 kg/m^2, those younger than 30 years of age, or exclusively male participants.

2.3. Data Extraction

The data extraction tool, designed within Covidence, captured pertinent information including author names, publication year, study country, sample size, study design, disease indication, intervention duration, and study duration. Mean baseline differences in BMI and age were recorded, along with between-group differences (intervention vs. control). Additionally, the percentage of female and Black participants was extracted. Interventions were categorized as lifestyle, bariatric surgery, pharmacotherapy, or control, with controls comprising placebo or comparison groups.

Outcome results (e.g., weight, inflammatory markers, hormones) were tabulated pre- and post-intervention, including standard deviations (SD) and participant numbers. For consistency, mean and standard deviations were calculated for studies that only reported confidence intervals (CI) or standard error means (SEM). To ensure comprehensive coverage, additional tracking measures were instituted to limit the number of studies per outcome (e.g., maximum of 10 studies per intervention group for a single outcome). Data extraction was conducted independently by A.D.C. and E.G. following the quality assessment protocol. Any missing data were evaluated for relevance to outcomes, and authors were contacted to obtain necessary information.

2.4. Quality Assessment

To evaluate potential bias in randomized controlled trials, the Cochrane Risk of Bias 2.0 tool was employed, focusing on five domains: bias stemming from the randomization process, bias due to deviations from intended interventions, bias due to missing outcome data, bias in the measurement of outcomes, and bias in the selection of the reported result [32]. Nonrandomized studies, including case-control and cohort studies, were assessed using the Newcastle-Ottawa Scale (NOS) quality assessment tool [33]. Nonrandomized, or unclearly documented, studies were assessed for selection, comparability, and exposure. Studies that received $\geq$6 out of 9 stars were included in the analysis. Risk of bias and quality assessment analysis was conducted independently by A.D.C. and E.G. A third reviewer (V.B.-J.) was consulted to resolve any discrepancies.

2.5. Statistical Analysis

A comprehensive synthesis of the effect that reduced body weight has on inflammatory indicators and hormones related to the risk of EC was conducted among the different intervention groups and within the intervention groups. Averages were adjusted from the pooled data to estimate the portion of males included in the datasets to approximate female-only data (adj. avg.).

For the meta-analysis, the random-effects model was utilized. In the random-effects model, it is presumed that the true effect differs among studies. The overall effect is determined by computing the weighted average of the observed effects from various studies. Generally, this model yields a cautious estimate, acknowledging the natural variation in the true effects found across the studies. The analysis of the extracted data for the controlled studies was performed using Cochrane Review Manager Web (RevMan Web), where outcome effect measures are reported as mean differences (MDs) with confidence intervals (CIs) [34].

2.6. Heterogeneity

Heterogeneity was evaluated using Cochran's Q test, with a significance threshold set at less than 0.10. Additionally, the I^2 statistic was employed to quantify the total observed variation across the studies. I^2 value exceeding 75% was indicative of high heterogeneity, while a value below 25% suggested low heterogeneity. A value of 0% indicated no heterogeneity.

Sensitivity analysis was performed by subgrouping identified covariates within the analysis, such as the menopausal status of women, the percentage of men included in the sample, the presence of inadequate control groups, and study duration. This approach allowed for an exploration of the impact of these covariates on the overall results and provided insights into potential sources of heterogeneity.

3. Results

Upon review of the online databases, a total of 18,193 abstracts were identified, with 10,963 from PubMed, 3211 from Scopus, 2509 from Embase, and 1510 from unspecified sources. Following the removal of duplicates, 10,020 unique studies remained for screening. Subsequently, upon evaluation of titles and abstracts, 9018 studies were deemed irrelevant and excluded from further consideration. This process resulted in 1002 studies being selected for full-text review to assess eligibility, leading to the exclusion of 964 studies. Thirty-eight studies remained for quality assessment and data extraction. Twenty-four studies utilizing control groups were included for the meta-analysis. A visual representation of this selection process is provided in Figure 2.

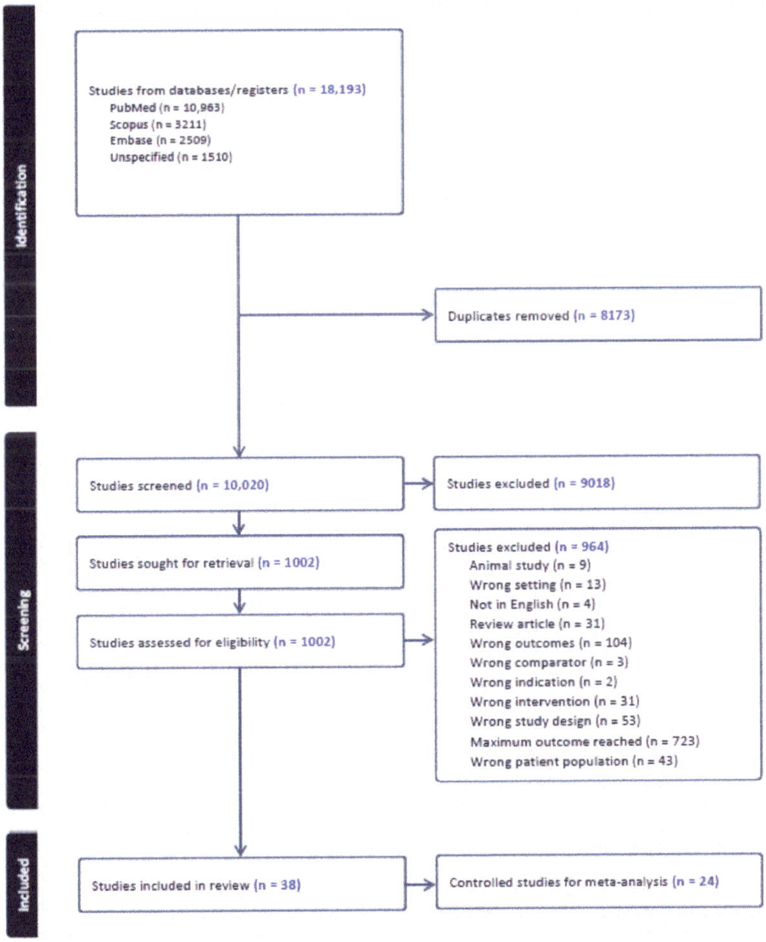

Figure 2. Flow diagram of the literature search.

3.1. Study Characteristics

The 38 unique eligible studies included (40 total studies analyzed for results of interventions due to 2 studies [35,36] incorporating both a lifestyle intervention group and bariatric surgery group as part of the study) a total of 9844 participants (male and female). The average age was 45.6 years and average BMI was 37.1 kg/m^2. Females accounted for 81.8% (n = 8062) of the participants, with 43% (n = 3445) of the females considered postmenopausal (>50 years of age). Among the 38 eligible studies with available data on race and ethnicity, 18.9% (n = 1518) identified as Black and 76.9% (n = 6720) as White (Table 1).

Table 1. Characteristics of included studies that examined weight loss strategies for obesity.

Study Name (Year) Country	Study Design	Intervention	Population (Intervention/Control)	Biomarkers Measured Pre- and Post-Intervention	BMI (Mean)	% Female	% Black	% White	Age (Mean)	Intervention Length
Intervention: Lifestyle										
* Abbenhardt (2013) USA [37]	RCT	Diet + exercise Exercise Diet Control	Women aged 50 to 75 (n = 116/117/118/87)	BMI; leptin; adiponectin	30.9	100	8	92	57.9	12 months
* Abulmeaty (2023) Saudi Arabia [35]	CT	Diet + exercise Control	Adults aged 18 to 60 (n = 14/24)	BMI; CRP; IL-6; TNF-α	40.9	45	NR	87	35	6 months
* Aldubayan (2022) Denmark [38]	RCT	Diet + exercise Control	Adults (n = 49/51)	BMI; weight; leptin; CRP; IL-6; adiponectin; TNF-α	32.2	69	NR	NR	45.3	2.5 months
* Babatunde (2020) USA [39]	RCT	Diet + PA Control	Adult women (n = 176/161)	BMI; weight; CRP; IL-6	39.1	100	100	0	49.5	12 months
* Campbell (2012) USA [40]	RCT	Diet + exercise Exercise Diet Control	Women aged 50 to 75 (n = 117/117/118/87)	BMI; weight; estradiol; estrone; testosterone	30.9	100	8	92	57.9	12 months
Claessens (2009) Netherlands [41]	RT	CR 6 weeks: HC maintenance HPC maintenance HPW maintenance	Adults aged 30 to 60 (n = 20/20/20)	BMI; weight; leptin; adiponectin	32.9	52	NR	NR	45.4	4.5 months
* Duggan (2019) USA [42]	RCT	Diet + exercise Control	Women aged 50 to 75 (n = 151/270)	BMI; estradiol; estrone; testosterone	30	100	13	85	58.5	30 months
Gomez-Huelgas (2019) Spain [12]	Cohort	Diet + Exercise	Adult women (n = 115)	BMI; weight; CRP; IL-6; adiponectin; resistin; TNF-α	36.3	100	NR	NR	44.5	24 months
* Lopez-Legarrea (2014) Spain [43]	RCT	RESMENA Diet Control	Adults (n = 48/48)	BMI; weight; TNF-α; IL-6; CRP	35.9	50	NR	NR	50	2 months
* Lorenzo (2022) Spain [36]	RCT	VLCKD RESMENA Diet Control	Adults (n = 20/20/32)	BMI; IL-6	35.7	63	NR	NR	40.2	6 months
Moszak (2018) Poland [44]	Cohort	PA + hypocaloric diet	Adults (n = 24)	BMI; weight; CRP	39.7	58	NR	NR	46	<1 month
Pinto (2020) UK [45]	RT	CER IER	Adults aged 35 to 75 (n = 22/21)	BMI; weight.; leptin; adiponectin	31	72	10	90	53	1 month
* Porter Starr (2019) USA [46]	HCT	HP TWL (control)	Adults aged 65 or older (n = 25/14)	BMI; weight; adiponectin; leptin; IL-6; CRP; TNF-α	37	31	11	73	68.3	6 months
* Shah (2011) USA [47]	RCT	Diet + exercise Exercise Diet Control	Adults aged 65 or older (n = 28/26/26/27)	BMI; leptin; estradiol	37.8	63	NR	NR	69.8	12 months
Stolzenberg-Solomon (2012) USA [48]	RT	PC IT	Women aged 50 or older (n = 105/173)	BMI; weight; estradiol; estrone; testosterone	33.4	100	37	62	59.3	18 months
Swora-Cwynar (2016) Poland [49]	RT	LC IM	Women aged 18 to 40 (n = 39/38)	BMI; weight; estradiol; testosterone	38.1	100	NR	NR	31.4	3 months
Study name (year) country	**Study design**	**Intervention details**	**Population (intervention/control)**	**Biomarkers measured Pre-and post-intervention**	**BMI (mean)**	**% Female**	**% Black**	**% White**	**Age (mean)**	**Intervention length**
Intervention: Bariatric Surgery										
* Abulmeaty (2023) Saudi Arabia [35]	CT	SG Control	Adults aged 18 to 60 (n = 18/24)	BMI; CRP; IL-6; TNF-α	40.9	45	NR	NR	35	6 months
Ceccarini (2019) Italy [50]	CT	GB Lean control Surgical control	Adults aged 24 to 59 (n = 51/41/9)	BMI; leptin	48.1	64	NR	NR	47.7	18 months
* Dalmas (2011) France [51]	CT	RYGB Control	Adult women (n = 51/14)	BMI; leptin; adiponectin; TNF-α; IL-6; CRP; VEGF	36	100	NR	NR	41.2	24 months
* Freitas (2018) Brazil [52]	RCT	GB Control	Adults aged 18 to 65 (n = 55/14)	BMI; weight; leptin; adiponectin; TNF-α	47.1	85	22	78	41.3	6 months
Jacobsen (2012) Denmark [53]	Cohort	GB	Adults aged 20 to 60 (n = 8)	BMI; weight; leptin	46.7	75	NR	NR	35.5	<1 month
Kim (2023) Korea [54]	Cohort	GB	Adults aged 20 to 65 (n = 63)	BMI; weight; leptin; adiponectin; Resistin	38.9	70	NR	NR	37.5	12 months
Kjaer (2017) Denmark [55]	Cohort	RYGB	Women aged less than 50 (n = 31)	BMI; weight; CRP; estradiol; estrone; testosterone	44.1	100	NR	NR	34	12 months
* Lima (2013) Brazil [56]	RCT	RYGB Control	Women aged less than 50 (n = 10/10)	BMI; weight; leptin; adiponectin; resistin; IL-6; TNF-α; CRP	45.65	100	NR	NR	35.9	15 months

Table 1. *Cont.*

Study name (year) country	Study design	Intervention details	Population (intervention/control)	Biomarkers measured Pre-and post-intervention	BMI (mean)	% Female	% Black	% White	Age (mean)	Intervention length
* Lorenzo (2022) Spain [36]	RCT	Laparoscopic Control	Adults (n = 39/32)	BMI; IL-6	45.6	63	NR	NR	40.2	6 months
MacKintosh (2019) UK [57]	Cohort	GB SG	Adult women (n = 72)	BMI; weight; leptin; CRP; IL-6; adiponectin; estradiol; progesterone; testosterone	52.1	100	NR	NR	42	12 months
* Moreira (2023) Brazil [58]	RCT	GB Control	Women aged 18 to 65 (n = 64/11)	BMI; weight; leptin; adiponectin;	47	100	12	88	42.3	6 months
* Moriconi (2022) Italy [59]	CT	RYGB Control	Adults aged 18 to 65 (n = 50/11)	BMI; leptin; adiponectin	47	100	12	NR	42.3	6 months
Nikolic (2011) Croatia [60]	Cohort	IGB	Adults aged 20 to 60 (n = 43)	BMI; weight; leptin	41.1	82	0	NR	35	12 months
Sarwer (2018) USA [61]	Cohort	GB	Adult women (n = 106)	BMI; weight; estradiol; testosterone	44.5	100	3	97	41	48 months
* Tussing-Humphreys (2011) USA [62]	CT	GB Control	Adult women (n = 20/20)	BMI; CRP; IL-6, TNF-α	46.6	100	48	52	36.3	6 months
* Wojciechowska-Kulik (2020) Poland [63]	CT	IGB Control	Adults (n = 30/18)	BMI; weight; leptin; adiponectin; CRP	40.9	57	NR	NR	41.2	6 months
Intervention: Pharmacotherapy										
* Dushay (2012) USA [64]	RCT	Exenatide Placebo	Women aged 18 to 70 (n = 21/21)	BMI: weight; leptin; adiponectin	33.1	100	NR	NR	48	9 months
* Gadde (2011) USA [65]	RCT	Phen/top 7.5/46 mg Phen/top 15/92 mg Placebo	Adults aged 18 to 70 (n = 498/995/994)	BMI; weight; CRP; adiponectin	36.5	70	11	86	51.1	14 months
* Garvey (2022) USA [66]	RCT	Semaglutide Placebo	Adults aged 18 or older (n = 152/152)	BMI; weight; CRP	38.6	78	4	96	47.4	26 months
* Iepsen (2015) UK [67]	RCT	Liraglutide Placebo	Adults aged 18 to 65 (n = 27/25)	BMI; weight; leptin	30.8	85	NR	NR	46	12 months
* Joo (2011) Korea [68]	RCT	Diacerein Placebo	Adults aged 20 or older (n = 12/7)	BMI; weight; CRP; adiponectin; TNF-α	31	28	NR	NR	38	3 months
* Kim (2015) Australia [69]	RCT	Beloranib Placebo	Adults aged 18 to 65 (n = 109/38)	BMI; weight; leptin; CRP; adiponectin;	37.6	93	NR	NR	48.3	3 months
Sari (2010) Turkey [70]	RT	Sibutramine Sibutramine + metformin	Adults aged 18 to 65 (n = 36/34)	BMI; weight; leptin; CRP	39.8	100	NR	NR	46.9	12 months
* Smith (2010) USA [71]	RCT	Lorcaserin Placebo	Adults aged 18 to 75 (n = 1595/1587)	BMI; weight; CRP	36.2	83.5	19	81	44.1	12 months

textbfCER: continuous energy restriction; **CR**: calorie restriction; **CT**: controlled trial; **GB**: gastric bypass; **HC**: high carbohydrates; **HCT**: historically controlled trial; **HP**: high protein; **HPC**: high-protein casein; **HPW**: high-protein whey; **IER**: intermittent energy restriction; **IGB**: intragastric balloon; **IM**: isocaloric + metformin; **IT**: interactive technology; **LC**: low calorie; **NR**: not reported; **PA**: physical activity; **PC**: personal contact; **Phen/top**: phentermine plus topiramate; **RCT**: randomized controlled trial; **RT**: randomized trial; **SG**: sleeve gastroplasty; **RYGB**: Roux-en-y gastric bypass; **TWL**: traditional weight loss; **VEGF**: vascular endothelial growth factor; **VLCKD**: very low-calorie ketogenic diet. * Included in meta-analysis.

3.1.1. Systematic Review

Among the 38 unique studies, investigators examined the effects of lifestyle interventions in 16 studies [12,35–49], bariatric surgery in 16 studies [35,36,50–63], and pharmacotherapy in 8 studies [67–71] (Table 1). Two studies [35,36] that evaluated both lifestyle and bariatric surgery were counted within each of those respective intervention groupings. In the lifestyle intervention group, interventions predominantly consisted of dietary modifications, exercise programs, or a combination of both, often compared to control groups. In the bariatric surgery group, fewer controls were observed, and the predominant procedure performed was gastric bypass (e.g., Roux-en-Y). In the pharmacotherapy group, medications examined for weight loss included exenatide, semgalutide, liraglutide, diacerein, beloranib, sibutramine, lorcaserin, and phentermine plus topiramate, with all studies incorporating control groups.

3.1.2. Meta-Analysis

Of the 38 eligible studies, 24 unique studies were controlled (10 in the lifestyle intervention group [35–40,42,43,46,47], 9 in the bariatric surgery group [35,36,50–52,58,59,62,63], and 7 in the pharmacotherapy group [64–69,71]). Two studies [35,36] that evaluated both lifestyle and bariatric surgery were counted (and included in meta-analysis) within each of those respecitve intervention groupings. One controlled study with a mismatched control group (e.g., normal weight group compared to an obese weight group) was analyzed as both included and excluded from the meta-analysis [63] to demonstrate differences in the effect of the results.

For the meta-analysis, the association of weight loss on changes to inflammatory markers CRP, IL-6, TNF-α, and hormones leptin, adiponectin, estradiol, estrone, and testosterone was compared between the three different intervention strategies and/or the percent of weight loss. When heterogeneity was considered significant, subgroup analysis was performed on sex percentages (100% female (yes or no)), menopause status (pre-menopause or post-menopause), sample size (<500 or ≥500), BMI (<35 or ≥35), and intervention duration (<6 months, ≥6 months, or ≥12 months).

3.2. Comparison of the Effects of Different Interventions on Weight Loss

Among the 29 eligible studies reporting weight changes (11 lifestyle [12,38–41,43–46,48,49], 10 bariatric surgery [52–58,60,61,63], 8 pharmacotherapy [64–71]), the overall average weight loss was 13.8% (adj. avg. = 12.0%). Clear distinctions were evident among the intervention groups, with the bariatric surgery group demonstrating strong reduction in body weight, averaging 25.8% (adj. avg. = 21.6%), followed by the pharmacotherapy group with a 7.6% (adj. avg. = 6.1%) reduction and lifestyle group with an average reduction of 5.9% (adj. avg. = 4.3%).

When performing a subgroup analysis of studies that included only female participants and adjusted for menopause status, the results showed three studies (n = 1053) involving postmenopausal women and nine studies (n = 534) involving premenopausal women. The postmenopausal group consisted solely of lifestyle intervention studies, with an average weight loss of 5% [39,40,48]. The premenopausal group included two lifestyle studies, five bariatric surgery studies, and two pharmacotherapy studies [12,49,55–58,61,64,70], with an average weight loss of 22%.

Meta-Analysis on the Effectiveness of Weight-loss Interventions on Reducing Weight

The first analysis evaluated the effectiveness of weight-loss interventions by examining the amount of weight participants lost. Of the 16 controlled studies reporting weight changes (n = 7425, females: 5931 or 81.5%, males: 1493 or 18.5%), the total mean difference was −4.38 (observed total effect size: 4.15; 95% CI: −6.45, −2.31; $p < 0.0001$), indicating significant effects of these interventions on weight reduction. There was significant overall heterogeneity ($I^2 = 100\%$, $p_{\text{heterogeneity}} < 0.00001$) observed, as illustrated in Figure 3 [38–40,43,46,52,56,58,63–69,71]. The analysis was repeated excluding Wojciechowska-Kulik et al. [64], who used normal weight controls. The overall effectiveness in weight loss remained statistically significant, and the weight loss effect for the bariatric surgery group demonstrated statistically significant results (observed total effect size: 2.77; 95% CI: −40.95, −6.98; $p = 0.006$).

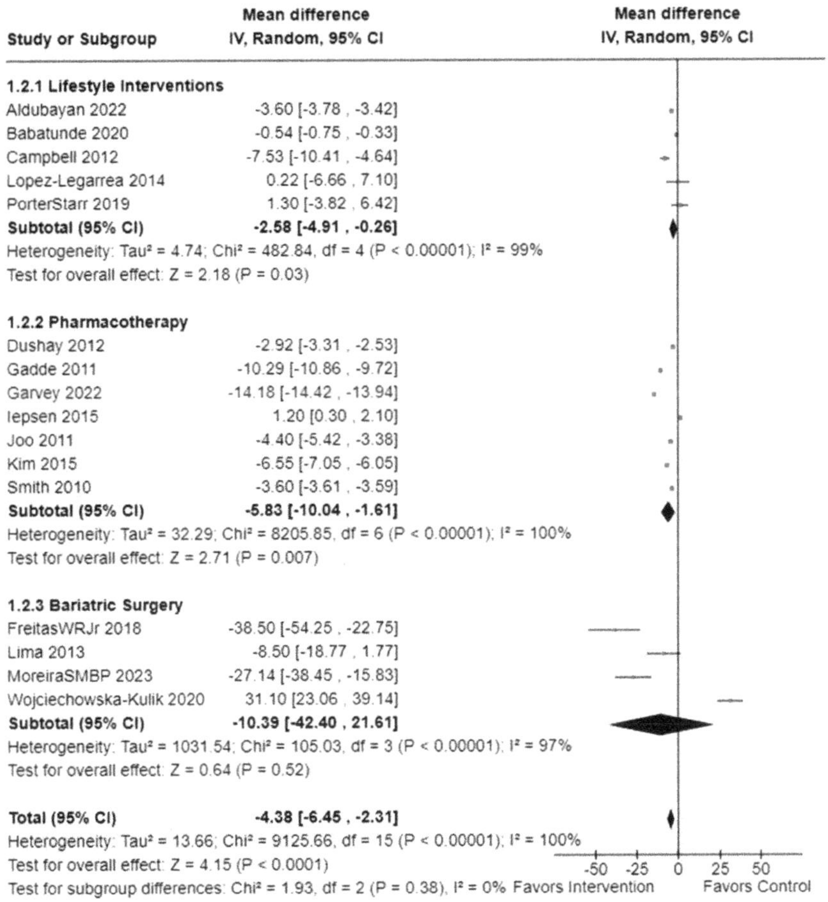

Figure 3. Forest plot illustrating the effects of interventions on weight loss (n = 7425). Black diamonds indicate total effect observed within groups and between groups. Red squares indicate the observed effect per individual study. Value of zero indicates no significant effect where a negative value indicates a positive effect of intervention on weight loss. Results not adjusted for female participants. **IV**: inverse variance; **CI**: confidence interval [XXX].

When performing subgroup analysis of the controlled studies that included only female participants and adjusted for menopause status, the results showed two studies (n = 775) involving postmenopausal women and three studies (n = 123) involving premenopausal women. The total mean difference of weight loss for the postmenopausal group was 3.88 (observed total effect size: 1.11; 95% CI: −10.72, 2.96; $p = 0.27$) and 12.01 for the premenopausal group (observed total effect size: 1.73; 95% CI: −25.62, 1.60; $p = 0.08$), indicating menopause status was not statistically for weight loss. There was no heterogeneity between the subgroups ($I^2 = 8.6\%$, $p_{\text{heterogeneity}} = 0.30$) but high heterogeneity overall ($I^2 = 97\%$, $p_{\text{heterogeneity}} < 0.00001$) [37,39,40,42,56,58,59].

3.3. Comparison of Intervention-Associated Weight Loss on Inflammatory Markers

3.3.1. CRP

Twenty-one eligible studies (20 unique; Abulmeaty et al. [35] counted for lifestyle and bariatric surgery) assessed changes in circulating CRP levels following weight loss interventions (7 lifestyle [12,35,38,39,43,44,46], 8 bariatric surgery [35,51,55–57,59,62,63], 6 pharmacotherapy [65,66,68–71]). On average, CRP decreased by 33.5% (adj. avg. = 28.7%) from baseline due to weight loss across these studies. When comparing CRP reductions across the three types of weight loss strategies, bariatric surgery resulted in the most substantial reduction in CRP levels (47%, adj. avg. = 43%), with the greatest effect demonstrated by Lima et al. [56] who reported an 89% decrease among female participants 12 months post-surgery. Weight loss from pharmacotherapy also demonstrated notable reductions (30.1%, adj. avg. = 22%), with Garvey et al. [66] reporting a 56.7% (adj. avg. = 43.9%) reduction following the administration of semaglutide for 104 weeks. In contrast, the lifestyle intervention studies showed a smaller effect of weight loss on CRP reductions (18.7%, adj. avg. = 15.7%). For example, Moszak et al. [44] showed little effect in reducing CRP levels with a hypocaloric diet with physical activity (2.4% (adj. avg. = 1.4%); however, the short duration of the intervention should be considered when assessing its effect.

Meta-Analysis on the Effectiveness of Weight Loss on CRP

For the meta-analysis, changes in CRP were assessed for their association with the percentage of weight loss achieved from the interventions. Of the 11 controlled studies reporting weight changes (n = 6762, females: 5287 or 73.4%, males: 1474 or 26.5%), the total mean difference was −0.41 (observed total effect size: 1.20; 95% CI: −1.08, 0.26; $p = 0.23$). There was significant overall heterogeneity ($I^2 = 100\%$, $p_{\text{heterogeneity}} < 0.00001$) as well as significant heterogeneity between the groups ($I^2 = 98.7\%$, $p_{\text{heterogeneity}} < 0.00001$), as illustrated in Figure 4. [38,39,43,46,56,63,65,66,68,69,71].

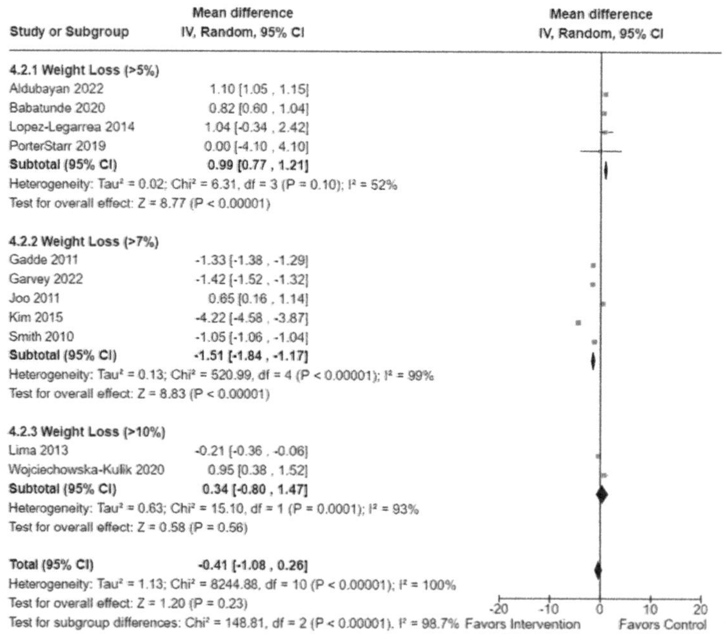

Figure 4. Forest plot illustrating effectiveness of weight loss on reducing CRP (n = 6762). Black diamonds indicate total effect observed within groups and between groups. Red squares indicate the

observed effect per individual study. Value of zero indicates no significant effect where a negative value indicates a positive effect of weight loss on circulating CRP. Results not adjusted for female participants. **IV**: inverse variance; **CI**: confidence interval. The analysis was repeated excluding Wojciechowska-Kulik et al. [63], who used normal weight controls, and Lima et al. [56] as there was no power to assess effect size; therefore, the > 10% weight loss group was removed from the repeat analysis. The overall effectiveness of weight loss on CRP remained the same as well as the heterogeneity overall and between group differences remained high (I^2 = 99.3%, $p_{\text{heterogeneity}}$ < 0.00001).

3.3.2. TNF-α

Nine eligible studies (eight unique; Abulmeaty et al. [35] counted for lifestyle and bariatric surgery) investigated changes in circulating TNF-α levels following weight loss interventions (5 lifestyle [12,35,38,43,46], 3 bariatric surgery [35,52,62], 1 pharmacotherapy [68]), revealing an overall average decrease of 13% (adj. avg. = 5.3%) from baseline. When comparing TNF-α reductions across the three types of weight loss strategies, a bariatric surgery study conducted by Abulmeaty et al. [35] observed a substantial reduction 12 months post-gastric bypass surgery (52.3%, adj. avg. = 23.5%). In examining the lifestyle intervention studies, two out of five also showed no effect of weight loss on reducing TNF-α levels post-intervention [38,46]. Additionally, only one pharmacotherapy study investigated weight loss on TNF-α levels, reporting a 42.5% (adj. avg. = 11.9%) reduction at the end of 3 months of administering diacerein [68]. Lima et al. [56] was excluded from the analysis as sensitivity testing revealed abnormal values that were outside of the limits.

Meta-Analysis on the Effectiveness of Weight Loss on TNF-α

The meta-analysis revealed that between group comparisons could not be performed as there was only one study in the >7% weight loss group and one study in the > 10% weight loss group. However, out of the five controlled studies that assessed TNF-α levels (n = 332, females: 216 or 62.2%; males: 115 or 37.8%), the total mean difference was –0.03 (observed total effect size: 0.13; 95% CI: −0.47, 0.41; p = 0.89), demonstrating no statistically significant effect. There was significant overall heterogeneity (I^2 = 96%, $p_{\text{heterogeneity}}$ < 0.00001), as illustrated in Figure 5 [38,43,46,52,68].

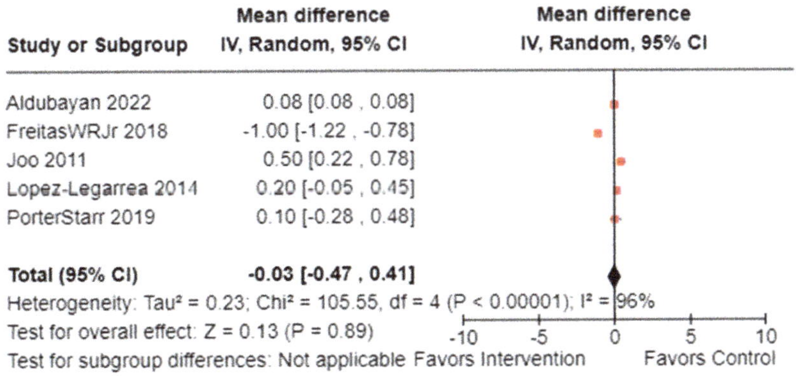

Figure 5. Forest plot illustrating effectiveness of weight loss on reducing TNF-α (n = 332). Black diamonds indicate total effect observed within groups and between groups. Red squares indicate the observed effect per individual study. Value of zero indicates no significant effect where a negative value indicates apositive effect of weight loss on circulating TNF-α. Results not adjusted for female participants. **IV**: inverse variance; **CI**: confidence interval.

3.3.3. IL-6

Ten eligible studies (eight unique; Abulmeaty et al. [35] and Lorenzo et al. [36] counted for lifestyle and bariatric surgery) (7 lifestyle [12,35,36,38,39,43,46], 5 bariatric surgery

[35,36,51,56,57]) examined circulating IL-6 levels pre- and post-intervention, revealing weight loss caused an average reduction of 41.9% (adj. avg. = 38.2%). When comparing IL-6 reductions across the three types of weight loss strategies, four out of the five bariatric studies [35,36,51,56,57] reported that weight loss reduced IL-6 levels by an average of 46.3% (adj. avg. = 41.4%); however, MacKintosh et al. [57] reported a 95% decrease from baseline 12 months post-surgery. In the lifestyle intervention studies, the overall reported reduction of IL-6 levels was 33.5% (adj. avg. = 28.2%). The greatest effect in this group was observed by Babatunde et al. [39], who reported an 81.3% reduction in IL-6 after weight loss from a 12-month diet plus physical activity intervention, specifically in Black females with obesity. IL-6 levels were not documented in any of the eligible pharmacotherapy studies. Sensitivity analysis revealed that Tussing-Humphreys et al. [62]. had abnormal values reported and was removed from this analysis.

Meta-Analysis on the Effectiveness of Weight Loss on IL-6

When analyzing the association of weight loss on IL-6 levels on the studies that reported weight changes, the overall findings were statistically significant. Four controlled studies (n = 922, females: 704 or 75.7%, males: 217 or 24.2%) showed a total mean difference of −0.25 (observed total effect size: 6.17; 95% CI: −0.33, −0.17; $p < 0.00001$). There was no subgroup analysis performed, as only one study with > 10% weight loss had reported data, while the >7% wight loss group had none. The heterogeneity was statistically significant among the > 5% weight loss group (I^2 =74%, $p_{\text{heterogeneity}}$ = 0.010).

3.4. Comparison of Intervention-Associated Weight Loss on Hormones

3.4.1. Leptin

Twenty eligible studies (6 lifestyle [37,38,41,45–47], 10 bariatric surgery [50–54,56–58,60,63], four pharmacotherapy [64,67,69,70]) examined pre- and post-intervention leptin levels, revealing an overall average reduction of 40% (adj. avg. = 33.2%). When comparing leptin reductions across the three types of weight loss strategies, three of ten bariatric surgery [54,56,58] studies demonstrated substantial improvements in leptin levels post-intervention exceeding 70% (adj. avg. = 64.6%). The pharmacotherapy [64,67,69,70] and lifestyle intervention [37,38,41,45–47] studies both reported reductions averaging 27.4% (adj. avg. = 21.4%). Notably, Lima et al. [56] and Moreira et al. [58] both reported significant leptin reductions at 78% (adj. avg. = 78%) after implementing a low-caloric liquid diet and exercise regimen for 6 weeks. Kim et al. [69], Sari et al. [70], and Iepsen et al. [67] reported 30% reductions after administering different weight loss medications.

Meta-Analysis on the Effectiveness of Weight loss on Leptin

When analyzing the association of weight loss on leptin levels for the studies that reported weight changes, the overall findings were statistically significant. Eight controlled studies (n = 492, females: 405 or 85.3%, males: 86 or 16.5%) showed a total mean difference of −1.55 (observed total effect size: 12.57; 95% CI: −1.79, −1.31; $p < 0.00001$); however, there was significant overall heterogeneity (I^2 = 100%, $p_{\text{heterogeneity}} < 0.00001$). Significant heterogeneity was also present for between group comparisons (I^2 = 99.8%, $p_{\text{heterogeneity}} < 0.00001$) as illustrated in Figure 6 [38,46,52,56,63,64,67,69].

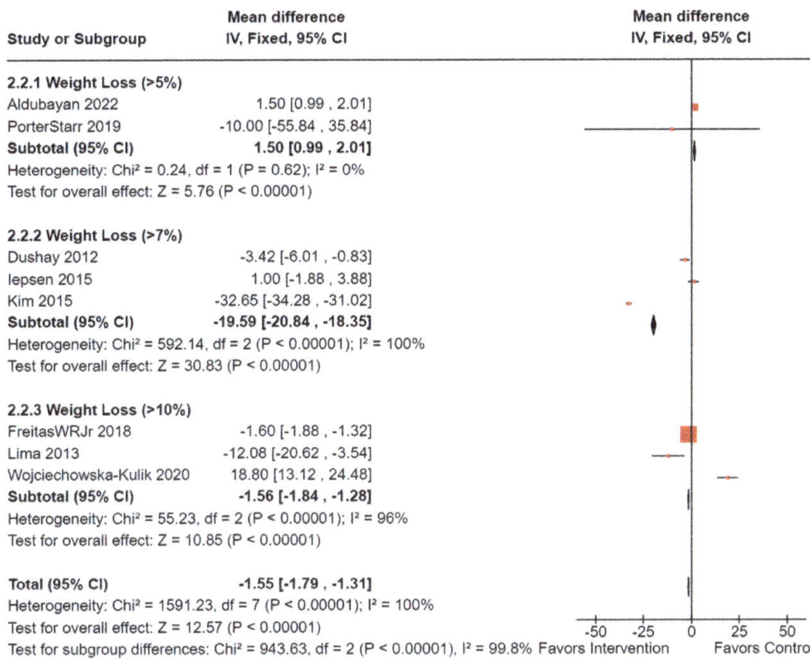

Figure 6. Forest plot illustrating effectiveness of weight loss on leptin levels (n = 492). Black diamonds indicate total effect observed within groups and between groups. Red squares indicate the observed effect per individual study. Value of zero indicates no significant effect where a negative value indicates a positive effect of weight loss on circulating leptin. Results not adjusted for female participants. **IV**: inverse variance; **CI**: confidence interval. The analysis was repeated excluding Wojciechowska-Kulik et al. [63], who used normal weight controls. The overall effectiveness of weight loss on leptin remained the same as well as heterogeneity for between group differences.

3.4.2. Adiponectin

Fifteen eligible studies conducted pre- and post-intervention assessments of circulating adiponectin (6 lifestyle [12,37,38,41,45,46], 5 bariatric surgery [51,52,56,57,63], 4 pharmacotherapy [64,65,68,69]), the average overall increase was 22.8% (adj. avg. = 20.9%). When comparing increases in adiponectin levels across the three types of weight loss strategies, the bariatric surgery group had an increase of 54.7% (adj. avg. 51.3%) compared to an increase of 19.9% (adj. avg. 17.4%) observed in the pharmacotherapy group. The lifestyle group had mixed results. Three out of the six studies showed a decrease in adiponectin (4.5% adj. avg. 3.7%), while the other three studies showed an increase of 2.2% (adj. avg. 1.9%). Two studies were removed from the analysis after sensitivity testing revealed Kim et al. [54] and Moreira et al. [58] reported values outside of the limit.

Meta-Analysis on the Effectiveness of Weight Loss on Adiponectin

Among the studies that reported weight loss changes and circulating adiponectin levels, nine controlled studies (n = 2946, females: 2107 or 75.7%, males: 838 or 24.2%) showed a total mean difference of 0.45 (observed total effect size: 1.02; 95% CI: −0.42, 1.32; $p = 0.31$) indicating no statistically significant effect. Furthermore, there was significant overall heterogeneity ($I^2 =$ 100%, $p_{\text{heterogeneity}} < 0.00001$). Significant heterogeneity was also observed between the groups ($I^2 = 89\%$, $p_{\text{heterogeneity}} = 0.0001$) (Figure 7) [38,46,52,56,63–65,68,69].

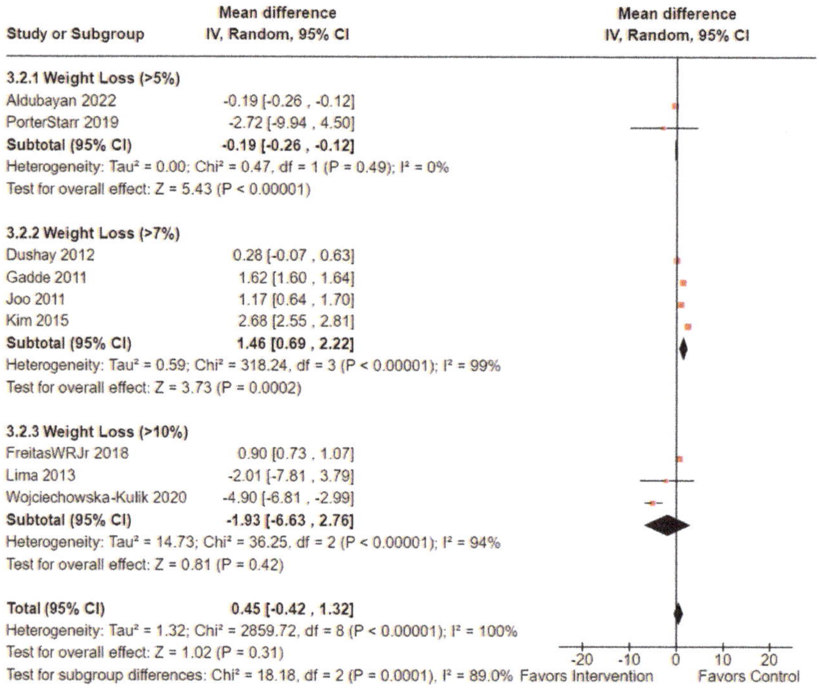

Figure 7. Forest plot illustrating effectiveness of weight loss on adiponectin levels (n = 2946). Black diamonds indicate total effect observed within groups and between groups. Red squares indicate the observed effect per individual study. Value of zero indicates no significant effect where a negative value indicates a positive effect of weight loss on increasing adiponectin. Results not adjusted for female participants. **IV**: inverse variance; **CI**: confidence interval. The analysis was repeated excluding Wojciechowska-Kulik et al. [63], who used normal weight controls. The overall effectiveness of weight loss on adiponectin remained the same, and heterogeneity remained significant for between group differences (I^2 = 98.7%, $p_{\text{heterogeneity}} < 0.00001$).

3.4.3. Sex Hormones

Seven eligible studies (5 lifestyle [40,42,47–49], 2 bariatric surgery [55,57]) reported changes in levels of sex hormones following weight loss interventions. No eligible pharmacotherapy studies assessed the weight loss effects on estradiol, estrone, or testosterone levels, precluding comment on the effectiveness of this intervention on these outcomes. After conducting sensitivity analysis, Sarwer et al. [61] was removed from the analysis for estradiol due to abnormal values that were outside the limit.

As endometroid EC is a hormone-driven cancer that is a risk factor for postmenopausal women with obesity, it is important to assess weight loss on circulating sex hormones in both pre-and postmenopausal status due to drastic hormone differences between the groups.

The subgroup analysis of six studies that included only female participants and adjusted for menopause status presented three studies (n = 1138) involving postmenopausal women [49,55,57] and three studies (n = 120) involving premenopausal women [40,42,48]. The postmenopausal group consisted solely of lifestyle intervention studies, whereas the premenopausal group was solely bariatric surgery studies. The average reduction of estradiol in postmenopausal women was 16.4%. However, estrone is the predominant form of estrogen in postmenopausal women, and the average reduction in estrone was 8.3%. Testosterone, which can be converted into estrone or estradiol, was reduced by 4.8% in postmenopausal women. More bariatric surgery studies need to assess the weight loss effects on sex hormones in this high-risk group.

Estradiol

Seven eligible studies (5 lifestyle [40,42,47–49] and 2 bariatric surgery [55,57]) reported pre-and post-intervention changes in circulating estradiol levels. When comparing changes in estradiol levels across the two weight loss groups, an average decrease was observed of 6.4% (adj. avg. = 5.8%) from baseline. A lifestyle study conducted by Shah et al. [47] reported changes in estradiol in both men and women (6.4% compared to 10.3%). One out of the two bariatric surgery studies showed a 32% decrease 12 months after surgery, while one lifestyle group showed a 21.3% decrease at 30 months post-intervention. As two studies (two lifestyle, two bariatric surgery) showed increases in estradiol, this indicates more research is needed to confirm a significant effect between the two interventions [49,57].

Estrone

Four eligible studies (3 lifestyle [40,42,48] and 1 bariatric surgery [55]) assessed circulating estrone levels pre- and post-intervention, showing reductions averaging 14.7%. When comparing changes across the two weight loss groups, the average reduction in estrone in the lifestyle group was 8.3%, while Kjaer et al. [55] reported a 33.8% reduction 12 months following bariatric surgery among premenopausal participants.

Testosterone

Seven eligible studies (4 lifestyle [40,42,48,49] and 3bariatric surgery [55,57,61]) reported pre- and post-intervention changes in circulating testosterone levels, with an average decrease of 18.1%. When comparing changes across the two weight loss groups, the bariatric surgery group showed an average reduction in testosterone from baseline of 34.3%. Sarwer et al. [61] specifically addressed the impact of bariatric surgery on sex hormones in women with obesity, showing testosterone levels decreased by 53.6% post-surgery. The two other bariatric surgery studies investigating testosterone levels showed an average decrease of 24.6% [55,57], whereas weight loss from lifestyle interventions resulted in a 5.9% decrease, and Duggan et al. [42] showed no change.

Meta-Analysis on the Effectiveness of Weight Loss on Sex Hormones

There were only two controlled studies (lifestyle interventions only) reporting changes in sex hormones (n = 860, 100% postmenopausal women). Among the sex hormones assessed for a weight loss effect, estrone is the only hormone that demonstrated a positive effect from weight loss [40,42].

When analyzing the association of weight loss on estradiol levels for two studies that reported changes, the overall findings were not statistically significant, with a total mean difference of −3.91 (observed total effect size: 1.57; 95% CI: −8.77, 0.96; p = 0.12). There was significant heterogeneity among the two studies (I^2 =99%, $p_{\text{heterogeneity}} < 0.00001$).

When analyzing the association of weight loss on estrone levels for two studies that reported changes, the overall findings were statistically significant, with a total mean difference of −2.88 (observed total effect size: 8.07; 95% CI: −3.58, −2.18; $p < 0.00001$). There was no significant heterogeneity among the two studies (I^2 =0%, $p_{\text{heterogeneity}} = 0.58$).

When analyzing the association of weight loss on testosterone levels for two studies that reported changes, the overall findings were not statistically significant with total mean difference of –1.12 (observed total effect size: 0.89; 95% CI: −3.60, 1.36; p = 0.38). There was significant heterogeneity among the two studies (I^2 =89%, $p_{\text{heterogeneity}} = 0.003$).

3.5. Intervention Effects on Key Tumor Growth Factors

It is worth noting that one bariatric surgery study [51]), which consisted of 100% female participants, assessed pre- and post-intervention changes of VEGF. This growth factor is increased in later stages of tumor growth, including endometroid EC. Reducing levels of this protein can be an indication of a positive effect from an intervention. Dalmas et al. [51] reported a 28.1% reduction from baseline.

3.6. Heterogeneity

Among the 24 controlled studies evaluated for the meta-analysis, heterogeneity was pronounced in all total effects ($I^2 > 75\%$), with reduced heterogeneity observed for between group comparisons. Therefore, the random-effects model was used for all analyses. Sources for heterogeneity were analyzed using sensitivity analysis and subgroup analysis. The results of the subgroup analysis to assess sources for possible heterogeneity are available in Supplementary Table S2.

3.7. Bias

There were 20 unique randomized controlled studies (Lorenzo et al. [36] was counted in lifestyle and bariatric surgery) (9 lifestyle [36–40,42,43,46,47], 4 bariatric surgery [36,52,56,58], 7 pharmacotherapy [64–69,71]) and 5 randomized studies (4 lifestyle [41,45,48,49], 0 bariatric surgery, 1 pharmacotherapy [70]) not utilizing controls assessed using the Cochrane Risk of Bias 2.0 (RoB 2.0) tool as shown in Figure 8. For non-randomized studies, the Newcastle-Ottawa Scale (NOS) was used to determine the quality of the studies for inclusion in the meta-analysis. The reasons for researchers not randomizing were not clarified. In the context of dietary interventions, the choice not to randomize might be due to factors such as limitations in sample size or the inherent characteristics of the intervention. Moreover, bariatric studies seldom use randomization as all participants are preselected candidates for surgery with predetermined schedules.

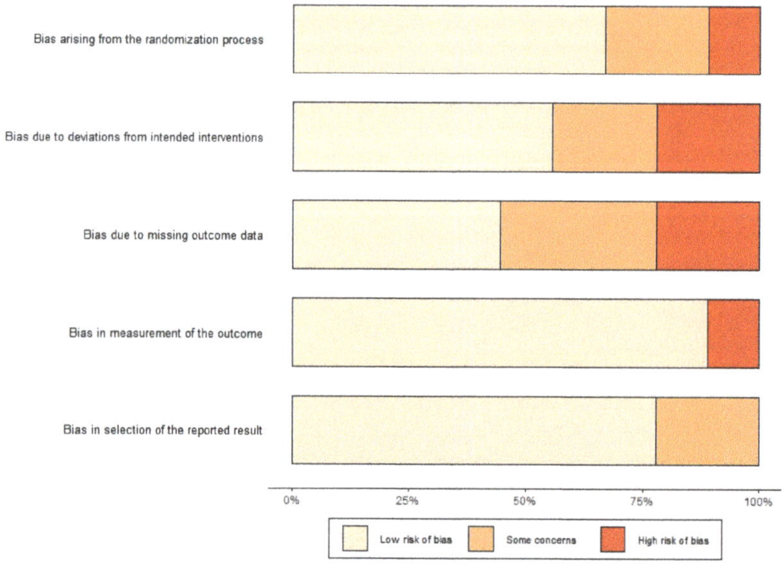

Figure 8. Graph displaying the RoB 2.0 five domains and risk of bias per domain for all included randomized trials (n = 25 unique studies).

3.8. Quality

Table 2 summarizes the quality of the 14 eligible studies using the NOS scale. A study can receive up to nine stars over three categories: selection, comparability, and outcomes/exposures. For selection, four studies gained four stars, six gained three stars, and four studies gained two stars. For comparability, the controlled studies received either one or two whereas the uncontrolled studies received zero stars. For outcomes and exposures, 13 studies received three stars, one study received two stars, and zero studies received one star. Overall, the 14 studies were deemed good quality (6–9/9 stars). No study was considered fair or low quality for this analysis.

Table 2. Quality assessment using the Newscastle-Ottawa Scale (NOS). "*" is an awarded star.

Study	Selection	Comparability	Outcomes/Exposure	Total
Abulmeaty, 2023 [35]	****	*	***	8
Ceccarini, 2019 [50]	***	*	**	6
Dalmas, 2011 [51]	**	*	***	6
Gomez-Huelgas, 2019 [12]	***		***	6
Jacobsen, 2012 [53]	****		***	7
Kim, 2023 [54]	****		***	7
Kjaer, 2017 [55]	***		***	6
MacKintosh, 2019 [57]	***		***	6
Moriconi, 2022 [59]	****	**	***	9
Moszak, 2018 [44]	***		***	6
Nikolic, 2011 [60]	***		***	6
Sarwer, 2018 [61]	***		***	6
Tussing-Humphreys, 2011 [62]	**	*	***	6
Wojciechowska-Kulik (2020) [63]	**	*	***	6

4. Discussion

EC is classified into various histologic types, each with distinct features and prognostic implications [72]. The most common type, endometrioid carcinoma, is linked to excess estrogen and generally has a favorable prognosis. Reducing body weight can significantly lower estrogen levels, thereby decreasing the risk and improving outcomes for estrogen-positive endometrioid carcinoma. Serous carcinoma (aggressive, not estrogen-dependent), clear cell carcinoma (rare, poor prognosis), mucinous carcinoma (rare, mucus-secreting), and carcinosarcoma (highly malignant, mixed tissue) are the other histologic types that influence prognosis and treatment strategies [72,73].

Obesity significantly increases the risk of women developing estrogen-positive endometrioid EC. This risk is particularly elevated in postmenopausal women due to increased aromatase activity in adipocytes, which converts androgens to estrogen. Excess estrogen overstimulates various pro-growth pathways, such as PI3K/Akt/mTOR, in many cells, including adipocytes and epithelial [74].

The overgrowth of cells leads to overcrowding, causing cells to become necrotic. Necrotic cells release signals, called cytokines, that alert the immune system to aid in the cleanup of cellular debris [75]. The infiltration of immune cells, particularly macrophages, leads to a state of chronic inflammation, disrupting homeostasis. This inflammatory state is characterized by the release of proinflammatory cytokines, such as TNF-α, IL-6, and CRP, which further perpetuate inflammation and, in obesity, aid in insulin resistance [76].

Inflammation from obesity is well documented as a contributing factor to several chronic, life-threatening diseases, including cardiovascular diseases, type 2 diabetes, and non-alcoholic fatty liver disease (NAFLD) [25,77,78]. In the context of cancer, chronic inflammation creates a microenvironment that promotes tumorigenesis. Elevated levels of proinflammatory cytokines lead to increased cell proliferation, angiogenesis, and inhibition of apoptosis [78]. Furthermore, the imbalance between proinflammatory adipokines, like leptin, and anti-inflammatory adipokines, like adiponectin, exacerbates metabolic disturbances, increasing the risk of developing endometroid EC [79]. Addressing obesity and its related inflammation through weight loss can be crucial in reducing the incidence of endometroid EC.

This systematic review and meta-analysis examined the impact of weight loss strategies on reducing body weight, which can reduce body fat, waist circumference, and BMI, to reduce the risk of developing endometrioid EC. Reducing these anthropometric outcomes can help reduce the incidence of inflammatory biomarkers that can promote endometroid EC development. Reducing this risk is particularly important for postmenopausal women, who are at greater risk of developing obesity, heightening their risk in developing endometroid EC [17]. However, weight loss is not the only determinant in reducing endometroid EC. Decreases in endometrial stimulation by estrogen also need to be assessed.

For this assessment, three weight loss strategies—bariatric surgery, pharmacotherapy, and lifestyle interventions—were analyzed to determine whether weight loss reduced circulating levels of CRP, TNF-α, IL-6, leptin, and estrogens, all of which are considered markers of increased endometrial proliferation and inflammation due to obesity. The results suggest that bariatric surgery has a 25.8% reduction in weight higher than lifestyle or pharmacotherapy interventions. The meta-analysis suggests that any intervention is significant at reducing body weight ($p < 0.0001$). It is important to note that bariatric surgery may not be an option for everyone. Surgery requires a lengthy recovery time and an increased risk for other health complications, and it can take from months to years to achieve the long-term benefits. For this reason, other alternatives need to be considered to prevent hormone-dependent endometrial hyperplasia, such as nutritional, behavioral, and estrogen-progesterone therapies.

The most notable biomarkers of inflammation linked to obesity and endometroid EC—CRP, IL-6, TNF-α, and leptin—were assessed to determine whether weight loss reduces their circulating levels across the three interventions. Overall, the results suggest that weight loss reduces CRP levels by 33.5%, with weight loss >7% having the most significant effect ($p < 0.00001$). The results also suggest that weight loss also improves IL-6 levels, with an average reduction of 41.9%. These results were confirmed by the meta-analysis showing a significant on IL-6 levels effect due to weight loss ($p < 0.00001$). Regarding TNF-α, the meta-analysis showed no significant overall effect; however, weight reductions of 10% or greater did suggest a significant effect ($p < 0.00001$) confirming the results of the systematic review showed an average decrease of 13%.

Leptin, an adipokine secreted by adipocytes, is significantly involved in the progression of estrogen-dependent cancers, such as EC. Elevated leptin levels, common in obesity, are linked to several oncogenic mechanisms. Leptin promotes angiogenesis by stimulating the production of vascular endothelial growth factor (VEGF), which supports tumor growth by enhancing blood supply [80]. It also activates pro-growth signaling pathways like JAK/STAT and PI3K/AKT, leading to increased cancer cell proliferation and migration [81]. Furthermore, leptin fosters a proinflammatory tumor microenvironment by increasing cytokine secretion and immune cell infiltration, which can enhance estrogen receptor activity and tumor progression [82]. Additionally, leptin inhibits apoptosis by inhibiting the AMPK pathway, allowing for uncontrolled cell growth [80].

Weight loss is an opportunistic approach to improve the ratio of circulating leptin and adipokine levels, thereby reducing factors that promote metabolic disorders and increase the risk of EC. Results from the meta-analysis suggest that weight loss has a significant effect on reducing leptin levels ($p < 0.00001$) with any percentage of weight loss improving circulating levels. The same suggestion holds true for weight loss up to 7% improving levels of anti-inflammatory adiponectin; however, weight loss over 10% did not have as significant of an effect ($p = 0.42$) suggesting weight loss overall may not improve circulating adiponectin ($p = 0.31$).

Cytokines and adipokines are not the only indicators of inflammation and imbalanced homeostasis related to obesity and increasing the risk of endometroid EC. Various sex hormones also play roles in pro- and anti-inflammatory processes that obesity can exacerbate, leading to cancer development, such as estrogen.

In postmenopausal women, high levels of estrogen are a profound contributor to the increased risk of obesity contributing to endometroid EC. Estrogen drives many cell proliferation pathways, including PKA/CREB and PI3K/Akt/mTOR that are known to be upregulated in EC [81]. Weight loss can directly reduce estrogen levels by reducing body fat percentages and visceral adiposity, as demonstrated by Campbell et al. [40] and Duggan et al. [42]. Estrone, a weaker form of estrogen, is more prominent in postmenopausal women, who are at higher risk for EC. Therefore, reducing visceral adiposity can affect circulating estrone levels [83].

In the meta-analysis, only two studies assessed the impact of weight loss on circulating sex hormones, all within the lifestyle intervention group. This highlights the need for more controlled studies measuring sex hormones as outcomes in obesity and weight loss research. Overall, the reduction in estradiol from weight loss was not statistically significant ($p = 0.12$). Estrone was significantly reduced by weight loss in both studies that examined this outcome ($p < 0.00001$). And testosterone, which can be converted into estradiol or estrone by aromatase, had similar findings to estradiol: no overall effect of weight loss ($p = 0.38$) [40,42]. Due to the limited number of controlled studies and varied heterogeneity, a definitive conclusion on the effectiveness of weight loss on reducing sex hormones to decrease the risk of EC remains to be elucidated.

Overall, the conclusions from this review and meta-analysis suggest that weight loss interventions, in general, promote significant weight reductions that can reduce proinflammatory markers and hormones that are implicated in the risks of developing endometroid EC. However, the outcomes from this review must be interpreted with caution due to the large amount of heterogeneity observed among the included studies. For example, the pharmacotherapy group sparsely reported, or did not report, any outcomes related to TNF-α, IL-6, leptin, adiponectin, and sex hormones affecting between group comparisons and overall effects. In addition, the high heterogeneity among the ages and sex of the participants was a limiting factor in correlating the weight-loss effect on the specific risk to postmenopausal women. As premenopausal women are protected by progesterone produced by the ovaries, some of the effects observed in women may be related to intact ovaries as well as other lifestyle conditions and underlying diseases.

An important observation worth addressing in this review is the underrepresentation of Black participants across the weight loss studies examined, as demonstrated by the fact that only 18.9% of the participants identified were Black, while White participants represented 76.9%. Out of the 38 unique studies, only 14 included Black participants (7 lifestyle [37,39,40,42,45,46,48], 5 bariatric surgery [52,58,59,61,62], 3 pharmacotherapy [65,66,71]), while the other studies did not disclose race or only enrolled a specific race that excluded Black participants. Studies conducted in the US, the UK, and Brazil mostly reported race, while studies in other countries typically did not.

The lack of diversity in study samples not only hampers the generalizability of research findings but also perpetuates health disparities, particularly in conditions such as obesity and EC, where Black individuals experience disproportionate burdens [84]. The understanding of weight loss interventions to reduce the risk of cancer or other co-morbidities is critical, especially with the high rate of mortality from endometrioid EC observed in Black women [2,4]. Only one eligible study included in this review, Babatunde et al. [40], consisted of 100% Black, female participants with obesity (n = 336) to address the lack of research available in this population. Interestingly, in this randomized controlled study, a 12-month diet and exercise intervention had little to no effect on weight loss (0.3%) but did reduce proinflammatory markers IL-6 (81.3%) and CRP (15.7%) [39]. This data warrant that weight loss is not a "one size fits all" solution, and culture, race, and ethnicity may influence response. Not addressing race as a factor in research undermines efforts to develop effective interventions and treatments tailored to the needs of diverse populations and hinders the identification of culturally appropriate strategies for weight management and prevention of obesity-related diseases among Black individuals.

The strengths of this study include a comprehensive examination of the effects of weight loss from different types of interventions on key biomarkers related to obesity and the risk of endometroid EC. Another strength is the large pool of participants and the high proportion of included females with the comparison of weight loss on sex hormones. This review also identifies significant gaps in current research, such as the underrepresentation of cytokine, adipokine, and sex hormone outcomes in many of the weight loss interventions but especially in pharmacotherapy. Additionally, it highlights the lack of diversity in study sampling, which impacts the generalizability of weight loss strategies, particularly in addressing health disparities among participants with obesity who are at greater risk of developing endometroid EC.

There are also several limitations of this study that have been identified. The limited number of controlled studies, especially in the context of sex hormones and pharmacotherapy, restricts the ability to draw definitive conclusions. Additionally, there is variable efficacy among different interventions, with lifestyle modifications showing limited effectiveness in some cases, suggesting the need for more robust and sustained strategies. Furthermore, the included studies varied in sample size, intervention duration, and follow-up periods, which can impact the overall conclusions and consistency of the findings.

5. Conclusions

This review highlights the intricate relationship between obesity and EC pathogenesis, emphasizing the need for comprehensive and inclusive approaches to improve outcomes across diverse populations. Future pharmacotherapy research is crucial to investigate the effects of weight loss on common proinflammatory indicators and hormones linked to obesity and EC, given their pivotal roles in cancer-related pro-growth pathways. The lack of assessment of weight loss effects on these indicators emphasizes a significant gap in current knowledge.

Various weight loss strategies, including bariatric surgery, lifestyle interventions, and pharmacotherapy, have been explored to evaluate their impact on EC risk. Bariatric surgery, known for its efficacy in weight reduction, is promising in lowering the risk of endometroid EC due to decreases in body weight that can lead to reductions in fat mass and BMI, which are favorable hormonal alterations. However, most women with obesity do not have access to bariatric surgery due to cost, adverse effects, or exclusion criteria. In contrast, lifestyle interventions, which promote gradual weight loss, and less invasive approaches have had smaller reductions in the percent of weight lost (5% or less) that indicate more innovative solutions are warranted.

Pharmacological approaches with weight loss drugs represent a less invasive alternative to bariatric surgery, with increased weight loss over lifestyle interventions, but they are expensive, have unwanted gastrointestinal side effects, and currently, there is little research on their effectiveness in reducing endometroid EC. This represents a critical need to understand the potential of pharmacologic weight-loss interventions more fully, such as incretin mimetics, for EC risk reduction.

This study significantly contributes to understanding the relationship between obesity, weight loss, and the risk of endometroid EC. However, several limitations warrant consideration. The high heterogeneity among included studies, influenced by variations in sample sizes, intervention durations, and follow-up time points, presents a challenge to interpretation. The limited number of studies reporting on inflammatory and hormone outcomes, particularly in the pharmacotherapy group, may impact overall conclusions. Additionally, the inclusion of both males and females necessitates estimating averages for females, potentially affecting the precision of results. Nonetheless, a notable strength of the study is the extensive participant pool, with a high proportion of females, allowing for relevant comparisons within a high-risk population for risk of endometrioid EC. Moreover, the study is groundbreaking in its comprehensive analysis of major weight loss interventions alongside the latest pharmacotherapies, providing valuable insights into their relative efficacy.

Future research should focus on standardizing intervention methods and controlling for confounding variables to better understand the relationship between weight loss and circulating inflammatory biomarker levels. This review provides valuable insights into the potential advantages of weight loss therapies as supplementary elements in the holistic treatment of EC. By outlining interventions and their outcomes, this endeavor seeks to enhance clinical practices, optimize therapeutic approaches, and incorporate customized interventions that address various factors essential for promoting equitable healthcare and improving outcomes for individuals impacted by EC.

Supplementary Materials: The following are available online at https://www.mdpi.com/article/10.3390/cancers16122197/s1, Table S1: List of terms used for literature searches; Table S2: Subgroup analysis for sources of heterogeneity based on random-effects model for association between weight loss and levels of CRP, IL-6, TNF-α, leptin, and adiponectin. Estradiol, estrone, and testosterone were assessed between all studies that reported values (no groupings).

Author Contributions: Conceptualization: A.D.C., V.L.B.-J. and S.D.H. Data extractions: A.D.C. and E.G. Analysis of data: A.D.C. and E.G. Writing draft: A.D.C. Revising and completing manuscript: A.D.C., E.G., V.L.B.-J. and S.D.H. All authors have read and agreed to the published version of the manuscript.

Funding: This research received no external funding.

Data Availability Statement: The datasets utilized and analyzed for this review are available from the corresponding author upon reasonable request.

Conflicts of Interest: The authors declare no conflicts of interest.

References

1. Siegel, R.L.; Miller, K.D.; Wagle, N.S.; Jemal, A. Cancer statistics, 2023. *CA Cancer J. Clin.* **2023**, *73*, 17–48. [CrossRef]
2. Cote, M.L.; Ruterbusch, J.J.; Olson, S.H.; Lu, K.; Ali-Fehmi, R. The Growing Burden of Endometrial Cancer: A Major Racial Disparity Affecting Black Women. *Cancer Epidemiol. Biomark. Prev.* **2015**, *24*, 1407–1415. [CrossRef]
3. Clarke, M.A.; Devesa, S.S.; Harvey, S.V.; Wentzensen, N. Hysterectomy-Corrected Uterine Corpus Cancer Incidence Trends and Differences in Relative Survival Reveal Racial Disparities and Rising Rates of Nonendometrioid Cancers. *J. Clin. Oncol.* **2019**, *37*, 1895–1908. [CrossRef]
4. Dubil, E.A.; Tian, C.; Wang, G.; Tarney, C.M.; Bateman, N.W.; Levine, D.A.; Conrads, T.P.; Hamilton, C.A.; Maxwell, G.L.; Darcy, K.M. Racial disparities in molecular subtypes of endometrial cancer. *Gynecol. Oncol.* **2018**, *149*, 106–116. [CrossRef]
5. Morrow, C.; Bundy, B.N.; Kurman, R.J.; Creasman, W.T.; Heller, P.; Homesley, H.D.; Graham, J.E. Relationship between surgical-pathological risk factors and outcome in clinical stage I and II carcinoma of the endometrium: A Gynecologic Oncology Group study. *Gynecol. Oncol.* **1991**, *40*, 55–65. [CrossRef]
6. Papatla, K.; Huang, M.; Slomovitz, B. The obese endometrial cancer patient: How do we effectively improve morbidity and mortality in this patient population? *Ann. Oncol.* **2016**, *27*, 1988–1994. [CrossRef]
7. Arem, H.; Irwin, M.L. Obesity and endometrial cancer survival: A systematic review. *Int. J. Obes.* **2013**, *37*, 634–639. [CrossRef]
8. Kokts-Porietis, R.L.; Elmrayed, S.; Brenner, D.R.; Friedenreich, C.M. Obesity and mortality among endometrial cancer survivors: A systematic review and meta-analysis. *Obes. Rev.* **2021**, *22*, e13337. [CrossRef]
9. Winder, A.A.; Kularatna, M.; Maccormick, A.D. Does Bariatric Surgery Affect the Incidence of Endometrial Cancer Development? A Systematic Review. *Obes. Surg.* **2018**, *28*, 1433–1440. [CrossRef]
10. Rock, C.L.; Pande, C.; Flatt, S.W.; Ying, C.; Pakiz, B.; Parker, B.A.; Williams, K.; Bardwell, W.A.; Heath, D.D.; Nichols, J.F. Favorable changes in serum estrogens and other biologic factors after weight loss in breast cancer survivors who are overweight or obese. *Clin. Breast Cancer* **2013**, *13*, 188–195. [CrossRef]
11. Jensterle, M.; Janez, A.; Fliers, E.; Devries, J.H.; Vrtacnik-Bokal, E.; Siegelaar, S.E. The role of glucagon-like peptide-1 in reproduction: From physiology to therapeutic perspective. *Hum. Reprod. Update* **2019**, *25*, 504–517. [CrossRef]
12. Gomez-Huelgas, R.; Ruiz-Nava, J.; Santamaria-Fernandez, S.; Vargas-Candela, A.; Alarcon-Martin, A.V.; Tinahones, F.J.; Bernal-Lopez, M.R. Impact of Intensive Lifestyle Modification on Levels of Adipokines and Inflammatory Biomarkers in Metabolically Healthy Obese Women. *Mediat. Inflamm.* **2019**, *2019*, 1–9. [CrossRef]
13. Kleist, B.; Wahrburg, U.; Stehle, P.; Schomaker, R.; Greiwing, A.; Stoffel-Wagner, B.; Egert, S. Moderate Walking Enhances the Effects of an Energy-Restricted Diet on Fat Mass Loss and Serum Insulin in Overweight and Obese Adults in a 12-Week Randomized Controlled Trial. *J. Nutr.* **2017**, *147*, 1875–1884. [CrossRef] [PubMed]
14. Jakicic, J.M.; Marcus, B.H.; Lang, W.; Janney, C. Effect of exercise on 24-month weight loss maintenance in overweight women. *Arch. Intern. Med.* **2008**, *168*, 1550–1559; discussion 9–60. [CrossRef] [PubMed]
15. von Gruenigen, V.E.; Courneya, K.S.; Gibbons, H.E.; Kavanagh, M.B.; Waggoner, S.E.; Lerner, E. Feasibility and effectiveness of a lifestyle intervention program in obese endometrial cancer patients: A randomized trial. *Gynecol. Oncol.* **2008**, *109*, 19–26. [CrossRef]
16. Linkov, F.; Maxwell, G.L.; Felix, A.S.; Lin, Y.; Lenzner, D.; Bovbjerg, D.H.; Lokshin, A.; Hennon, M.; Jakicic, J.M.; Goodpaster, B.H.; et al. Longitudinal evaluation of cancer-associated biomarkers before and after weight loss in RENEW study participants: Implications for cancer risk reduction. *Gynecol. Oncol.* **2012**, *125*, 114–119. [CrossRef]

17. Madeddu, C.; Sanna, E.; Gramignano, G.; Tanca, L.; Cherchi, M.C.; Mola, B.; Petrillo, M.; Macciò, A. Correlation of Leptin, Proinflammatory Cytokines and Oxidative Stress with Tumor Size and Disease Stage of Endometrioid (Type I) Endometrial Cancer and Review of the Underlying Mechanisms. *Cancers* **2022**, *14*, 268. [CrossRef] [PubMed]
18. Rundle-Thiele, D.; Shrestha, S.; Janda, M. Prevention of endometrial cancer through lifestyle Interventions: A systematic review and synthesis. *Gynecol. Oncol. Rep.* **2022**, *39*, 100900. [CrossRef] [PubMed]
19. Thompson, H.J.; Sedlacek, S.M.; Wolfe, P.; Paul, D.; Lakoski, S.G.; Playdon, M.C.; McGinley, J.N.; Matthews, S.B. Impact of Weight Loss on Plasma Leptin and Adiponectin in Overweight-to-Obese Post Menopausal Breast Cancer Survivors. *Nutrients* **2015**, *7*, 5156–5176. [CrossRef]
20. Fabian, C.J.; Kimler, B.F.; Donnelly, J.E.; Sullivan, D.K.; Klemp, J.R.; Petroff, B.K.; Phillips, T.A.; Metheny, T.; Aversman, S.; Yeh, H.-W. Favorable modulation of benign breast tissue and serum risk biomarkers is associated with > 10 % weight loss in postmenopausal women. *Breast Cancer Res. Treat.* **2013**, *142*, 119–132. [CrossRef]
21. Christou, N.V.; Lieberman, M.; Sampalis, F.; Sampalis, J.S. Bariatric surgery reduces cancer risk in morbidly obese patients. *Surg. Obes. Relat. Dis.* **2008**, *4*, 691–695. [CrossRef]
22. Linkov, F.; Elishaev, E.; Gloyeske, N.; Edwards, R.; Althouse, A.D.; Geller, M.A.; Svendsen, C.; Argenta, P.A. Bariatric surgery-induced weight loss changes immune markers in the endometrium of morbidly obese women. *Surg. Obes. Relat. Dis.* **2014**, *10*, 921–926. [CrossRef]
23. Ward, K.K.; Roncancio, A.M.; Shah, N.R.; Davis, M.-A.; Saenz, C.C.; McHale, M.T.; Plaxe, S.C. Bariatric surgery decreases the risk of uterine malignancy. *Gynecol. Oncol.* **2014**, *133*, 63–66. [CrossRef]
24. McCawley, G.M.; Ferriss, J.S.; Geffel, D.; Northup, C.J.; Modesitt, S.C. Cancer in obese women: Potential protective impact of bariatric surgery. *J. Am. Coll. Surg.* **2009**, *208*, 1093–1098. [CrossRef]
25. Knudsen, L.B.; Lau, J. The Discovery and Development of Liraglutide and Semaglutide. *Front. Endocrinol. (Lausanne)* **2019**, *10*, 155. [CrossRef]
26. Frías, J.P.; Davies, M.J.; Rosenstock, J.; Pérez Manghi, F.C.; Fernández Landó, L.; Bergman, B.K.; Liu, B.; Cui, X.; Brown, K. Tirzepatide versus Semaglutide Once Weekly in Patients with Type 2 Diabetes. *N. Engl. J. Med.* **2021**, *385*, 503–515. [CrossRef]
27. Coskun, T.; Sloop, K.W.; Loghin, C.; Alsina-Fernandez, J.; Urva, S.; Bokvist, K.B.; Cui, X.; Briere, D.A.; Cabrera, O.; Roell, W.C.; et al. LY3298176, a novel dual GIP and GLP-1 receptor agonist for the treatment of type 2 diabetes mellitus: From discovery to clinical proof of concept. *Mol. Metab.* **2018**, *18*, 3–14. [CrossRef]
28. Jastreboff, A.M.; Aronne, L.J.; Stefanski, A. Tirzepatide Once Weekly for the Treatment of Obesity. *Reply N. Engl. J. Med.* **2022**, *387*, 1434–1435. [CrossRef]
29. Rubino, D.M.; Greenway, F.L.; Khalid, U.; O'Neil, P.M.; Rosenstock, J.; Sørrig, R.; Wadden, T.A.; Wizert, A.; Garvey, W.T.; Arauz-Pacheco, C.; et al. Effect of Weekly Subcutaneous Semaglutide vs Daily Liraglutide on Body Weight in Adults with Overweight or Obesity without Diabetes: The STEP 8 Randomized Clinical Trial. *JAMA* **2022**, *327*, 138–150. [CrossRef]
30. Page, M.J.; McKenzie, J.E.; Bossuyt, P.M.; Boutron, I.; Hoffmann, T.C.; Mulrow, C.D.; Shamseer, L.; Tetzlaff, J.M.; Akl, E.A.; Brennan, S.E.; et al. The PRISMA 2020 statement: An updated guideline for reporting systematic reviews. *BMJ* **2021**, *372*, n71. [CrossRef]
31. Covidence Systematic Review Software. [Internet]. Innovation, V.H. Available online: http://www.covidence.org (accessed on 23 September 2023).
32. Flemyng, E.; Dwan, K.; Moore, T.H.; Page, M.J.; Higgins, J.P. Risk of Bias 2 in Cochrane Reviews: A phased approach for the introduction of new methodology. *Cochrane Database Syst. Rev.* **2020**, *10*, ED000148. [CrossRef]
33. Wells, G.A.; Shea, B.; O'Connell, D.; Peterson, J.; Welch, V.; Losos, M.; Tugwell, P. The Newcastle-Ottawa Scale (NOS) for Assessing the Quality of Nonrandomised Studies in Meta-Analyses 2013. Available online: http://www.evidencebasedpublichealth.de/download/Newcastle_Ottowa_Scale_Pope_Bruce.pdf (accessed on 23 September 2023).
34. The Cochrane Collaboration. *Review Manager Web (RevMan Web)*; The Cochrane Collaboration: London, UK, 2020.
35. Abulmeaty, M.M.A.; Ghneim, H.K.; Alkhathaami, A.; Alnumair, K.; Al Zaben, M.; Razak, S.; Al-Sheikh, Y.A. Inflammatory Cytokines, Redox Status, and Cardiovascular Diseases Risk after Weight Loss via Bariatric Surgery and Lifestyle Intervention. *Medicina* **2023**, *59*, 751. [CrossRef]
36. Lorenzo, P.M.; Sajoux, I.; Izquierdo, A.G.; Gomez-Arbelaez, D.; Zulet, M.A.; Abete, I.; Castro, A.I.; Baltar, J.; Portillo, M.P.; Tinahones, F.J.; et al. Immunomodulatory effect of a very-low-calorie ketogenic diet compared with bariatric surgery and a low-calorie diet in patients with excessive body weight. *Clin. Nutr.* **2022**, *41*, 1566–1577. [CrossRef] [PubMed]
37. Abbenhardt, C.; McTiernan, A.; Alfano, C.M.; Wener, M.H.; Campbell, K.L.; Duggan, C.; Foster-Schubert, K.E.; Kong, A.; Toriola, A.T.; Potter, J.D.; et al. Effects of individual and combined dietary weight loss and exercise interventions in postmenopausal women on adiponectin and leptin levels. *J. Intern. Med.* **2013**, *274*, 163–175. [CrossRef]
38. Aldubayan, M.A.; Pigsborg, K.; Gormsen, S.M.; Serra, F.; Palou, M.; Galmés, S.; Palou-March, A.; Favari, C.; Wetzels, M.; Calleja, A.; et al. A double-blinded, randomized, parallel intervention to evaluate biomarker-based nutrition plans for weight loss: The PREVENTOMICS study. *Clin. Nutr.* **2022**, *41*, 1834–1844. [CrossRef]
39. Babatunde, O.A.; Adams, S.A.; Truman, S.; Sercy, E.; Murphy, A.E.; Khan, S.; Hurley, T.G.; Wirth, M.D.; Choi, S.K.; Johnson, H.; et al. The impact of a randomized dietary and physical activity intervention on chronic inflammation among obese African-American women. *Women Health* **2020**, *60*, 792–805. [CrossRef]

40. Campbell, K.L.; Foster-Schubert, K.E.; Alfano, C.M.; Wang, C.-C.; Wang, C.-Y.; Duggan, C.R.; Mason, C.; Imayama, I.; Kong, A.; Xiao, L.; et al. Reduced-calorie dietary weight loss, exercise, and sex hormones in postmenopausal women: Randomized controlled trial. *J. Clin. Oncol.* **2012**, *30*, 2314–2326. [CrossRef] [PubMed]
41. Claessens, M.; van Baak, M.A.; Monsheimer, S.; Saris, W.H. The effect of a low-fat, high-protein or high-carbohydrate ad libitum diet on weight loss maintenance and metabolic risk factors. *Int. J. Obes. (Lond.)* **2009**, *33*, 296–304. [PubMed]
42. Duggan, C.; Tapsoba, J.D.; Stanczyk, F.; Wang, C.Y.; Schubert, K.F.; McTiernan, A. Long-term weight loss maintenance, sex steroid hormones, and sex hormone-binding globulin. *Menopause* **2019**, *26*, 417–422.
43. Lopez-Legarrea, P.; de la Iglesia, R.; Abete, I.; Navas-Carretero, S.; Martinez, J.A.; Zulet, M.A. The protein type within a hypocaloric diet affects obesity-related inflammation: The RESMENA project. *Nutrition* **2014**, *30*, 424–429. [CrossRef]
44. Moszak, M.; Klupczynska, A.; Kanikowska, A.; Kokot, Z.; Zawada, A.; Grzymisławska, M.; Grzymisławski, M. The influence of a 3-week body mass reduction program on the metabolic parameters and free amino acid profiles in adult Polish people with obesity. *Adv. Clin. Exp. Med.* **2018**, *27*, 749–757. [CrossRef]
45. Pinto, A.M.; Bordoli, C.; Buckner, L.P.; Kim, C.; Kaplan, P.C.; Del Arenal, I.M.; Jeffcock, E.J.; Hall, W.L. Intermittent energy restriction is comparable to continuous energy restriction for cardiometabolic health in adults with central obesity: A randomized controlled trial; the Met-IER study. *Clin. Nutr.* **2020**, *39*, 1753–1763. [CrossRef]
46. Porter Starr, K.N.; Orenduff, M.; McDonald, S.R.; Mulder, H.; Sloane, R.; Pieper, C.F.; Bales, C.W. Influence of Weight Reduction and Enhanced Protein Intake on Biomarkers of Inflammation in Older Adults with Obesity. *J. Nutr. Gerontol. Geriatr.* **2019**, *38*, 33–49. [CrossRef]
47. Shah, K.; Armamento-Villareal, R.; Parimi, N.; Chode, S.; Sinacore, D.R.; Hilton, T.N.; Napoli, N.; Qualls, C.; Villareal, D.T. Exercise training in obese older adults prevents increase in bone turnover and attenuates decrease in hip bone mineral density induced by weight loss despite decline in bone-active hormones. *J. Bone Miner. Res.* **2011**, *26*, 2851–2859. [CrossRef] [PubMed]
48. Stolzenberg-Solomon, R.Z.; Falk, R.T.; Stanczyk, F.; Hoover, R.N.; Appel, L.J.; Ard, J.D.; Batch, B.C.; Coughlin, J.; Han, X.; Lien, L.F.; et al. Sex hormone changes during weight loss and maintenance in overweight and obese postmenopausal African-American and non-African-American women. *Breast Cancer Res.* **2012**, *14*, R141. [CrossRef]
49. Swora-Cwynar, E.; Kujawska-Łuczak, M.; Suliburska, J.; Reguła, J.; Kargulewicz, A.; Kręgielska-Narożna, M.; Marcinkowska, E.; Kanikowska, A.; Bielas, M.; Grzymisławski, M.; et al. The effects of a low-calorie diet or an isocaloric diet combined with metformin on sex hormones In obese women of child-bearing age. *Acta Sci. Pol. Technol. Aliment.* **2016**, *15*, 213–220. [CrossRef]
50. Ceccarini, G.; Pelosini, C.; Ferrari, F.; Magno, S.; Vitti, J.; Salvetti, G.; Moretto, C.; Marioni, A.; Buccianti, P.; Piaggi, P.; et al. Serum IGF-binding protein 2 (IGFBP-2) concentrations change early after gastric bypass bariatric surgery revealing a possible marker of leptin sensitivity in obese subjects. *Endocrine* **2019**, *65*, 86–93. [CrossRef] [PubMed]
51. Dalmas, E.; Rouault, C.; Abdennour, M.; Rovere, C.; Rizkalla, S.; Bar-Hen, A.; Nahon, J.-L.; Bouillot, J.-L.; Guerre-Millo, M.; Clément, K.; et al. Variations in circulating inflammatory factors are related to changes in calorie and carbohydrate intakes early in the course of surgery-induced weight reduction. *Am. J. Clin. Nutr.* **2011**, *94*, 450–458. [CrossRef] [PubMed]
52. Freitas, W.R., Jr.; Oliveira, L.V.F.; Perez, E.A.; Ilias, E.J.; Lottenberg, C.P.; Silva, A.S.; Urbano, J.J.; Oliveira, M.C.; Vieira, R.P.; Ribeiro-Alves, M.; et al. Systemic Inflammation in Severe Obese Patients Undergoing Surgery for Obesity and Weight-Related Diseases. *Obes. Surg.* **2018**, *28*, 1931–1942. [CrossRef]
53. Jacobsen, S.H.; Olesen, S.C.; Dirksen, C.; Jørgensen, N.B.; Bojsen-Møller, K.N.; Kielgast, U.; Worm, D.; Almdal, T.; Naver, L.S.; Hvolris, L.E.; et al. Changes in gastrointestinal hormone responses, insulin sensitivity, and beta-cell function within 2 weeks after gastric bypass in non-diabetic subjects. *Obes. Surg.* **2012**, *22*, 1084–1096. [CrossRef]
54. Kim, S.R.; Ene, G.E.V.; Simpson, A.; Gesink, D.; Ferguson, S.E. Acceptability of bariatric surgery in people with endometrial cancer and atypical hyperplasia: A qualitative study. *Gynecol. Oncol.* **2023**, *169*, 12–16. [CrossRef] [PubMed]
55. Kjær, M.M.; Madsbad, S.; Hougaard, D.M.; Cohen, A.S.; Nilas, L. The impact of gastric bypass surgery on sex hormones and menstrual cycles in premenopausal women. *Gynecol. Endocrinol.* **2017**, *33*, 160–163. [CrossRef] [PubMed]
56. Lima, M.M.O.; Pareja, J.C.; Alegre, S.M.; Geloneze, S.R.; Kahn, S.E.; Astiarraga, B.D.; Chaim, A.; Baracat, J.; Geloneze, B. Visceral fat resection in humans: Effect on insulin sensitivity, beta-cell function, adipokines, and inflammatory markers. *Obesity* **2013**, *21*, E182–E189. [CrossRef] [PubMed]
57. MacKintosh, M.L.; Derbyshire, A.E.; McVey, R.J.; Bolton, J.; Nickkho-Amiry, M.; Higgins, C.L.; Kamieniorz, M.; Pemberton, P.W.; Kirmani, B.H.; Ahmed, B.; et al. The impact of obesity and bariatric surgery on circulating and tissue biomarkers of endometrial cancer risk. *Int. J. Cancer* **2018**, *144*, 641–650. [CrossRef] [PubMed]
58. Moreira, S.M.B.P.; Bachi, A.L.L.; Jirjos, E.I.; Malheiros, C.A.; Vencio, S.; Alves, V.L.S.; Sousa, A.R.T.; Felipe, L.A.; Perez, E.A.; Lino, M.E.M.; et al. Roux-en-Y Gastric Bypass Improves Adiponectin to Leptin Ratio and Inflammatory Profile in Severely Obese Women with and without Metabolic Syndrome: A Randomized Controlled Trial. *Nutrients* **2023**, *15*, 3312. [CrossRef] [PubMed]
59. Moriconi, D.; Antonioli, L.; Masi, S.; Bellini, R.; Pellegrini, C.; Rebelos, E.; Taddei, S.; Nannipieri, M. Glomerular hyperfiltration in morbid obesity: Role of the inflammasome signalling. *Nephrology* **2022**, *27*, 673–680. [CrossRef]
60. Nikolic, M.; Boban, M.; Ljubicic, N.; Supanc, V.; Mirosevic, G.; Pezo Nikolic, B.; Krpan, R.; Posavec, L.; Zjacic-Rotkvic, V.; Bekavac-Beslin, M.; et al. Morbidly obese are ghrelin and leptin hyporesponders with lesser intragastric balloon treatment efficiency: Ghrelin and leptin changes in relation to obesity treatment. *Obes. Surg.* **2011**, *21*, 1597–1604. [CrossRef] [PubMed]

61. Sarwer, D.B.; Wadden, T.A.; Spitzer, J.C.; Mitchell, J.E.; Lancaster, K.; Courcoulas, A.; Gourash, W.; Rosen, R.C.; Christian, N.J. 4-Year Changes in Sex Hormones, Sexual Functioning, and Psychosocial Status in Women Who Underwent Bariatric Surgery. *Obes. Surg.* **2018**, *28*, 892–899. [CrossRef]
62. Tussing-Humphreys, L.; Pini, M.; Ponemone, V.; Braunschweig, C.; Fantuzzi, G. Suppressed cytokine production in whole blood cultures may be related to iron status and hepcidin and is partially corrected following weight reduction in morbidly obese pre-menopausal women. *Cytokine* **2011**, *53*, 201–206. [CrossRef] [PubMed]
63. Wojciechowska-Kulik, A.; Blus, E.; Kowalczyk, Z.; Baj, Z.; Majewska, E. The Effect of Noninvasive Bariatric Surgery on the Levels of Certain Adipokines and Atherosclerosis Risk Factors in Patients with Metabolic Syndrome. *J. Am. Coll. Nutr.* **2020**, *39*, 481–487. [CrossRef]
64. Dushay, J.; Gao, C.; Gopalakrishnan, G.S.; Crawley, M.; Mitten, E.K.; Wilker, E.; Mullington, J.; Maratos-Flier, E. Short-term exenatide treatment leads to significant weight loss in a subset of obese women without diabetes. *Diabetes Care* **2012**, *35*, 4–11. [CrossRef]
65. Gadde, K.M.; Allison, D.B.; Ryan, D.H.; Peterson, C.A.; Troupin, B.; Schwiers, M.L.; Day, W.W. Effects of low-dose, controlled-release, phentermine plus topiramate combination on weight and associated comorbidities in overweight and obese adults (CONQUER): A randomised, placebo-controlled, phase 3 trial. *Lancet* **2011**, *377*, 1341–1352. [CrossRef] [PubMed]
66. Garvey, W.T.; Batterham, R.L.; Bhatta, M.; Buscemi, S.; Christensen, L.N.; Frias, J.P.; Jódar, E.; Kandler, K.; Rigas, G.; Wadden, T.A.; et al. Two-year effects of semaglutide in adults with overweight or obesity: The STEP 5 trial. *Nat. Med.* **2022**, *28*, 2083–2091. [CrossRef]
67. Iepsen, E.W.; Lundgren, J.; Dirksen, C.; Jensen, J.-E.; Pedersen, O.; Hansen, T.; Madsbad, S.; Holst, J.J.; Torekov, S.S. Treatment with a GLP-1 receptor agonist diminishes the decrease in free plasma leptin during maintenance of weight loss. *Int. J. Obes. (Lond.)* **2015**, *39*, 834–841. [CrossRef]
68. Joo, N.S.; Kim, S.M.; Kim, K.M.; Kim, C.W.; Kim, B.T.; Lee, D.J. Changes of body weight and inflammatory markers after 12-week intervention trial: Results of a double-blind, placebo-control pilot study. *Yonsei Med. J.* **2011**, *52*, 242–248. [CrossRef] [PubMed]
69. Kim, D.D.; Krishnarajah, J.; Lillioja, S.; de Looze, F.; Marjason, J.; Proietto, J.; Shakib, S.; Stuckey, B.G.A.; Vath, J.E.; Hughes, T.E. Efficacy and safety of beloranib for weight loss in obese adults: A randomized controlled trial. *Diabetes Obes. Metab.* **2015**, *17*, 566–572. [CrossRef]
70. Sari, R.; Eray, E.; Ozdem, S.; Akbas, H.; Coban, E. Comparison of the effects of sibutramine versus sibutramine plus metformin in obese women. *Clin. Exp. Med.* **2010**, *10*, 179–184. [CrossRef]
71. Smith, S.R.; Weissman, N.J.; Anderson, C.M.; Sanchez, M.; Chuang, E.; Stubbe, S.; Bays, H.; Shanahan, W.R.; Behavioral Modification and Lorcaserin for Overweight and Obesity Management (BLOOM) Study Group. Multicenter, placebo-controlled trial of lorcaserin for weight management. *N. Engl. J. Med.* **2010**, *363*, 245–256. [CrossRef]
72. Zheng, W. Molecular Classification of Endometrial Cancer and the 2023 FIGO Staging: Exploring the Challenges and Opportunities for Pathologists. *Cancers* **2023**, *15*, 4101. [CrossRef] [PubMed]
73. Santoro, A.; Angelico, G.; Travaglino, A.; Inzani, F.; Arciuolo, D.; Valente, M.; D'alessandris, N.; Scaglione, G.; Fiorentino, V.; Raffone, A. New Pathological and Clinical Insights in Endometrial Cancer in View of the Updated ESGO/ESTRO/ESP Guidelines. *Cancers* **2021**, *13*, 2623. [CrossRef]
74. Schüler-Toprak, S.; Skrzypczak, M.; Gründker, C.; Ortmann, O.; Treeck, O. Role of Estrogen Receptor β, G-Protein Coupled Estrogen Receptor and Estrogen-Related Receptors in Endometrial and Ovarian Cancer. *Cancers* **2023**, *15*, 2845. [CrossRef]
75. Maliniak, M.L.; Cheriyan, A.M.; Sherman, M.E.; Liu, Y.; Gogineni, K.; Liu, J.; He, J.; Krishnamurti, U.; Miller-Kleinhenz, J.; Ashiqueali, R.; et al. Detection of crown-like structures in breast adipose tissue and clinical outcomes among African-American and White women with breast cancer. *Breast Cancer Res.* **2020**, *22*, 65. [CrossRef] [PubMed]
76. I'yasova, D.; Colbert, L.H.; Harris, T.B.; Newman, A.B.; Bauer, D.C.; Satterfield, S.; Kritchevsky, S.B. Circulating levels of inflammatory markers and cancer risk in the health aging and body composition cohort. *Cancer Epidemiol. Biomarkers Prev.* **2005**, *14*, 2413–2418.
77. Rustgi, V.K.; Li, Y.; Gupta, K.; Minacapelli, C.D.; Bhurwal, A.; Catalano, C.; Elsaid, M.I. Bariatric Surgery Reduces Cancer Risk in Adults with Nonalcoholic Fatty Liver Disease and Severe Obesity. *Gastroenterology* **2021**, *161*, 171–184.e10. [CrossRef] [PubMed]
78. O'Flanagan, C.H.; Bowers, L.W.; Hursting, S.D. A weighty problem: Metabolic perturbations and the obesity-cancer link. *Horm. Mol. Biol. Clin. Investig.* **2015**, *23*, 47–57. [CrossRef] [PubMed]
79. Garikapati, K.K.; Ammu, V.; Krishnamurthy, P.T.; Chintamaneni, P.K.; Pindiprolu, S. Type-II endometrial cancer: Role of adipokines. *Arch. Gynecol. Obstet.* **2019**, *300*, 239–249. [CrossRef] [PubMed]
80. Boron, D.; Nowakowski, R.; Grabarek, B.O.; Zmarzly, N.; Oplawski, M. Expression Pattern of Leptin and Its Receptors in Endometrioid Endometrial Cancer. *J. Clin. Med.* **2021**, *10*, 2787. [CrossRef]
81. Tao, Z.; Cheng, Z. Hormonal regulation of metabolism-recent lessons learned from insulin and estrogen. *Clin. Sci. (Lond.)* **2023**, *137*, 415–434. [CrossRef]
82. Yang, X.; Wang, J. The Role of Metabolic Syndrome in Endometrial Cancer: A Review. *Front. Oncol.* **2019**, *9*, 744. [CrossRef] [PubMed]

83. Barros, R.P.; Gustafsson, J.A. Estrogen receptors and the metabolic network. *Cell Metab.* **2011**, *14*, 289–299. [CrossRef]
84. Allard, J.E.; Maxwell, G.L. Race disparities between black and white women in the incidence, treatment, and prognosis of endometrial cancer. *Cancer Control.* **2009**, *16*, 53–56. [CrossRef] [PubMed]

cancers

MDPI

Article

Enhancing Progestin Therapy with a Glucagon-Like Peptide 1 Agonist for the Conservative Management of Endometrial Cancer

Andrea R. Hagemann [1], Ian S. Hagemann [1,2], David G. Mutch [1], Eric J. Devor [3], Paige K. Malmrose [4], Yuping Zhang [3], Abigail M. Morrison [3], Kristina W. Thiel [3,*,†] and Kimberly K. Leslie [5,*,†]

1 Division of Gynecologic Oncology, Department of Obstetrics and Gynecology, Washington University, and Alvin J. Siteman Cancer Center, St. Louis, MO 63110, USA; hagemanna@wustl.edu (A.R.H.); hagemani@wustl.edu (I.S.H.); mutchd@wustl.edu (D.G.M.)
2 Department of Pathology and Immunology, Washington University, St. Louis, MO 63110, USA
3 Department of Obstetrics and Gynecology and the Holden Comprehensive Cancer Center, University of Iowa, Iowa City, IA 52242, USA; eric-devor@uiowa.edu (E.J.D.); yuping-zhang@uiowa.edu (Y.Z.); abby-morrison@uiowa.edu (A.M.M.)
4 Department of Pathology, University of Illinois at Chicago, Chicago, IL 60612, USA; pmalmr2@uic.edu
5 Department of Internal Medicine and the University of New Mexico Comprehensive Cancer Center, University of New Mexico Health Sciences Center, Albuquerque, NM 87131, USA
* Correspondence: kristina-thiel@uiowa.edu (K.W.T.); kkleslie@salud.unm.edu (K.K.L.)
† These authors contributed equally to this work.

Simple Summary: Since obesity is a major risk factor for endometrial cancer, we explored a dual treatment strategy by combining a weight loss drug (semaglutide—a glucagon-like peptide-1 (GLP-1) agonist) with a progestin. Progestins are a safe and highly effective non-surgical treatment strategy for early-stage/grade endometrial cancer, but the response is often not sustained. Using cell lines and patient-derived preclinical models of endometrial cancer, we discovered an unexpected interplay between the GLP-1 receptor and progesterone receptor. Specifically, GLP-1 agonists increased expression of the progesterone receptor, and the combination of the two agents resulted in more pronounced cell death as compared to a single agent alone. As more patients are exploring GLP-1 agonists for weight management, these data make an important contribution to the field by suggesting that such drugs can also increase the efficacy of hormone therapy in endometrial cancer.

Abstract: Objective: Obesity is a major risk factor for endometrial cancer. In addition to hormone therapy with progestins, glucagon like peptide-1 receptor (GLP-1R) agonists such as semaglutide may be helpful to achieve weight loss during conservative treatment of endometrial hyperplasia or cancer. Methods: We theorized that the combination of semaglutide and the progestin levonorgestrel would be useful as a novel treatment or prevention regimen and tested this hypothesis using endometrial cancer cell lines and patient-derived organoids (PDOs). Results: Hec50, KLE, and Ishikawa endometrial cancer cells express GLP-1R, as determined by both qPCR and Western blotting, and GLP-1R agonist treatment induces GLP-1R mRNA transcription through positive feedback mechanisms in cell models. PDOs from six individuals with grade 1 endometrial carcinomas were treated with progesterone, levonorgestrel, semaglutide, or levonorgestrel + semaglutide. Multiple models demonstrated a significant reduction in viability in response to combinatorial treatment, and the effect was noted in models from both PR high- and PR low-expressing tumors. Most interesting was the induction not only of the membrane GLP-1R with treatment, but also the significant upregulation of nuclear and membrane progesterone receptors—PR and PGRMC1/2, respectively—indicating a potential positive feedback loop between semaglutide and progestins such as levonorgestrel. Conclusion: In summary, we identify synergistic molecular cross-talk between the GLP-1R and steroid hormone receptor pathways, with the potential to enhance the anticancer activity of levonorgestrel when combined with semaglutide.

Keywords: endometrial cancer; hormone therapy; glucagon-like peptide 1 receptor agonists; conservative management; progesterone

1. Introduction

Although obesity is a strong risk factor for many cancers, endometrial cancer (EC) is the malignancy most commonly associated with obesity [1]. Up to 90% of the 65,000 women diagnosed with endometrial cancer each year in the US are diagnosed as overweight or obese, and up to 60% of endometrial cancer cases are attributed to obesity [2]. Furthermore, predictive models show that anovulation, nulliparity, polycystic ovarian syndrome, insulin resistance, and diabetes—all correlates of obesity—increase the risk of atypical endometrial hyperplasia (AEH) in premenopausal women by six- to ten-fold [3,4]. Thus, as the obesity rate in the US has increased, so too has the rate of endometrial cancer [5].

For postmenopausal women or women not desiring fertility preservation, hysterectomy is highly effective in preventing endometrial cancer development in patients with atypical endometrial hyperplasia (AEH) and is frequently curable for women with early-stage/grade endometrial cancer (EC). However, for premenopausal women with AEH and/or early-stage/grade EC, removing the uterus would preclude pregnancy, whereas delaying hysterectomy could allow cancer to develop. Thus, such patients are often offered progestin therapy, usually via a levonorgestrel intrauterine device (LNG-IUD) or oral progestational agents. These patients often choose progestin therapy because of their desire for fertility. Although progestins are effective, the risk of relapse is as high as 50% within 6 months [6,7]. Thus, better strategies to prevent endometrial cancer in premenopausal women with AEH who desire pregnancy are needed.

In a normal cycling female, the endometrial lining of the uterus is dynamically regulated by a balance of estrogen and progesterone throughout the menstrual cycle. In the normal endometrium, estrogen provides a pro-proliferative signal, whereas progesterone counters the effects of estrogen to promote differentiation of the endometrial epithelial cells [4]. An estimated 80% of EC diagnoses are attributed to an imbalance in estrogen and progesterone—either too much estrogen or insufficient progesterone. Estrogen acting through estrogen receptors (ERs) α and β plays a key role in EC through transcriptional regulation of pro-growth signaling pathways, including recruitment of specific co-activators to estrogen-responsive elements. In addition, estrogen has non-genomic actions that contribute to oncogenesis, such as activation of the PI3K and MAPK pathways that are often aberrantly activated in EC [8]. In contrast, progesterone-mediated activation of the progesterone receptor (PR) serves as a tumor suppressive signal in the endometrium, including induction of differentiation, cell cycle arrest, and, potentially, apoptosis along with inhibition of inflammation and invasion [9–17]. Thus, targeting this hormone imbalance is a potentially safe strategy to prevent and/or treat EC.

Synthetic progesterone mimetics, termed progestins, have been explored for treatment of EC (reviewed in references [9,18,19]). Like progesterone, progestins are hypothesized to promote differentiation of tumor cells and potentially block cell division and/or induce apoptosis. There is a trend towards improved response to progestins with expression of PR. For example, the overall response rate in patients with expression of PR by immunohistochemistry (IHC) is 72%, whereas patients with no detectable expression of PR in the tumor epithelial cells have a response rate of 12% [20]. Unfortunately, a number of patients experience relapse after their initial response to progestin, indicating that further optimization of progestin therapy is necessary to attain a durable response in EC.

Progestins are synthesized from progesterone and androgen backbones. For example, levonorgestrel is a derivative of 17α-ethynyl-19-nortestosterone-18-methylestrane, and levonorgestrel binds to both PR and other steroid hormone receptors, including the androgen receptor (AR). Classic nuclear receptors are well-described transcription factors that, upon ligand binding, dimerize, translocate into the nucleus, and bind to hormone-responsive gene promoters. In addition to ligand binding, nuclear receptors can be activated by phosphorylation in a ligand-dependent or -independent manner. However, in addition to nuclear receptors, progesterone and progestins also activate more recently identified membrane receptors PGRMC1 and 2, which are activated in the cell membrane by ligand binding and enhance downstream signaling through molecules such as Src and Erk [21]. There is a potential synergy between Src and Erk signaling induced by PGRMC1 and 2, as these signaling molecules both phosphorylate PRB, which is the PR receptor isoform that resides in the cytoplasm in its inactive, unphosphorylated state. However, upon phosphorylation, PRB is bound by ligand and translates into the nucleus, thereby allowing PR gene transcription. Importantly, GLP-1R agonists also bind to membrane receptors that enhance Src and Erk signaling, and GLP-1R is reported to interact with and co-activate with PGRMC1, suggesting a synergistic effect on PR-mediated EC cell growth inhibition. Hence, in this study, we theorized that the combination of a GLP-1R agonist and a progestin would be useful as a novel treatment or prevention regimen and tested this hypothesis in preclinical models using EC cell lines and patient-derived organoids.

2. Methods

Drugs: Semaglutide, liragutide, and levonorgestrel were purchased from Selleck Chemicals and reconstituted in DMSO.

Cell lines: KLE endometrial cancer cells were purchased from the American Type Culture Collection (CRL-1622), while Ishikawa and Hec50 endometrial cancer cells were gifts from Dr. Erlio Gurpide (New York University). The origins and validation of these cell lines as models of endometrial cancer have been extensively reported [22,23]. Cell line identities were verified via CODIS DNA profiling at BioSynthesis, Inc., Lewisville, TX, USA. Ishikawa and Hec50co cells were cultured in DMEM supplemented with 10% fetal bovine serum (FBS) and 1% antibiotic [penicillin (p)/streptomycin (s)] (all from Thermo Fisher Scientific, Inc., Waltham, MA, USA). KLE cells were cultured in RPMI 1640 (Thermo Fisher Scientific, Inc.) supplemented with 10% FBS and 1% p/s.

Western blotting: Expression of GLP-1R was assessed in 20 μg of protein collected from whole cell lysates of Hec50, KLE, and Ishikawa cells by Western blotting. Protein concentrations were determined using the BCA Protein Assay Kit (Thermo Fisher Scientific, Inc.) per the manufacturer's instructions. Proteins were separated by SDS-PAGE with 10% acrylamide followed by transfer to nitrocellulose membrane. After blocking in 5% milk in Tris-buffered saline buffer (TBST), membranes were exposed to primary antibodies against GLP-1R (# bs-1559R, Bioss Antibodies, Woburn, MA, USA, 1:1000 dilution in 5% BSA in TBST, incubated overnight at 4 °C) or β-actin (#A1978, Sigma-Aldrich, St. Louis, MO, USA, 1:5000 dilution in 5% milk in TBST, incubated 1 hr. at room temperature), followed by exposure to either anti-rabbit-HRP for GLP-1R or anti-mouse HRP for β-actin (each secondary antibody was at a 1:10,000 dilution in 5% BSA in TBST for 1 h). Bands were visualized using HRP-conjugated chemiluminescence reagent (SuperSignal West Pico PLUS, Thermo Fisher Scientific, Inc.). Expression of GLP-1R in each cell line by densitometry (NIH Image J version 1.54m) was calculated relative to Ishikawa cells after normalization to β-actin loading control.

RNA extraction and quantitative polymerase chain reaction: Total cellular RNA was isolated using the RNeasy Plus Kit (QIAGEN, Germantown, MD, USA), and 500 ng of RNA was subjected to reverse transcription using SuperScript III Reverse Transcriptase and an oligo (dT) primer (Thermo Fisher Scientific, Inc.) to yield cDNA, followed by amplification with Power SYBR Green (Thermo Fisher Scientific, Inc.) using the Applied Biosystems Model 7900HT in the Genomics Division of the Iowa Institute of Human Genetics (IIHG). The following genes were assessed by quantitative PCR (qPCR): GLP1R, PGRMC1, PGRMC2, FOXO1, ER, AR, GR, MR, and PR, with 18S rRNA used as the housekeeping gene. The PrimerQuest primer design tool (IDT, Coralville, IA, USA) was used to design primer sequences. The sequences of forward and reverse primers are as follows:

Gene		**5′-SEQUENCE-3′**	**Tm**
GLP1R	For	CAGGCTCGTTCGTGAATGT	55.4
	Rev	GCGATAACCAGAGCAGAGAAG	55.3
PGRMC1	For	AGCAGGAGACTCTGAGTGACTG	58.0
	Rev	CCTCATCTGCGTACACAGTGGG	56.7
PGRMC2	For	CGTGACCAAAGGCAGCAAGTTC	59.1
	Rev	CTCTAAGTGCATCTTTATCTAGGC	52.8
FOXO1	For	CTACGAGTGGATGGTCAAGAGC	57.1
	Rev	CCAGTTCCTTCATTCTGCACACG	58.4
ERα	For	TGGGCTTACTGACCAACCTG	57.1
	Rev	CCTGATCATGGAGGGTCAAA	54.4
AR	For	ATGGTCCCTGGCAGTCTCCAAA	60.7
	Rev	ATGGTGAGCAGAGTGCCCTATC	58.8
GR	For	GGAATAGGTGCCAAGGATCTGG	57.4
	Rev	GCTTACATCTGGTCTCATGCTGG	57.7
MR	For	AAATCACACGGCGACCTGTCGT	61.4
	Rev	ATGGCATCCTGAAGCCTCATCC	59.4
PGR	For	ATCCTACAAACACGTCAGTGGGCA	60.5
	Rev	ACTGGGTTTGACTTCGTAGCCCTT	60.3
18S rRNA	For	AACCTTTCGATGGTAGTCGCCG	59.4
	Rev	CCTTGGATGTGGTAGCCGTTT	57.6

Data were calculated as the fold change in expression relative to untreated samples at the same time point using the ΔΔCt method.

Immunohistochemistry (IHC): ER and PR expression in primary tissues from which the patient-derived organoid models were generated was assessed by IHC by the Comparative Pathology Laboratory at the University of Iowa (Iowa City, IA, USA). IHC was performed on sequential sections using the following primary antibodies from Dako (Agilent, Santa Clara, CA, USA): PR (PgR636); ER (M7047). Secondary antibody was goat-anti-mouse for both PR and ER. Stained sections were incubated in DAB chromogen substrate and counterstained with hematoxylin prior to mounting with coverslips. Two independent reviewers applied the modified H-score (range of 0–300) to score staining. Specifically, the modified H-score is determined by multiplying the intensity of tumor staining (0–3) by the percent of cells with staining (0–100%). Positive control tissue was scored as a 3+ for intensity and a 100% for percentage. To ensure tissue integrity, the presence of signal was assessed in the stroma and myometrium. Modified H-scores from each reviewer (N = 2 reviewers) were averaged for each slide to yield the average modified H-score. Representative images were acquired with an EVOS inverted bright field microscope at 20× magnification.

Patient-derived organoids: Patient-derived organoids (PDOs) were isolated in a study approved by the University of Iowa's IRB, protocol #201809807, from patients with grade 1 endometrial cancer undergoing surgical resection and established based upon our previously reported methods [17,24,25]. Organoids with a diameter of >50 μm typically developed within 3–14 days after tumor processing. To assess drug sensitivity, organoid harvesting solution (Cultrex, R&D Systems, Minneapolis, MN, USA) was used to collect organoids. After digestion into single-cell suspensions using TrypLE Express (Gibco, Waltham, MA, USA), cells were resuspended in AdDE+++ medium supplemented with 10% Matrigel. Next, 10,000 cells were seeded into each well of a 96-well ultra-low attachment white plate (U-bottom). Organoids were treated with 100 nM of progesterone (P4), 100 nM of levonorgestrel (L), 100 nM of semaglutide (S), or L+S for 72 h in organoid culture media (29, 30) containing 1 nM of estradiol as an additive to the media. Viability was assessed using CellTiter-Glo 3D reagent (Promega, Madison, WI, USA) per the manufacturer's protocol; luminescence was determined with a Gen5 Microplate Reader (BioTek, Agilent, Winooski, VT, USA). All the tests were conducted in triplicate. Data were normalized to control, which was set at 100% viability.

Statistical analyses: All data were analyzed with GraphPad Prism version 10. To assess statistical significance, we used either Student's *t*-test or one-way ANOVA with Tukey's multiple comparison test.

3. Results

3.1. GLP-1R Expression in Endometrial Cancer Cell Models

As previously reported, GLP-1Rs are expressed in the endometrium, both in malignant and non-malignant tissues. Kanda et al. reported expression of GLP-1R in Ishikawa endometrial cancer cells, as well as in over 100 tissue samples from endometrial tumors [26]. We confirmed that Hec50, KLE, and Ishikawa endometrial cancer cell models also express GLP-1Rs by Western blotting (Figures 1A and S1). Of interest, expression of GLP-1R was consistent over multiple cell lines representing endometrioid (Ishikawa, KLE) and serous (Hec50) models [22,23].

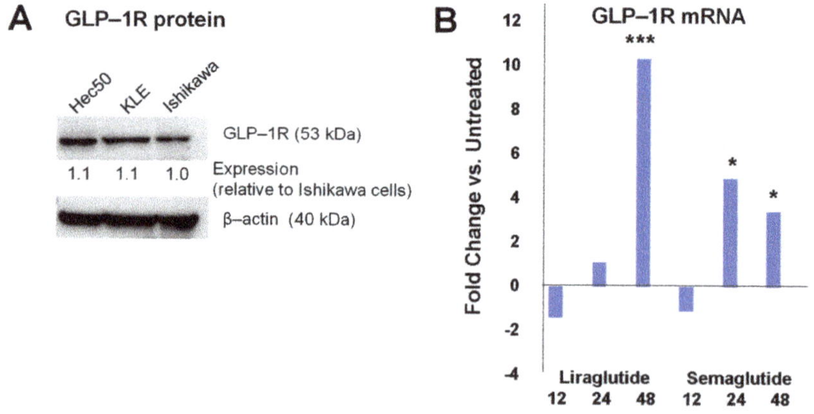

Figure 1. GLP-1Rs are expressed in endometrial cancer cells, and transcription is induced in response to GLP-1R agonist treatment. (**A**): Western blotting confirms protein expression of GLP-1R in multiple endometrial cancer cell lines. Expression was determined by densitometry and calculated relative to Ishikawa cells after normalization to the β-actin loading control. (**B**): qPCR for GLP receptor transcripts in Ishikawa endometrial cancer cells demonstrates expression, as well as significant upregulation of this gene in response to agonists liraglutide and semaglutide at 24 and 48 h after treatment. * $p \leq 0.05$; *** $p \leq 0.001$ vs. untreated control at the same timepoint.

Since our goal is to assess the combination of a GLP-1R agonist with a progestin, in subsequent experiments, we focused on Ishikawa cells, which are endometrial cancer cells with PR expression [22,23]. GLP-1R transcription is significantly induced upon treatment of Ishikawa endometrial cancer cells with GLP-1R agonists liraglutide or semaglutide beginning at 24 h, with continued induction through 48 h (Figure 1B).

3.2. GLP-1R Agonists and Levonorgestrel Enhance Hormone Signaling Through the Upregulation of Steroid Hormone Receptors and Other Pathway Genes

As shown in Figure 1, the induction of GLP-1R in response to agonist binding predicts a feed-forward loop of activity for agents such as liraglutide and semaglutide in endometrial cancer cells. Furthermore, the impact of both semaglutide and levonorgestrel on the expression of genes comprising the hormonal response pathway is substantial (Figure 2). Levonorgestrel induces GLP-1R, PGRMC1, ER, AR, MR, and PR. In the case of ER and MR, the magnitude is 40- to 95-fold. The impact on ER, AR, MR, and PR signaling is potentially enhanced given the induction of those genes. This portends the synergistic activation of the pathway when both drugs are used together.

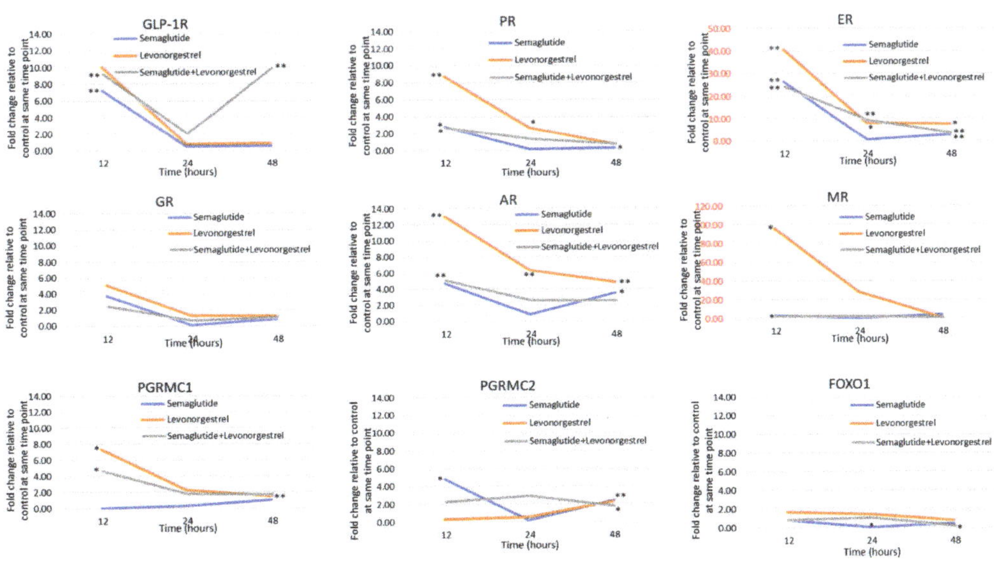

Figure 2. qPCR for the following mRNAs: GLP-1R, PGRMC1, PGRMC2, FOXO1, ER, AR, GR, MR, and PR following treatment of Ishikawa cells for 12, 24, and 48 h with semaglutide (blue line), levonorgestrel (orange line), or semaglutide + levonorgestrel (gray line). Fold change in expression was calculated relative to time-matched controls (DMSO vehicle) after correcting for housekeeping gene (18S rRNA) expression using the ΔΔCt method. * $p < 0.05$; ** $p < 0.01$ vs. time-matched control by *t*-test.

3.3. Effects of GLP-1 Agonist in Combination with Levonorgestrel on Growth of Patient-Derived Organoid Models of Endometrial Cancer

To validate the potential for benefit of the combination of levonorgestrel with semaglutide in endometrial cancer treatment, we next performed studies using patient-derived models of early-stage/grade endometrial cancer. These cases are typically associated with elevated BMI and historically recognized as "Type I" endometrial cancer [27,28]. Specifically, we studied the impact of levonorgestrel, semaglutide, or the combination on cell viability in a short-term experiment in six independent patient-derived organoids (PDOs) isolated after hysterectomy from women with early-stage, grade I endometrial cancer (Table 1).

Table 1. Clinical and demographic characteristics of endometrial cancer patients from which patient-derived organoid models were created. Disease status was determined at ~3 years after primary cytoreductive surgery. NED: no evidence of disease. Estrogen receptor (ER) and progesterone receptor (PR) were assessed by immunohistochemistry (IHC) by two independent reviewers and calculated as the modified H-score as described in Materials and Methods. H-scores are the average of two independent reviewers blinded to sample identify.

ID	Age	Race/Ethnicity	Diagnosis (Endometrial Adenocarcinoma)	Primary Adjuvant Treatment	Disease Status (as of 11/2024)	ER IHC	PR IHC
UBC 7282	68	White Non-Hispanic	Stage IA/Grade 1	Clinical surveillance	Dead (metastatic cancer; cannot rule out endometrial cancer as primary site of origin)	170	300
UBC 7661	69	White Non-Hispanic	Stage IA/Grade 1	Clinical surveillance	NED	185	165
UBC 7104	66	White Non-Hispanic	Stage IA/Grade 1	Clinical surveillance	NED	70	105
UBC 7325	34	White Non-Hispanic	Stage IA/Grade 1	Clinical surveillance	NED	200	100
UBC 7067	69	White Non-Hispanic	Stage 1B/Grade 1	Vaginal brachytherapy	NED	255	45
UBC 7200	45	White Non-Hispanic	Stage IA/Grade 1	Clinical surveillance	NED	240	30

We first established expression of ER and PR in the primary tumors from which the PDO models were created (matched specimen, Figure 3). Of note, there is heterogeneity in steroid hormone receptor expression both amongst tumors and intratumorally. PR staining is primarily nuclear in all patient samples, and in some models (e.g., UBC 7104), we observed robust stromal staining of PR in addition to expression in the glandular structures.

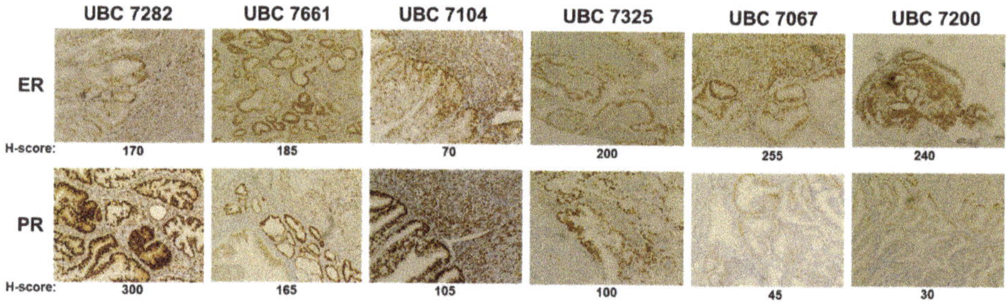

Figure 3. Heterogeneity in expression of estrogen receptor (ER) and progesterone receptor (PR) in early-stage/grade endometrial tumors. All images are at 20× magnification. The modified H-score for each specimen is noted below each image.

We next assessed the impact of progestin treatment in the absence or presence of a GLP-1R agonist on PDO growth, with cell viability as the endpoint. In most cases, there was a significant reduction in cell viability by 72 h in response to exposure to 100 nM of semaglutide + 100 nM of levonorgestrel (Figure 4). Responsiveness was noted in PDOs obtained from tumors expressing relatively high PR, as well as tumors with more modest PR expression as defined by the H-score. These data confirm the activity of semaglutide in endometrial cells, as well as its potential to further reduce cell viability in the presence of levonorgestrel. The activity of this combination in some models obtained from tumors with low PR expression is of interest. Given the induction of membrane and nuclear PR expression along with other steroid hormone receptors in response to semaglutide + levonorgestrel (Figure 2), hormone responsiveness is potentially magnified even in cases that initially express low PR levels and would be considered to be hormone therapy-unresponsive using a progestin alone.

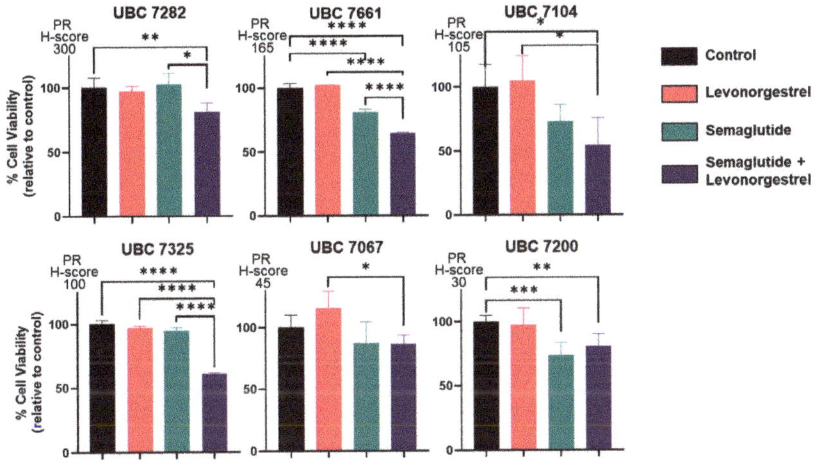

Figure 4. Semaglutide + levonorgestrel inhibits the growth of endometrial cancer patient-derived organoids (PDOs). PDOs from Grade 1 endometrial carcinomas were treated with progesterone (P4), levonorgestrel (L), semaglutide (S), or L+S. All drug concentrations are 100 nM. Cells were treated for 72 h. PR expression was determined by IHC of the primary tumor specimens from which the PDOs were generated (see Figure 3); models are arranged in order of highest to lowest PR expression. Control: vehicle (DMSO). * $p < 0.05$, ** $p < 0.01$, *** $p < 0.001$, or **** $p < 0.0001$ by one-way ANOVA.

4. Discussion

Obesity is a risk factor for many types of cancer and a key driver in the hormonal imbalance underlying development of EC. Indeed, obesity is not only associated with increased risk of developing cancer but also with higher cancer-related mortality and death due to other comorbidities [5,6]. We now appreciate that the increase in EC incidence and mortality is driven in part by the obesity epidemic, and the healthcare burden of EC will only continue to increase if we cannot simultaneously combat obesity. Studies into the mechanisms by which obesity increases risk of EC have identified abnormal hormone levels, in particular, of estrogen, as a key driver. Specifically, androgens are converted to estrone and estradiol in adipocytes by aromatase. There is a clear link between higher aromatase levels and increased BMI in postmenopausal women [7,29]; for example, as compared to postmenopausal women with BMI in the normal range, obese age-matched women have a 40% increase in circulating levels of both estrone and estradiol [7]. In addition to excess estrogen, a reduction in circulating progesterone also occurs in obesity and can contribute to the development and/or progression of EC, such as in obese anovulatory premenopausal patients. In this case, the lack of progesterone due to anovulation mimics polycystic ovarian syndrome (PCOS), a well-known risk factor for future EC development. In addition to a decrease in the ligand, changes in expression and activity of PR are also dysregulated in EC. For example, our group has identified several epigenetic modifications (e.g., miRNAs and DNA methylation) underlying loss of PR expression in EC [30].

Based on the link between obesity and AEH/EC, we hypothesized that the combination of a GLP-1R agonist and levonorgestrel would be useful as a novel treatment or prevention regimen and tested this hypothesis in preclinical models using EC cell lines and patient-derived organoids. In these studies, we demonstrated the expression of GLP-1R in our models, which is highly relevant to the clinical situation, because GLP-1R is known to be expressed in the normal and malignant endometrium. In line with our findings, Kanda et al. reported expression of GLP-1R in Ishikawa endometrial cancer cells, as well as in over 100 tissue samples from endometrial tumors [26]. We now report that GLP-1R agonists, when combined with progestins such as levonorgestrel, have the potential to enhance progestin therapy for AEH and EC as shown in EC cell and PDO models. The treatment-related transcriptional upregulation of receptors and components of the GLP-1R/PR signaling cascade sheds light on a possible mechanism of synergy as depicted in Figure 5. Specifically, the membrane progesterone receptor component PGRMC1 has been reported to cross-talk with and enhance GLP-1-induced insulin secretion in a rat insulinoma cell line. Moreover, PGRMC1 has been shown to increase cAMP production, as well as signaling by EGFR and subsequent downstream PIK3-induced insulin secretion. The role of the membrane progesterone receptor is less defined, though it has been shown to activate PI3K-mediated survival in breast cancer, as reviewed by Camacho-Arroyo and colleagues [31]. Thus, signaling downstream of GLP-1R has the potential to create synergy between progestins and GLP-1R agonists [32]. As diagrammed in Figure 5, our data indicate that in response to semaglutide and liraglutide, the expression of GLP-1R and steroid hormone receptors, including nuclear PR, is significantly induced in endometrial cancer cell models. We hypothesize that this induced receptor expression creates a feed-forward loop of activity for agents, such as liraglutide and semaglutide, that may enhance progestin therapy. In endometrial cancer organoids, the addition of levonorgestrel to semaglutide resulted in reduced cell viability compared to either agent alone. Importantly, the impact was noted in organoids from tumors expressing both high and low levels of nuclear PR, indicating that semaglutide treatment has the potential to enhance PR activity even in tumors considered to be less sensitive to hormone therapy.

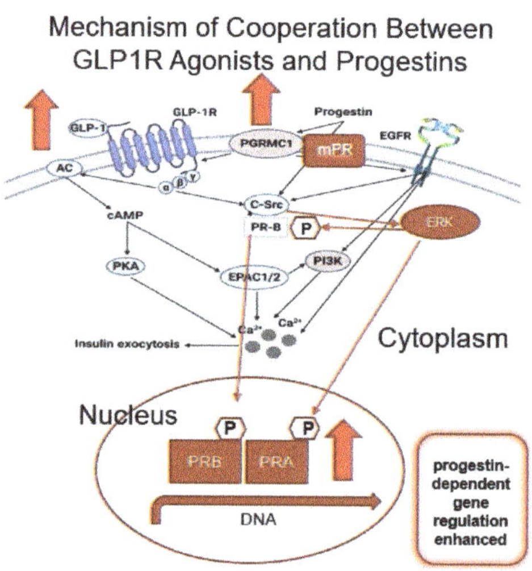

Figure 5. A proposed mechanism of cooperation between GLP-1R agonists and progestins. GLP-1R agonists and progestins bind to membrane receptors, including the 7-transmembrane GLP-1R and PGRMC1. It is proposed that PGRMC1 enhances the activation of GLP-1R, which signals downstream through C-Src and ERK to phosphorylate and activate PRB. That phosphorylation event induces PRB translocation to the nucleus and ligand binding, allowing activated PRB and PRA to enhance progestin-dependent gene regulation and endometrial differentiation. Activated GLP-1R also enhances cAMP effects, inducing insulin exocytosis. Red arrows denote key factors that are increased at the mRNA level by the combination of semaglutide and levonorgestrel in this study.

5. Conclusions

In summary, these studies highlight the potential usefulness of adding GLP-1R agonists to progestin therapy for AEH and EC. In addition to the obvious benefit of weight loss associated with GLP-1R agonist treatment, we show that GLP-1R agonists enhance the impact of progestin therapy, potentially through inducing hormone receptors and molecular cross-talk between GLP-1R, PGRMC1, and nuclear PRs. These findings support future clinical use of GLP-1R agonists in addition to progestins for women with AEH and EC. We propose that GLP- 1R agonists enhance progestin effectiveness, and progestins also enhance GLP-1R activation, resulting in the amplification of the effects of both agents.

Supplementary Materials: The following supporting information can be downloaded at: https://www.mdpi.com/article/10.3390/cancers17040598/s1, Figure S1: Western blot for GLP-1R in EC cell lines.

Author Contributions: Conceptualization, A.R.H., I.S.H., D.G.M., K.W.T., and K.K.L.; methodology, E.J.D.; validation, K.K.L.; formal analysis, E.J.D. and K.W.T.; investigation, E.J.D., P.K.M., Y.Z., and A.M.M.; resources, A.R.H., D.G.M., and K.K.L.; writing—original draft preparation, K.W.T. and K.K.L.; writing—review and editing, A.R.H., I.S.H., D.G.M., and A.M.M.; supervision, D.G.M., K.W.T., and K.K.L.; visualization, Y.Z.; supervision, K.W.T. and K.K.L.; funding acquisition, K.K.L. All authors have read and agreed to the published version of the manuscript.

Funding: This study was supported by grants from: the Department of Defense: CDMRP/GRANT CA2220729, Progestin Therapy for Endometrial Cancer (to K.K.L. and K.W.T.); the Department of Defense: CDMRP/GRANT OC190352, Converting Mutant p53 into an Actionable Target for the Treatment of Ovarian Cancer (to K.K.L.); the Department of Defense: CDMRP/GRANT 13461923, Treating Advanced Uterine Endometrial Cancers by Reactivating p53 (to K.K.L.); the National Institutes of Health: R01 5CA99908-22 Targeted Therapy for Endometrial Cancer (to K.K.L.); the National Institutes of Health: P50CA265793, Route 66 Endometrial Cancer SPORE (to K.K.L. and D.G.M.); the National Institutes of Health: the University of New Mexico Comprehensive Cancer Center P30CA118100; the National Institutes of Health K22 CA263783 Attacking Aggressive p53 Mutants in Gynecologic Cancer (to K.W.T.); the National Institutes of Health: the University of Iowa Comprehensive Cancer Center P30 CA086862.

Institutional Review Board Statement: The study was conducted in accordance with the Declaration of Helsinki and approved by the Institutional Review Board of University of Iowa (protocol code 201809807 approved on 5 January 2022).

Informed Consent Statement: Under IRB protocol 201809807, informed consent was obtained from all subjects involved in the study.

Data Availability Statement: All data for this manuscript are included in the primary manuscript or Supplemental Data.

Conflicts of Interest: The authors declare no conflicts of interest.

References

1. Lauby-Secretan, B.; Scoccianti, C.; Loomis, D.; Grosse, Y.; Bianchini, F.; Straif, K.; International Agency for Research on Cancer Handbook Working Group. Body fatness and cancer—Viewpoint of the IARC working group. *N. Engl. J. Med.* **2016**, *375*, 794–798. [CrossRef] [PubMed]
2. Renehan, A.G.; Tyson, M.; Egger, M.; Heller, R.F.; Zwahlen, M. Body-mass index and incidence of cancer: A systematic review and meta-analysis of prospective observational studies. *Lancet* **2008**, *371*, 569–578. [CrossRef] [PubMed]
3. Wise, M.R.; Gill, P.; Lensen, S.; Thompson, J.M.; Farquhar, C.M. Body mass index trumps age in decision for endometrial biopsy: Cohort study of symptomatic premenopausal women. *Am. J. Obstet. Gynecol.* **2016**, *215*, 598.e1–598.e8. [CrossRef]
4. Wise, M.R.; Jordan, V.; Lagas, A.; Showell, M.; Wong, N.; Lensen, S.; Farquhar, C.M. Obesity and endometrial hyperplasia and cancer in premenopausal women: A systematic review. *Am. J. Obstet. Gynecol.* **2016**, *214*, e681–e689. e617. [CrossRef] [PubMed]
5. Siegel, R.L.; Miller, K.D.; Fuchs, H.E.; Jemal, A. Cancer statistics, 2021. *CA Cancer. J. Clin.* **2021**, *71*, 7–33. [CrossRef] [PubMed]

6. Orbo, A.; Arnes, M.; Vereide, A.B.; Straume, B. Relapse risk of endometrial hyperplasia after treatment with the levonorgestrel-impregnated intrauterine system or oral progestogens. *BJOG: Int. J. Obstet. Gynaecol.* **2016**, *123*, 1512–1519. [CrossRef]
7. Chen, J.; Cao, D.; Yang, J.; Yu, M.; Zhou, H.; Cheng, N.; Wang, J.; Zhang, Y.; Peng, P.; Shen, K. Fertility-sparing treatment for endometrial cancer or atypical endometrial hyperplasia patients with obesity. *Front. Oncol.* **2022**, *12*, 812346. [CrossRef]
8. Thiebaut, C.; Vlaeminck-Guillem, V.; Tredan, O.; Poulard, C.; Le Romancer, M. Non-genomic signaling of steroid receptors in cancer. *Mol. Cell Endocrinol.* **2021**, *538*, 111453. [CrossRef]
9. Yang, S.; Thiel, K.W.; Leslie, K.K. Progesterone: The ultimate endometrial tumor suppressor. *Trends Endocrinol. Metab.* **2011**, *22*, 145–152. [CrossRef]
10. Carlson, M.J.; Thiel, K.W.; Yang, S.; Leslie, K.K. Catch it before it kills: Progesterone, obesity, and the prevention of endometrial cancer. *Discov. Med.* **2012**, *14*, 215–222.
11. Dai, D.; Litman, E.S.; Schonteich, E.; Leslie, K.K. Progesterone regulation of Activating pProtein-1 transcriptional activity: A possible mechanism of progesterone inhibition of endometrial cancer cell growth. *J. Steroid. Biochem. Mol. Biol.* **2003**, *87*, 123–131. [CrossRef] [PubMed]
12. Dai, D.; Wolf, D.M.; Litman, E.S.; White, M.J.; Leslie, K.K. Progesterone inhibits human endometrial cancer cell growth and invasiveness: Down-regulation of cellular adhesion molecules through progesterone B receptors. *Cancer Res.* **2002**, *62*, 881–886.
13. Davies, S.; Dai, D.; Feldman, I.; Pickett, G.; Leslie, K.K. Identification of a novel mechanism of NF-kappaB inactivation by progesterone through progesterone receptors in Hec50co poorly differentiated endometrial cancer cells: Induction of a20 and abin-2. *Gynecol. Oncol.* **2004**, *94*, 463–470. [CrossRef] [PubMed]
14. Kumar, N.S.; Richer, J.; Owen, G.; Litman, E.; Horwitz, K.B.; Leslie, K.K. Selective down-regulation of progesterone receptor isoform B in poorly differentiated human endometrial cancer cells: Implications for unopposed estrogen action. *Cancer Res.* **1998**, *58*, 1860–1865. [PubMed]
15. Leslie, K.K.; Thiel, K.W.; Yang, S. Endometrial cancer: Potential treatment and prevention with progestin-containing intrauterine devices. *Obstet. Gynecol.* **2012**, *119*, 419–420. [CrossRef]
16. Singh, M.; Zaino, R.J.; Filiaci, V.J.; Leslie, K.K. Relationship of estrogen and progesterone receptors to clinical outcome in metastatic endometrial carcinoma: A Gynecologic Oncology rGoup study. *Gynecol. Oncol.* **2007**, *106*, 325–333. [CrossRef]
17. Thiel, K.W.; Newtson, A.M.; Devor, E.J.; Zhang, Y.; Malmrose, P.K.; Bi, J.; Losh, H.A.; Davies, S.; Smith, L.E.; Padilla, J.; et al. Global expression analysis of endometrial cancer cells in response to progesterone identifies new therapeutic targets. *J. Steroid. Biochem. Mol. Biol.* **2023**, *234*, 106399. [CrossRef]
18. Yang, S.; Thiel, K.W.; De Geest, K.; Leslie, K.K. Endometrial cancer: Reviving progesterone therapy in the molecular age. *Discov. Med.* **2011**, *12*, 205–212.
19. Kim, J.J.; Chapman-Davis, E. Role of progesterone in endometrial cancer. *Semin. Reprod. Med.* **2010**, *28*, 81–90. [CrossRef] [PubMed]
20. Ehrlich, C.E.; Young, P.C.; Stehman, F.B.; Sutton, G.P.; Alford, W.M. Steroid receptors and clinical outcome in patients with adenocarcinoma of the endometrium. *Am. J. Obstet. Gynecol.* **1988**, *158*, 796–807. [CrossRef]
21. Garcia-Saenz, M.; Ibarra-Salce, R.; Pozos-Varela, F.J.; Mena-Ureta, T.S.; Flores-Villagomez, S.; Santana-Mata, M.; De Los Santos-Aguilar, R.G.; Uribe-Cortes, D.; Ferreira-Hermosillo, A. Understanding progestins: From basics to clinical applicability. *J. Clin. Med.* **2023**, *12*, 3388. [CrossRef]
22. Albitar, L.; Pickett, G.; Morgan, M.; Davies, S.; Leslie, K.K. Models representing Type I and Type II human endometrial cancers: Ishikawa H and Hec50co cells. *Gynecol. Oncol.* **2007**, *106*, 52–64. [CrossRef]
23. Devor, E.J.; Gonzalez-Bosquet, J.; Thiel, K.W.; Leslie, K.K. Genomic characterization of five commonly used endometrial cancer cell lines. *Int. J. Oncol.* **2020**, *57*, 1348–1357. [CrossRef]
24. Bi, J.; Newtson, A.M.; Zhang, Y.; Devor, E.J.; Samuelson, M.I.; Thiel, K.W.; Leslie, K.K. Successful patient-derived organoid culture of gynecologic cancers for disease modeling and drug sensitivity testing. *Cancers* **2021**, *13*, 2901. [CrossRef]
25. Bi, J.; Thiel, K.W.; Litman, J.M.; Zhang, Y.; Devor, E.J.; Newtson, A.M.; Schnieders, M.J.; Gonzalez Bosquet, J.; Leslie, K.K. Characterization of a TP53 somatic variant of unknown function from an ovarian cancer patient using organoid culture and computational modeling. *Clin. Obstet. Gynecol.* **2020**, *63*, 109–119. [CrossRef]
26. Kanda, R.; Hiraike, H.; Wada-Hiraike, O.; Ichinose, T.; Nagasaka, K.; Sasajima, Y.; Ryo, E.; Fujii, T.; Osuga, Y.; Ayabe, T. Expression of the glucagon-like peptide-1 receptor and its role in regulating autophagy in endometrial cancer. *BMC Cancer* **2018**, *18*, 657. [CrossRef]
27. Lindemann, K.; Vatten, L.J.; Ellstrom-Engh, M.; Eskild, A. The impact of BMI on subgroups of uterine cancer. *Br. J. Cancer* **2009**, *101*, 534–536. [CrossRef] [PubMed]
28. Onstad, M.A.; Schmandt, R.E.; Lu, K.H. Addressing the role of obesity in endometrial cancer risk, prevention, and treatment. *J. Clin. Oncol.* **2016**, *34*, 4225–4230. [CrossRef]
29. Park, J.Y.; Seong, S.J.; Kim, T.J.; Kim, J.W.; Bae, D.S.; Nam, J.H. Significance of body weight change during fertility-sparing progestin therapy in young women with early endometrial cancer. *Gynecol. Oncol.* **2017**, *146*, 39–43. [CrossRef]

30. Cholakian, D.; Hacker, K.; Fader, A.N.; Gehrig, P.A.; Tanner, E.J., 3rd. Effect of oral versus intrauterine progestins on weight in women undergoing fertility preserving therapy for complex atypical hyperplasia or endometrial cancer. *Gynecol. Oncol.* **2016**, *140*, 234–238. [CrossRef] [PubMed]
31. Valadez-Cosmes, P.; Vazquez-Martinez, E.R.; Cerbon, M.; Camacho-Arroyo, I. Membrane progesterone receptors in reproduction and cancer. *Mol. Cell Endocrinol.* **2016**, *434*, 166–175. [CrossRef]
32. Zhang, M.; Robitaille, M.; Showalter, A.D.; Huang, X.; Liu, Y.; Bhattacharjee, A.; Willard, F.S.; Han, J.; Froese, S.; Wei, L.; et al. Progesterone receptor membrane component 1 is a functional part of the glucagon-like peptide-1 (GLP-1) receptor complex in pancreatic beta cells. *Mol. Cell Proteom.* **2014**, *13*, 3049–3062. [CrossRef] [PubMed]

cancers

MDPI

Article

Five-Year Relative Survival Rates of Women Diagnosed with Uterine Cancer by County-Level Socioeconomic Status Overall and across Histology and Race/Ethnicity

Akemi T. Wijayabahu [1,*], Jennifer K. McGee-Avila [1], Meredith S. Shiels [1], Alfonsus Adrian H. Harsono [2], Rebecca C. Arend [2] and Megan A. Clarke [1]

1 Division of Cancer Epidemiology and Genetics, National Cancer Institute, Rockville, MD 20850, USA; jennifer.mcgee-avila@nih.gov (J.K.M.-A.); shielsms@mail.nih.gov (M.S.S.); megan.clarke@nih.gov (M.A.C.)
2 Division of Gynecologic Oncology, Department of Obstetrics and Gynecology, University of Alabama at Birmingham, Birmingham, AL 35233, USA; aharsono@uabmc.edu (A.A.H.H.); rarend@uabmc.edu (R.C.A.)
* Correspondence: akemi.wijayabahu@nih.gov

Simple Summary: Persistent survival disparities by racial/ethnic groups are expected to worsen with rising incidence rates of aggressive uterine cancer subtypes. Understanding the impact of area-based socioeconomic factors on survival outcomes may help to better understand these disparities. This research aims to understand how living in counties with lower educational attainment, higher poverty, higher unemployment, lower median household income, and population density in urban areas may impact survival rates of women diagnosed with uterine cancer across racial/ethnic groups. Our findings show that lower county-level socioeconomic characteristics are linked with worse survival rates, mainly impacting women diagnosed with aggressive histologic subtypes and underrepresented racial/ethnic groups. Regardless of tumor and socioeconomic characteristics, non-Hispanic (NH) Black women consistently experience the poorest survival outcomes compared to other racial/ethnic groups included in this study. Racial/ethnic disparities in survival were observed even in the most affluent counties, suggesting that other factors beyond county-level socioeconomic status are at play.

Abstract: Understanding socioeconomic factors contributing to uterine cancer survival disparities is crucial, especially given the increasing incidence of uterine cancer, which disproportionately impacts racial/ethnic groups. We investigated the impact of county-level socioeconomic factors on five-year survival rates of uterine cancer overall and by histology across race/ethnicity. We included 333,013 women aged $\geq$ 30 years with microscopically confirmed uterine cancers (2000–2018) from the Surveillance, Epidemiology, and End Results 22 database followed through 2019. Age-standardized five-year relative survival rates were compared within race/ethnicity and histology, examining the differences across tertiles of county-level percent (%) <high-school education, %<150 percent poverty, %unemployment, median household income, and %urbanicity. Overall age-adjusted five-year relative survival was 77.7%. Rates were lowest among those residing in the least advantaged counties (tertile 3) and highest among the most advantaged (tertile 1): education (74.7% vs. 80.2%), poverty (72.9% vs. 79.8%), unemployment (75.7% vs. 80.5%), and income (73.3% tertile 1 vs. 78.1% tertile 3). Impact of county-level socioeconomic characteristics on survival across histology was minimal. We observed considerable survival disparities among NH-Black and NH-Native American/Alaskan Native women, regardless of tumor and socioeconomic characteristics. These findings add to our understanding of how county-level socioeconomic characteristics affect uterine cancer survival inequalities among racial/ethnic groups.

Keywords: socioeconomic status; uterine cancer; race/ethnicity; histologic subtype; five-year survival; race/ethnic disparity

1. Introduction

Rising incidence and mortality rates of uterine cancer pose a significant public health challenge in the United States (U.S.), with an estimated 67,880 cases and 13,250 deaths projected for 2024 [1]. While the overall five-year relative survival rate for uterine cancer is favorable at 83.3%, survival varies significantly by stage at diagnosis (according to Summary Stage, localized: 95.4%, regional: 69.9%, distant: 18.1%) and histologic subtype (endometrioid: 91.8%, non-endometrioid: 57.5%, sarcoma: 52.8%) [2]. Further, while for most cancers five-year relative survival rates have improved over time, uterine cancer survival has decreased [1], likely owing to an increasing incidence of non-endometrioid subtypes within the last two decades, which have poorer prognosis and are more common in non-Hispanic (NH) Black women. Studies dating back to 1987 consistently highlighted racial differences in survival outcomes among Black women with uterine cancer compared to White women, particularly among women who did not receive recommended treatment [3]. More recent studies have shown that survival differences across racial/ethnic groups persist even after controlling for tumor characteristics, time to treatment, and type of treatment [4–6]. Therefore, it is imperative to investigate and understand other factors associated with these survival disparities, particularly in light of the increasing incidence rates of aggressive histologic subtypes [2,7].

Lower socioeconomic status, whether defined individually or as a group (e.g., based on the lived-in area, including county-level socioeconomic status as an indication of socioeconomic deprivation), is typically marked by lower educational attainment, higher unemployment, and lower income levels [8]. Living in areas with socioeconomic deprivation has been linked with poor survival outcomes in women with uterine cancer [8–10]. Lower area-based socioeconomic status has also been associated with delayed healthcare utilization and access, resulting in advanced stage at uterine cancer diagnosis [11]. Black and Hispanic women are more likely to present with advanced disease compared to White women [12]. Socioeconomic deprivation, often manifested through residential segregation, is linked to the differential distribution in lifestyle factors (e.g., adiposity, parity) [13,14], chronic comorbidities [15], and other potential risk exposures (e.g., air pollutants) [16–18] among racial/ethnic groups. Moreover, inequality in access to healthcare further exacerbates these disparities, and may contribute to the observed differences in survival across racial and ethnic groups [19,20].

To our knowledge, prior studies have not compared uterine cancer survival rates across histologic subtypes, socioeconomic factors, and racial/ethnic groups using a large, population-based database. It is important to investigate the intersectionality of these factors to gain a deeper understanding of healthcare outcome disparities. Thus, our objective was to compute age-standardized five-year relative survival rates for women diagnosed with uterine cancer, by county-level socioeconomic factors (education, poverty, unemployment, income, urbanization) overall, and by histologic subtype across racial/ethnic groups.

2. Materials and Methods

2.1. Data Source

We utilized the Surveillance, Epidemiology, and End Results (SEER) Research Plus Limited-Field Database 22 (SEER 22), excluding Illinois and Massachusetts, which aggregates incident uterine cancer cases from 20 population-based cancer registries across the United States and linked with the American Census Survey (ACS) database [21]. These registries collectively cover approximately 41.9% of the total U.S. population based on the 2020 U.S. Census [21]. All data utilized in this study are de-identified and are publicly available; thus, the study was exempt from institutional review board approval and the need for informed consent.

2.2. Case Selection and Study Design

In this retrospective cohort study of uterine cancer cases from the SEER database, we included malignant corpus uteri and uterine corpus, NOS (International Classification of Diseases for Oncology Site, Third Edition (ICD-O-3) codes: C540–C549, and C559) diagnosed between 2000 and 2018 in women aged 30 years and older. The rationale for the age restriction is to balance between minimizing the contribution of hereditary cancers and capturing a wider age range comparable to previous studies [2,22]. Cases were restricted to microscopically confirmed first primary uterine cancers. We included racial/ethnic categories provided in SEER, including NH-White, NH-Black, NH-American Indian or Alaska Native (hereafter referred to as Native American/Alaska Native, NA/AN), and Hispanic women, excluding unknown races (n = 1435). We also excluded cases identified with autopsy only or death certificates only, and those with no survival data. Case selection is depicted in the flow chart shown in Figure 1.

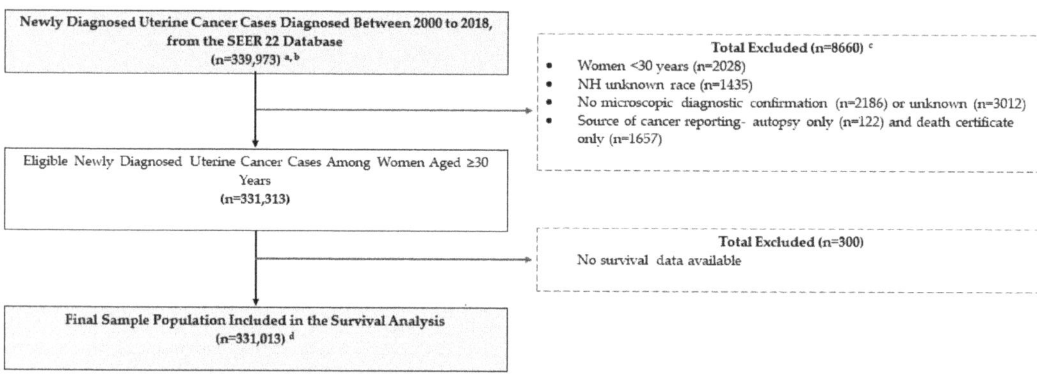

Figure 1. Flow chart of study population selection: women diagnosed with uterine cancer among U.S. women aged $\geq$ 30 years reported in the SEER 22 database. Abbreviations: SEER (Surveillance, Epidemiology, and End Results). Notes: [a] We utilized the Surveillance, Epidemiology, and End Results (SEER 22) database, excluding Illinois and Massachusetts, which combines 20 population-based cancer registries and links to American Community Survey data. [b] All cases included are first matching primary record of malignant uterine cancers (corpus uteri and uteri, not otherwise specified). [c] Some cases have one or more overlapping exclusion criteria. Thus, the total number excluded is less than the sum of individual categories of exclusion. [d] Final sample included in our survival analysis. Survival calculated cases diagnosed between 2000 and 2018, and survival data available from 2000 through 2019.

2.3. Demographic and Clinical Characteristics

We included race and ethnicity categories from the SEER database [23] (NH-White, NH-Black, NH-Asian/PI, NH-NA/AN, and Hispanic women). Race and ethnicity data in cancer registries are obtained from patient medical records, which are primarily based on self-reported data, caretaker reports, and physical appearance [24]. We grouped histologic subtypes as endometrioid, non-endometrioid, sarcomas, and other cancers, as shown in Table S1. We included stage at diagnosis (localized, which corresponds to FIGO Stage I, regional to FIGO Stage II–III, and distant to FIGO Stage IV) using SEER summary stage from the diagnosis period 2004 to 2018, because stage information was not available prior to 2004 [25,26]. In addition to the n = 54,918 women with cancers diagnosed before 2004, we set n = 10,978 cases classified as unknown/not staged, and n = 10 tumors classified as in situ according to the SEER summary stage diagnosed from 2004 to 2018 as missing for analyses evaluating rates by stage only.

2.4. County-Level Socioeconomic Characteristics

County-level socioeconomic characteristics including percent with less than a high-school education (<HS education), percentage of persons below 150 percent poverty line (<150% poverty), percentage of unemployment (unemployment), and median household income in U.S. dollars (USD) were selected from the 2015–2019 ACS (American Community Survey, conducted by the U. S. Census Bureau) cycle, linked with the SEER 22 database. We selected 2015–2019 ACS cycle data, the most recent county-level socioeconomic estimates that overlaps with our study population. The socioeconomic characteristics from the 2015–2019 ACS cycle are highly correlated with those from the 2000s, 2007–2011, and 2010–2014 ACS cycles. We chose these county-level socioeconomic characteristics because these area-based socioeconomic indicators effectively detect health outcome differences and monitor disparities within healthcare systems [27]. For each variable, we ranked the counties from lowest to highest values, and created tertiles based on this ranked distribution, with cutoff values identified for each tertile [28]. Tertile cutoff values for percentage of people with <HS education are <9.48% (Tertile 1), 9.48–14.74% (Tertile 2), and >14.74% (Tertile 3); percentage of people < 150% poverty level < 20.92% (Tertile 1), 20.92–28.12% (Tertile 2) and >28.12% (Tertile 3); percent unemployment < 4.08% (Tertile 1), 4.08–5.86% (Tertile 2), >5.86% (T3); and median household income < USD 47,060 (Tertile 3), USD 47,060–USD 56,590 (Tertile 2), >USD 56,590 (Tertile 1). Tertile distributions correspond to the county-level socioeconomic measure in the U.S. population and not the distribution of women with uterine cancer included in our study population. For any given socioeconomic measure, except for the median household income variable, tertile 1 corresponds to the first 33% of counties with the highest socioeconomic status, while tertiles 2 and 3 correspond to medium and the lowest socioeconomic status. Median household income is inverse, where tertile 3 corresponds to the highest income.

Additionally, we included the %urban (=total urban population/total population) measure from the 2010 U.S. Census, linked with the SEER 22 database, which denotes the percentage of individuals residing in urban areas within each county. We categorized the %urban measure into tertiles based on the distribution of counties as follows: <24.53%, 24.53–57.02%, and >57.02%.

2.5. Statistical Analysis

Five-year relative survival rates with a 95% CI were calculated using SEER*Stat survival session. Expected survival rates considered age-specific all-cause mortality rates within the U.S. reference population, utilizing actuarial and Ederer II methods. These computations account for survival variations due to racial/ethnic composition, geographic area, and socioeconomic status [29]. We included women diagnosed with uterine cancer between 2000 and 2018 and assessed survival from 2000 to 2019. To facilitate comparisons between participant characteristics, we displayed age-standardized five-year survival estimates in the main tables and figures, and include non-age standardized relative survival estimates in the Supplementary Materials.

We computed age-standardized five-year relative survival estimates for socioeconomic characteristics for the overall study population, and we performed stratified analyses by histology and race/ethnicity. We did not report survival rates or specific case/death counts for groups with fewer than 16 cases and/or 16 deaths overall [30]. Due to the small sample size across multiple strata, we did not include in the figures the survival rates for the %urban measure across racial/ethnic groups, nor the survival rates for non-endometrioid histology among NH-NA/AN women. Additionally, we restricted our stratified analysis by histology to the two most common histologic subtypes (endometrioid and non-endometrioid) due to the small sample sizes within stratum-specific categories for rare sarcomas and other heterogeneous histologic types.

Statistically significant group differences in the five-year relative survival estimates were determined using p-values calculated from the z-test generated by the SEER*Stat program. We calculated survival rate ratios (SRRs) to assess the magnitude of impact of the different levels of socioeconomic characteristics with respect to the reference group on five-year relative survival estimates. All analyses were conducted using the SEER*Stat software version 8.4.3. All statistical tests were two sided, and statistical significance was assessed at an α level of $p < 0.05$.

Post hoc analyses: We conducted two post hoc analyses. In our first post hoc analysis, we excluded all women with the "other uterine cancer" histologic subtype, which includes heterogeneous cancers (n = 7543, 3%), and other histologic subtypes with possible cancer site misclassification: ICD-O-3 histologic codes 8210/3, 8263/3, 8262/3, 8261/3, 8211/3, 8490/3, and 8141/3 from endometrioid cancer estimates (n = 1530, 0.6%), and 8981/3 from non-endometrioid estimates (n = 16, 0.03%). In our second post hoc analysis, we evaluated the impact of county-level socioeconomic characteristics and survival rates by U.S. geographic regions [31]. We classified SEER registries into three categories, including Northeast (Connecticut, New Jersey, New York), South (Georgia, Kentucky, Louisiana and Texas), Combined Midwest (Iowa), and West (Alaska, California, Hawaii, Idaho, New Mexico, Utah, Washington), and evaluated the five-year relative survival across socioeconomic characteristics.

3. Results

3.1. Demographic, Clinical, and Socioeconomic Characteristics of the Study Population

Population estimates and five-year relative survival rates are shown in Table 1 and Table S2. Of the 333,013 eligible women diagnosed with uterine cancer between 2000 and 2018, 59% were of ages 50 to 69 years, 69% were NH-White, 75% had endometrioid histology, and 70% presented with localized disease. Counties with mid (Tertile 2) and highest (Tertile 3) proportions of the population with <HS education accounted for 35% to 36% of women with uterine cancer in this study, respectively. Around half of the women with uterine cancer included in this study lived in counties with the lowest percentage of the population living at < 150% poverty line and within mid-level %unemployment. The majority of the study population was from counties with a median household income above USD 56,590 and highest %urban. Most NH-Black, Hispanic, and NH-NA/AN women with uterine cancer resided in counties with lower socioeconomic status compared to other NH-White and NH-Asian/PI women (Figure S1). The majority of women across all racial/ethnic groups lived in counties with higher %urban (Figure S1e).

Table 1. Age standardized five-year relative survival rates (2000–2019) of women diagnosed with microscopically confirmed uterine cancer overall and across demographic, clinical, and county-level socioeconomic characteristics, among U.S. women aged $\geq$ 30 Years, diagnosed between the years 2000 and 2018, in SEER 22.

Population Characteristics		Cases, N (%)	Deaths, N	Survival Rate (95% CI) [a]	SRR	p-Value [b]
Overall, All Women		333,013	79,782	77.7 (77.5, 77.9)	-	-
Race/Ethnicity						
	NH-White	229,273 (68.85)	51,882	81.2 (80.9, 81.4)	1.00	Reference
	NH-Black	34,200 (10.27)	14,096	57.7 (57.0, 58.4)	0.71	<0.001
	NH-Asian/PI	21,209 (6.37)	3890	75.8 (74.8, 76.8)	0.93	0.0908
	NH-NA/AN	1615 (0.48)	338	77.1 (72.6, 81.0)	0.95	0.0673
	Hispanic	44,716 (13.43)	9576	73.1 (72.4, 73.8)	0.90	<0.001
Histology						
	Endometrioid	247,240 (74.69)	39,919	87.1 (86.8, 87.3)	1.00	Reference
	Non-Endometrioid	61,620 (18.62)	28,525	55.7 (55.3, 56.2)	0.63	<0.001
	Sarcoma	14,610 (4.41)	7060	44.4 (43.2, 45.7)	0.51	<0.001
	Other	7543 (2.28)	4278	42.2 (41.0, 43.5)	0.48	<0.001

Table 1. *Cont.*

Population Characteristics		Cases, N (%)	Deaths, N	Survival Rate (95% CI) [a]	SRR	*p*-Value [b]
Stage at Diagnosis [c]						
	Localized	184,719 (69.68)	20,751	93.0 (92.7, 93.2)	1.00	Reference
	Regional	55,805 (21.05)	19,463	64.6 (64.1, 65.1)	0.69	<0.001
	Distant	24,583 (9.27)	19,475	16.6 (16.1, 17.2)	0.18	<0.001
%<HS Education [d]						
	Tertile 1	96,211 (29.07)	21,193	80.2 (79.8, 80.6)	1.00	Reference
	Tertile 2	119,901 (36.23)	28,580	78.3 (78.0, 78.7)	0.98	<0.001
	Tertile 3	114,873 (34.70)	30,000	74.7 (74.3, 75.1)	0.93	<0.001
%<150 Percent Poverty [d]						
	Tertile 1	149,375 (45.13)	33,111	79.8 (79.5, 80.2)	1.00	Reference
	Tertile 2	131,906 (39.85)	32,518	77.0 (76.6, 77.3)	0.96	<0.001
	Tertile 3	49,704 (15.02)	14,144	72.9 (72.3, 73.5)	0.91	<0.001
%Unemployment [d]						
	Tertile 1	45,981 (13.89)	10,014	80.5 (79.9, 81.1)	1.00	Reference
	Tertile 2	166,282 (50.24)	39,305	78.3 (78.0, 78.6)	0.97	<0.001
	Tertile 3	118,722 (35.87)	30,454	75.7 (75.3, 76.0)	0.94	<0.001
Median Household Income (USD) [d]						
	Tertile 1	28,623 (8.65)	8103	73.3 (72.5, 74.1)	1.00	Reference
	Tertile 2	43,064 (13.01)	10,620	78.3 (77.7, 78.9)	1.07	<0.001
	Tertile 3	259,298 (78.34)	61,050	78.1 (77.8, 78.3)	1.07	<0.001
% Urban [d]						
	Tertile 1	9019 (2.72)	2237	77.8 (76.4, 79.1)	1.00	Reference
	Tertile 2	30,149 (9.11)	7302	79.0 (78.3, 79.7)	1.02	0.93385
	Tertile 3	291,819 (88.17)	70,234	77.5 (77.3, 77.8)	1.00	0.11071

Notes. Percentages were calculated using non-missing data. Abbreviations: CI, confidence interval; NA/AN, Native American/ Alaska Native; NH, non-Hispanic; PI, Pacific Islander; SES, socioeconomic status; SRR, survival rate ratio. [a] First matching primary uterine cancer cases (diagnosed 2000–2018) were included, and survival assessed based on all-cause mortality estimated through 2019. [b] *p*-values (two tailed) indicates the statistical difference between two 5-year relative survival estimates computed from the z-statistic with the formula $p = 2 \times (1 - \Phi(|z|))$, where $\Phi(x)$ = the standard normal distribution's cumulative distribution function, and $|z|$ = absolute z-score. [c] Stage at diagnosis was not available prior to 2004, resulting in a total of n = 54,918 missing data points. Additionally, we set n = 10,978 (unknown/not-staged), and n = 10 (in situ) tumors as missing. [d] All county-level SES characteristics, except %urban (2010 U.S. Census County Attributes), were sourced from the ACS database (2015–2019) linked to SEER 22. The following ranges were used for SES tertile cutoffs: %<HS Education (1.12% to 73.56%; T1 < 9.48%, T2 9.48–14.74%, and T3 > 14.74%), %<150 Poverty (4.84% to 68.30%; T1 < 20.92%, T2 20.92–28.12%, and T3 > 28.12%), %Unemployed (0% to 27.15%; T1 < 4.07%, T2 4.07–5.86%, T3 > 5.86%), Median Household Income (USD 21,500 to USD 142,300; T1 < USD 47,060, T2 USD 47,060 to USD 56,590, T3 > USD 56,590), and %urban (0 to 100% urban; T1 < 24.53%, T2 24.53–57.02%, T3 > 57.02%).

3.2. Overall Five-Year Relative Survival across Demographics and Clinical Characteristics

The overall age-standardized five-year relative survival rate for women diagnosed with uterine cancer between 2000 and 2018 was 77.7% (95% CI 77.5%, 77.9%) (Table 1, non-age-standardized estimates are shown in Table S2). Overall, age-standardized five-year survival rates were lowest among NH-Black (57.7%) followed by Hispanic (73.1%), NH-Asian/PI (75.8%), NH-NA/AN (77.1%), and NH-White women (81.2%). For the defined histologic subtypes, survival was lowest among women diagnosed with sarcomas (44.4%), followed by non-endometrioid (55.7%) and endometrioid subtypes (87.1%). As expected, survival decreased with more advanced stage (localized 93.0%, regional 64.6%, and distant 16.6%). The observed survival differences based on tumor stage remained consistent across histologic subtypes and racial/ethnic groups (Figure S2).

3.3. *Five-Year Relative Survival across Socioeconomic Characteristics among All Women*

For all county-level measures of socioeconomic characteristics, survival rates were highest among those residing in the most advantaged counties compared with those living in the least advantaged counties. Except for the %urban category, all other socioeconomic categories followed a dose–response relationship, wherein survival rates declined sequentially with decreasing socioeconomic level (increasing disadvantage). County-level poverty (%<150 percent poverty) showed the greatest impact on survival rates based on absolute differences (79.8% in tertile 1 vs. 72.9% in tertile 3), and a similar impact was observed for %<HS education (80.2% vs. 74.7%), median household income (78.1% vs. 73.3%), and %unemployment (80.5% vs. 75.7%). We did not observe significant differences across tertiles of the %urban measure. The impact of socioeconomic status on survival was consistent across the stage of diagnosis, reflecting the overall patterns (Figure S3). Notably, the poor survival observed among NH-Black women persisted across all stages of diagnosis (Figure S3).

3.4. *Race/Ethnic Groups Specific Five-Year Relative Survival across County-Level Socioeconomic Characteristics, for All Histologic Subtypes Combined*

Age-standardized five-year survival rates declined with increasing county-level socioeconomic disadvantage (lower %<HS education, higher %<150 percent poverty, higher %unemployment, and lower median household income) among almost all racial/ethnic groups (Figure 2, with detailed results in Table S3). Although NH-Black women exhibited the worst survival rates across all socioeconomic characteristics, the impact of county-level socioeconomic characteristics appeared to be largest among NH-NA/AN women with uterine cancer, with the largest absolute differences from most to least advantaged observed for median household income (79.4% vs. 65.0%, $p = 0.022$), %<150 percent poverty (81.4% vs. 67.6%, $p = 0.026$) and %<HS education (80.1% vs. 71.0%, $p = 0.156$; not significant). Among other racial/ethnic groups, %<150 percent poverty and %<HS education were associated with the greatest differences in survival among NH-Black women (59.9% vs. 54.1% and 61.0% vs. 56.7%, respectively; both $p < 0.001$). Although NH-Asian/PI women residing in counties with higher %<150 percent poverty showed a dose–response relationship, the survival difference was significant only in the second tertile compared to the first tertile (76.7% vs. 74.9%, $p = 0.006$), and not the third tertile (76.7% vs. 71.2%, $p = 0.507$), likely due to only 5% of the population being in the third tertile (Figure S1). NH-Asian/PI women residing in counties within the second tertile of median household income had much lower survival (62.6%) compared to those in the first tertile (77.8%) and third tertile (76.1%), with a significant difference observed only between the first and second tertiles ($p = 0.027$). This disparity is likely due to population distribution, with 1% (n = 219) in the first tertile, 3% (n = 542) in the second tertile, and the majority in counties with median household income above USD 56,600 (Figure S1). The absolute differences in survival across county-level socioeconomic characteristics were small, though statistically significant, among NH-White and Hispanic women for nearly all measures (Figure 2). Survival estimates for %urban are not discussed further due to the majority residing in counties with higher %urban (Figure S1) and small sample sizes across multiple strata (Tables S3 and S4).

3.5. *Five-Year Relative Survival across County-Level Socioeconomic Characteristics Overall and by Histologic Subtype and Race/Ethnicity*

Approximately one-third (35%) of NH-Black women had the non-endometrioid subtype, compared to less than a fifth in other racial/ethnic groups (17% are non-endometrioid among NH-White, NH-Asian/PI, and Hispanic, and 15% in NH-NA/AN women) (Figure S4). Regardless, NH-Black women with both endometrioid and non-endometrioid subtypes exhibited the poorest survival rates across all socioeconomic characteristics (Figures 3 and 4). Detailed estimates are provided in the Supplementary Materials (Tables S3 and S4).

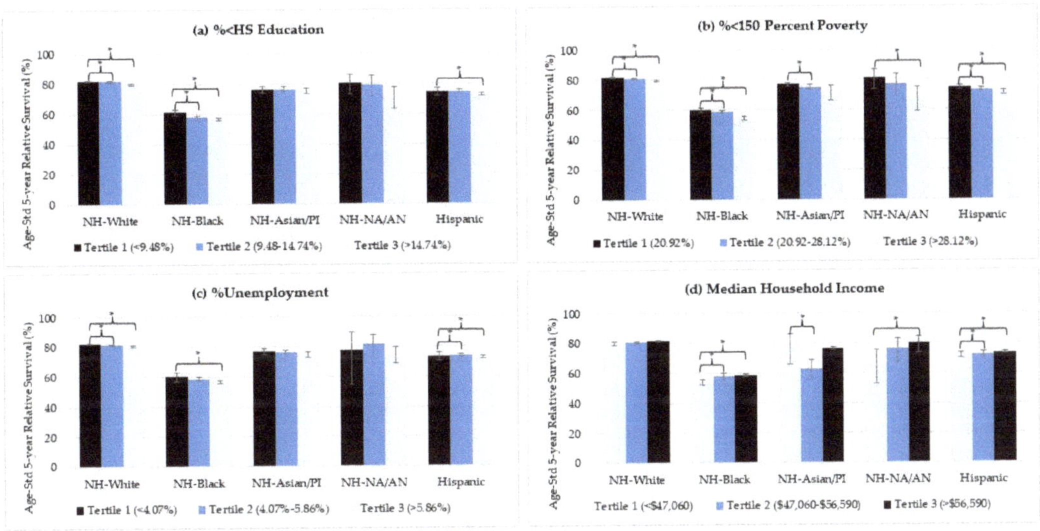

Figure 2. Overall race/ethnicity specific age-standardized 5-year survival rates across socioeconomic characteristics in women diagnosed with uterine cancer between 2000 and 2018, with survival assessed through 2019. Crossbars (*) indicate z-value-based statistical significance ($p < 0.05$), and error bars indicate 95% confidence intervals. Median household income is shown in U.S. dollars (USD denoted as $). Survival estimates for NH-NA/AN (%unemployment tertile 1) is based on small sample sizes (<100 cases and ≤20 deaths).

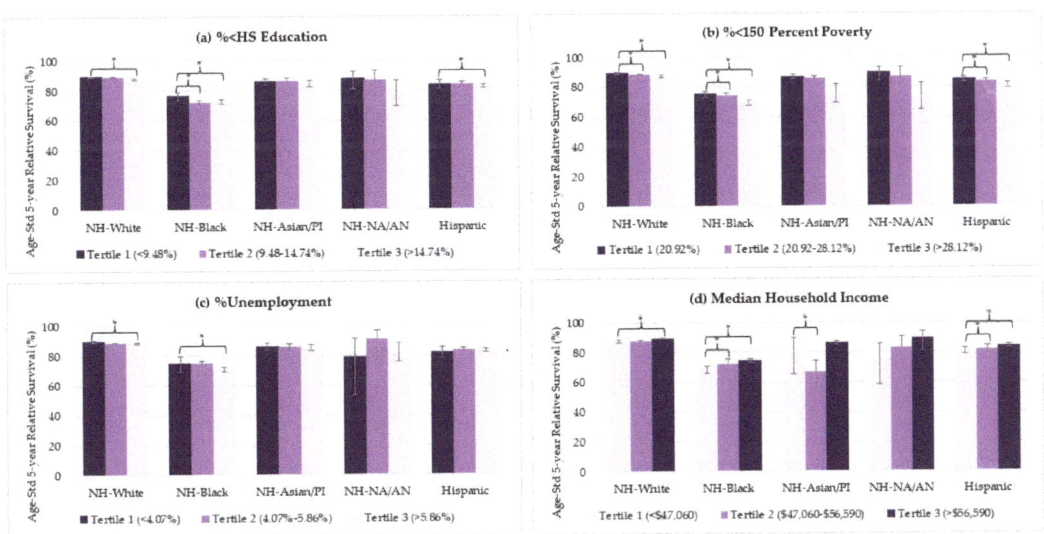

Figure 3. Race/ethnicity specific age-standardized 5-year survival rates across socioeconomic characteristics in women with endometrioid subtype of uterine cancer diagnosed between 2000 and 2018, with survival assessed through 2019. Crossbars (*) indicate z-value-based statistical significance ($p < 0.05$), and error bars indicate 95% confidence intervals. Median household income is shown in U.S. dollars (USD denoted as $). Survival estimates for NH-NA/AN (%unemployment tertile 1) and the NH-Asian/PI (median household income tertiles 1) are based on small sample sizes (<100 cases and/or ≤20 deaths).

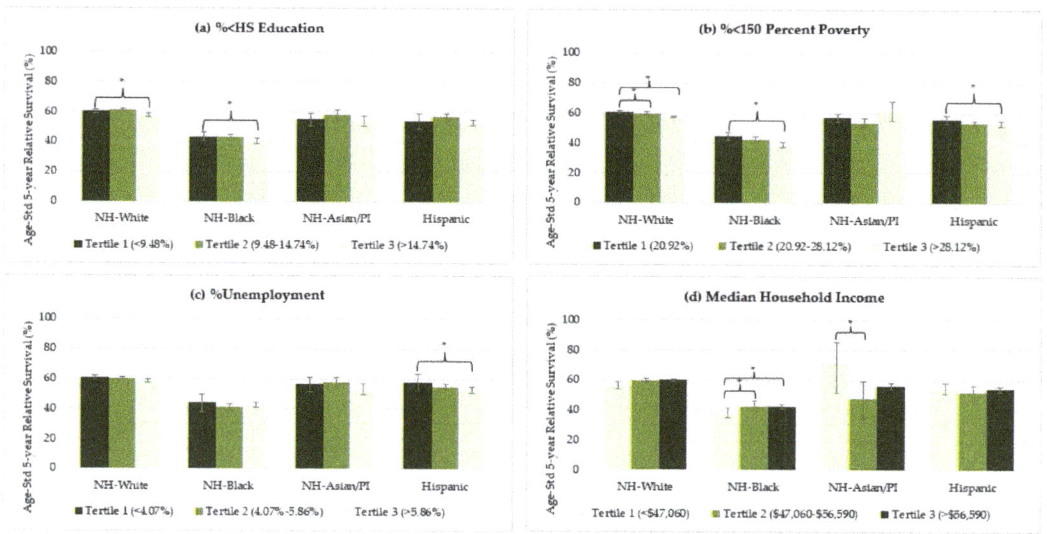

Figure 4. Race/ethnicity specific age-standardized 5-year survival rates across socioeconomic characteristics in women with non-endometrioid subtype of uterine cancer diagnosed between 2000 and 2018, with survival assessed through 2019. Crossbars (*) indicate z-value-based statistical significance ($p < 0.05$), and error bars indicate 95% confidence intervals. Median household income is shown in U.S. dollars (USD denoted as $). Survival estimates for NH-Asian/PI (median household income tertile 1) are based on small sample sizes (<100 cases and/or $\leq$20 deaths). Due to small sample size across multiple categories, NH-NA/AN estimates are not shown in the figure.

The following results present age-standardized five-year survival estimates, comparing the highest and lowest socioeconomic levels within each racial or ethnic group, rather than across different racial or ethnic groups.

Among NH-White women with the endometrioid subtype, there were small but significant survival differences between the highest (most advantaged) and lowest (least advantaged) socioeconomic levels for all four measures: %<HS education (89.3% vs. 87.8%, $p < 0.001$), %<150 percent poverty (89.4% vs. 87.5%, $p < 0.001$), %unemployment (89.5% vs. 88.5%, p 0.003), and median household income (89.2% vs. 87.7%, $p = 0.018$). For NH-White women with the non-endometrioid subtype, significant differences were observed for %<HS education (60.3% vs. 58.0%, $p = 0.001$) and %<150 percent poverty (60.7% vs. 57.4%, $p = 0.002$).

NH-Black women with the endometrioid subtype showed relatively larger differences between the highest and the lowest socioeconomic levels compared with other racial/ethnic groups included in this study: %<HS education (76.3% vs. 72.8%, $p < 0.001$), %<150 percent poverty (75.2% vs. 69.2%, $p < 0.001$), %unemployment (74.9% vs. 71.2%, $p = 0.005$), and median household income (74.2% vs. 68.6%, $p < 0.001$). For the non-endometrioid subtype, significant differences were observed in %<HS education (43.5% vs. 40.7%, $p = 0.020$), %<150 percent poverty (44.8% vs. 39.0%, $p < 0.001$), and median household income (42.7% vs. 38.4%, $p = 0.003$).

Hispanic women with the endometrioid subtype also showed statistically significant differences between the highest and lowest socioeconomic levels of %<HS education (83.8% vs. 82.7%, $p = 0.017$), %<150 percent poverty (84.7% vs. 80.7%, $p < 0.001$), and median household income characteristics (83.8% vs.80.7%, <0.001), but the magnitude of these differences was small. Similarly, there were small, but significant differences among Hispanic women with non-endometrioid cancer across the highest and lowest socioeconomic levels of %<150 percent poverty (56.0% vs. 53.3%, $p = 0.005$) and %unemployment (57.7% vs. 52.9%, $p = 0.021$).

Survival among NH-Asian/PI women in counties with second tertile median household income remained significantly lower compared to the first tertile (counties with median household income < USD 47,060) for both endometrioid and non-endometrioid subtypes, likely due to the population distribution (Figure S1).

Among NH-NA/AN women, certain socioeconomic characteristics seem to contribute to large absolute survival differences across socioeconomic levels (Table S3). However, due to the small sample size, these findings should be interpreted with caution (Table S4).

3.6. Findings from Post Hoc Analyses

Findings from our first post hoc analysis show that, given the rarity of the potentially misclassified and heterogeneous "other uterine cancers," the exclusion of these histologic subtypes did not impact our findings.

In our second post hoc analysis, we found significant survival differences across the highest and the lowest socioeconomic levels of %<HS education, <%150 percent poverty, %unemployment, and median household income in Northeast, South, and West/Midwest, with the largest differences between the socioeconomic levels observed in the Northeast region, followed by South (Table S5).

4. Discussion

4.1. Principal Findings

Our findings suggest that lower county-level socioeconomic status correlates with poorer five-year relative survival rates of uterine cancer across all racial/ethnic groups and histologic subtypes. In line with our findings, other studies have shown that lower area-based socioeconomic status, serving as a proxy for individual-level socioeconomic status, is associated with poorer survival outcomes in women with uterine cancer [8–10,12,32–34]. These associations persist even when considering stage, histology, and treatment factors, with the largest disparity observed among NH-Black women [4,5,12].

Within strata of the various county-level socioeconomic characteristic, wide gaps in survival persisted among NH-Black compared with all other racial and ethnic groups, particularly NH-White women. More importantly, worse survival rates were observed among NH-Black women even in the most advantaged counties with the highest education levels, lowest poverty and unemployment rates, and highest incomes. Collectively, these findings suggest that county-level socioeconomic factors including level of education, poverty, unemployment, and income play a minimal role in explaining uterine cancer survival disparities among NH-Black women.

While the proportion of women living in disadvantaged communities (communities with concentrated disadvantages according to socioeconomic determinants) tended to be highest among NH-Black, NH-NA/AN, and Hispanic women, the impact of county-level socioeconomic factors on survival rates among Hispanic women showed minimal absolute differences. Some studies show that living within ethnic enclaves may benefit the survival outcomes of people living with (any cancer, uterine and breast cancers) or without cancer of certain ethnic groups, but not all [35–38]. Data are limited and not consistent regarding the impact of ethnic enclaves on the survival outcomes of Hispanic/Latina women by socioeconomic status [35,37]. Further investigations are needed at the individual level, particularly among Hispanic women with uterine cancer.

Our post hoc analysis by geographic regions shows that the correlation between lower county-level socioeconomic status and poorer five-year relative survival rates of uterine cancer persists across different geographic regions. Additionally, our analysis revealed greater survival disparities across socioeconomic characteristics in the Northeast region, followed by the South region.

4.2. Socioeconomic Status and Survival Differences by Histology

While we did not observe strong differences in the influence of county-level socioeconomic characteristics on survival rates of women with uterine cancer by histology, it is worth noting that the impact of these measures appeared to be slightly greater for women with endometrioid histology. Because endometrioid cancers are treated surgically and highly curable, it is plausible that socioeconomic barriers related to treatment and healthcare access may have a greater influence on survival rates, whereas non-endometrioid cancers are less amenable to curative treatment and have worse outcomes in general.

NH-Black women consistently exhibit a survival gap compared to NH-White women, across both endometrioid and non-endometrioid subtypes. Despite a higher proportion being diagnosed with the aggressive non-endometrioid subtype, this disparity persists even after adjusting for clinical, socioeconomic, and treatment-related factors [5,39]. A mediation analysis found that clinicopathologic factors contributed most significantly to survival differences in NH-Black women with non-endometrioid uterine cancer compared to NH-White women (about 40%), followed by sociodemographic (about 9%) and treatment factors (about 7%) [39]. Despite adjustments, a persistent survival gap in NH-Black women remained, indicating unknown and unaccounted factors contribute to survival disparities in NH-Black women [39]. Histology differences in Black women with uterine cancer may reflect a complex interplay between multiple factors, possibly impacted by chronic lifestyle and socioeconomic stressors [40]. However, these complex associations, particularly by histology, remain poorly understood and are beyond the scope of this study.

4.3. Complex Interplay between Socioeconomic Characteristics, Race/Ethnicity, and Survival

We found that NH-Black women had worse survival rates compared to other racial/ethnic groups across all socioeconomic levels, regardless of histology and stage. The complex interplay between social determinants of health, systemic racism and discrimination, and race/ethnicity in regard to uterine cancer goes beyond our selected socioeconomic characteristics [20]. Instead, it encompasses the confluence of all these factors, particularly for Black women [15,33,40], which may pose challenges in interpreting county-level analyses but are crucial for comprehensive understanding and addressing disparities. For example, residential segregation [20,41] and the systematic distribution of risk factors [13,14] and healthcare resources [11] intentionally varies across race/ethnicity and socioeconomic levels, collectively influencing chronic comorbidities [15], and ultimately contributing to survival outcomes [11,13,14]. One study linked marginalization, including material deprivation factors like education, income, and unemployment, to advanced-stage diagnosis in women with uterine cancer, even after adjusting for age, obesity, comorbidities, and histology [19]. Another found either race/ethnicity or income, but not both, was associated with advanced-stage disease [33]. Both studies show a tight link (collinearity) between race and socioeconomic factors, highlighting the challenge in determining their independent impact [19,33].

With no standard screening protocols, uterine cancer diagnosis relies on symptomatic presentation (e.g., abnormal bleeding, abnormal imaging), individual health-seeking behaviors, and accurate, timely, and unbiased diagnostic evaluation by healthcare providers. Delays in any one of these steps may be associated with more advanced disease and worse outcomes [42]. NH-Black women in particular report delayed symptom appraisal, health-seeking behaviors, and encounter poor-quality healthcare such as less likelihood of receiving of guideline concordant diagnostic work-up and treatment, symptom dismissals, and delays in diagnosis and treatment [11,42–44]. Health literacy, often tied to education level, likely influences symptom recognition and, when combined with poverty and unemployment, may exacerbate barriers to accessing care. Racism, experienced through discrimination and residential segregation, often lead to socioeconomic disparities, limiting access to education and job opportunities [20]. These disparities can result in variations in neighborhood environments, such as access to healthy food, quality healthcare, and exposure to lifetime stressors [20,45,46], ultimately increasing the risk of morbidity and mortality [20,47]. These health inequalities are likely exacerbated by a lack of understanding of risk factors, early diagnostic factors, and prognostic factors associated with aggressive non-endometrioid subtypes, which are more common among NH-Black women [11,47,48].

4.4. Strengths, Limitations, and Future Directions

In our study, we offer a unique comparison of age-standardized five-year relative survival estimates of uterine cancer across various socioeconomic categories by race/ethnicity, encompassing major histologic subtypes. Such comprehensive analyses are scarce in the existing literature, underscoring the novelty and importance of our research. We calculated age-standardized survival estimates that facilitate comparisons across socioeconomic and racial/ethnic groups by mitigating differences in age distribution across groups [49], as well as the independent effects of age on outcomes [50]. We utilized data from 20 population-based cancer registries collectively covering 42% of the U.S. population based on the 2020 U.S. Census [21]. SEER provides a large and diverse sample of women diagnosed with uterine cancer with comprehensive histology, stage at diagnosis, and survival data. Moreover, through linkage with the ACS database and the U.S. Census, it offers information on county-level socioeconomic estimates including education, poverty, unemployment, income, and information on county-level urban population density.

While U.S.-level statistics generated using SEER data and the overall quality of the SEER database have been well established [1], we acknowledge certain limitations pertaining to its use in our current analysis [51]. Limitations of our study include the absence of key information on potential confounders (e.g., underreported/incomplete adjuvant therapy data, missing stage at diagnosis prior to 2004, and missing/incomplete data on tumor grade, comorbidities, and other risk factors) and possible misclassification of rare histologic subtypes of uterine cancer which would not have impacted our main findings. Moreover, individual migrations from registry catchment area to an outside area can result in loss to follow-up and misclassification of the county-level socioeconomic status, because socioeconomic characteristics were determined based on the lived-in area at the time of diagnosis [51]. Because each county-level socioeconomic characteristic is based on different combinations of counties that fall within a certain tertile level, we cannot rule out the influence of other geographic factors that may contribute to survival.

Our findings suggest a link between county-level socioeconomic factors and uterine cancer survival across racial/ethnic groups and minimal impact of socioeconomic factors on survival across histology; future investigations are warranted to include factors beyond traditionally considered socioeconomic indicators such as education, poverty, unemployment, and income, and inclusion of different research methods to identify root causes. It is important to note that our conclusions are limited to elucidating the link between county-level socioeconomic factors and overall survival rates in an ecological analysis. Racial/ethnic groups within the same geographic area may experience different socioeconomic conditions and individual-level challenges, which may have a larger impact on survival across racial/ethnic groups than county-level socioeconomic status alone. Moreover, a multivariable analysis that accounts for major confounders at individual level (e.g., treatment type and receipt of guideline concordant therapy, comorbidities, experiences of discrimination and other chronic lifetime stressors), in conjunction with a more refined geographical area-based socioeconomic status measure (defined at the zip code level) of, may offer a more precise understanding of the relationship between socioeconomic status and survival.

Considering the persistent survival disparities among underrepresented racial/ethnic groups, especially NH-Black women, which may worsen due to increasing incidence rates of aggressive histologic subtypes, it is crucial to initially identify and better understand modifiable socioeconomic determinants influencing these survival disparities.

5. Conclusions

We observed significant associations of lower county-level socioeconomic characteristics with reduced age-adjusted five-year relative survival rates, both overall and among different racial/ethnic groups. County-level socioeconomic characteristics had a minimal impact on survival rates across histologic subtypes. Regardless of the county-level socioeconomic level and tumor characteristics, NH-Black women consistently show worse survival outcomes compared to other racial/ ethnic groups studied in our analysis.

Our findings further our understanding of the relationship between selected county-level socioeconomic factors and survival inequalities among different racial/ethnic groups. Although reducing socioeconomic inequities could potentially have an impact on survival outcomes across racial and ethnic groups, it is clear from our findings and from other individual-level studies [8–10] that more research is needed to better understand reasons for persistent uterine cancer survival disparities among NH-Black women beyond these county-level socioeconomic characteristics. Future studies are warranted to investigate the relationship between social determinants of health that go beyond the traditional socioeconomic characteristics (county or individual level), including systemic racism, discrimination, exposure to chronic lifetime stressors and resilience, and survival outcomes, especially among NH-Black women [6–8].

Supplementary Materials: The following supporting information can be downloaded at: https://www.mdpi.com/article/10.3390/cancers16152747/s1, Figure S1. Proportion of eligible women diagnosed with uterine cancer by socioeconomic characteristics across race/ethnicity. Figure S2. Race/ethnicity specific age-standardized 5-year survival rates by histologic subtype and stage among women diagnosed with uterine cancer between 2000 and 2018, with survival assessed through 2019. Figure S3. Race/ethnicity specific age-standardized 5-year survival rates by socioeconomic characteristics and stage at diagnosis among women diagnosed with uterine cancer between 2000 and 2018, with survival assessed through 2019. Figure S4. Proportion of women diagnosed with uterine cancer by histologic subtypes and stage at diagnosis, across race/ethnicity. Table S1. Number of women diagnosed with uterine cancer cases by histologic subtype among U.S. women aged $\geq$ 30 years, SEER 22 (2000–2018). Table S2. Overall five-year relative survival rates (not standardized for age, 2000–2019) of women diagnosed with uterine cancer across demographic, clinical, and socioeconomic characteristics, among U.S. women aged $\geq$ 30 years, diagnosed between the years 2000 and 2018, SEER 22. Table S3. Race/ethnicity specific age-standardized five-year relative survival rates (2000–2019) of women diagnosed with uterine cancer across socioeconomic characteristics, overall and by major histologic subtypes (endometrioid and non-endometrioid cancer) among U.S. women aged $\geq$ 30 years, diagnosed between the years 2000 and 2018, SEER 22. Table S4. Number of women diagnosed with uterine cancer and deaths across race/ethnicity, socioeconomic characteristics, and histology, among U.S. women aged $\geq$ 30 years, diagnosed between the years 2000 and 2018, SEER 22. Table S5. Age-standardized five-year relative survival rates (2000–2019) of women diagnosed with uterine cancer by U.S. geographic region and socioeconomic characteristics among U.S. women aged $\geq$ 30 years, diagnosed between the years 2000 and 2018, SEER 22.

Author Contributions: Conceptualization, and methodology by both A.T.W. and M.A.C.; formal analysis, A.T.W.; writing—original draft preparation, A.T.W.; writing—review and editing, A.T.W., M.A.C., J.K.M.-A., M.S.S., A.A.H.H. and R.C.A.; supervision, M.A.C. All authors have read and agreed to the published version of the manuscript.

Funding: Supported by Intramural Research Program Grant No. Z01 CP010124-21 from the National Cancer Institute, National Institutes of Health.

Institutional Review Board Statement: All data utilized in this study are de-identified and are publicly available, thus the study was exempt from institutional review board approval.

Informed Consent Statement: All data utilized in this study are de-identified and are publicly available, thus the study was exempt from the need for informed consent.

Data Availability Statement: All data utilized in this study are de-identified and are publicly available. Data can be accessed at https://seer.cancer.gov/ (accessed on 8 April 2024).

Acknowledgments: We thank the SEER Program for providing data and software programs used in this study, supported by the National Cancer Institute's (NCI) Division of Cancer Control and Population Sciences (DCCPS).

Conflicts of Interest: The authors declare no conflicts of interest.

References

1. Siegel, R.L.; Giaquinto, A.N.; Jemal, A. Cancer statistics 2024. *CA Cancer J. Clin.* **2024**, *74*, 12–49. [CrossRef] [PubMed]
2. Clarke, M.A.; Devesa, S.S.; Harvey, S.V.; Wentzensen, N. Hysterectomy-Corrected Uterine Corpus Cancer Incidence Trends and Differences in Relative Survival Reveal Racial Disparities and Rising Rates of Nonendometrioid Cancers. *J. Clin. Oncol.* **2019**, *37*, 1895–1908. [CrossRef]
3. Bain, R.P.; Greenberg, R.S.; Chung, K.C. Racial differences in survival of women with endometrial cancer. *Am. J. Obstet. Gynecol.* **1987**, *157*, 914–923. [CrossRef]
4. Baskovic, M.; Lichtensztajn, D.Y.; Nguyen, T.; Karam, A.; English, D.P. Racial disparities in outcomes for high-grade uterine cancer: A California cancer registry study. *Cancer Med.* **2018**, *7*, 4485–4495. [CrossRef]
5. Bregar, A.J.; Rauh-Hain, J.A.; Spencer, R.; Clemmer, J.T.; Schorge, J.O.; Rice, L.W.; Del Carmen, M.G. Disparities in receipt of care for high-grade endometrial cancer: A National Cancer Data Base analysis. *Gynecol. Oncol.* **2017**, *145*, 114–121. [CrossRef] [PubMed]
6. Felix, A.S.; Cohn, D.E.; Brasky, T.M.; Zaino, R.; Park, K.; Mutch, D.G.; Creasman, W.T.; Thaker, P.H.; Walker, J.L.; Moore, R.G.; et al. Receipt of adjuvant endometrial cancer treatment according to race: An NRG Oncology/Gynecologic Oncology Group 210 Study. *Am. J. Obstet. Gynecol.* **2018**, *219*, 459. [CrossRef]
7. Giaquinto, A.N.; Miller, K.D.; Tossas, K.Y.; Winn, R.A.; Jemal, A.; Siegel, R.L. Cancer statistics for African American/Black People 2022. *CA Cancer J. Clin.* **2022**, *72*, 202–229. [CrossRef]
8. Donkers, H.; Bekkers, R.; Massuger, L.; Galaal, K. Systematic review on socioeconomic deprivation and survival in endometrial cancer. *Cancer Causes Control* **2019**, *30*, 1013–1022. [CrossRef] [PubMed]
9. Njoku, K.; Barr, C.E.; Hotchkies, L.; Quille, N.; Wan, Y.L.; Crosbie, E.J. Impact of socio-economic deprivation on endometrial cancer survival in the North West of England: A prospective database analysis. *Br. J. Obstet. Gynaecol.* **2021**, *128*, 1215–1224. [CrossRef]
10. Zadnik, V.; Žagar, T.; Tomšič, S.; Mihor, A.; Lokar, K. Cancer Patients' Survival According to Socioeconomic Environment in a High-Income Country with Universal Health Coverage. *Cancers* **2022**, *14*, 1620. [CrossRef]
11. Schlumbrecht, M.; Wright, K.; George, S. Unique Considerations in Early Detection, Risk, and Awareness of Endometrial Cancer in Black Women. *Cancer Control* **2023**, *30*, 10732748231202952. [CrossRef] [PubMed]
12. Whetstone, S.; Burke, W.; Sheth, S.S.; Brooks, R.; Cavens, A.; Huber-Keener, K.; Scott, D.M.; Worly, B.; Chelmow, D. Health Disparities in Uterine Cancer: Report from the Uterine Cancer Evidence Review Conference, USA. *Obstet. Gynecol.* **2022**, *139*, 645–659. [CrossRef] [PubMed]
13. Sarink, D.; Wilkens, L.R.; White, K.K.; Le Marchand, L.; Wu, A.H.; Setiawan, V.W.; Park, S.L.; Park, S.Y.; Killeen, J.L.; Merritt, M.A. Racial/ethnic differences in anthropometric and hormone-related factors and endometrial cancer risk: The Multiethnic Cohort Study. *Br. J. Cancer* **2021**, *124*, 1724–1733. [CrossRef]
14. Donkers, H.; Bekkers, R.; Massuger, L.F.; Galaal, K. Socioeconomic deprivation and survival in endometrial cancer: The effect of BMI. *Gynecol. Oncol.* **2020**, *156*, 178–184. [CrossRef]
15. Sonderlund, L.A.; Charifson, M.; Schoenthaler, A.; Carson, T.; Williams, N.J. Racialized economic segregation and health outcomes: A systematic review of studies that use the Index of Concentration at the Extremes for race, income, and their interaction. *PLoS ONE* **2022**, *17*, e0262962.
16. Adamkiewicz, G.; Zota, A.R.; Fabian, M.P.; Chahine, T.; Julien, R.; Spengler, J.D.; Levy, J.I. Moving environmental justice indoors: Understanding structural influences on residential exposure patterns in low-income communities. *Am. J. Public Health* **2011**, *101* (Suppl. S1), S238–S245. [CrossRef] [PubMed]
17. Collins, T.W.; Grineski, S.E.; Shaker, Y.; Mullen, C.J. Communities of color are disproportionately exposed to long-term and short-term PM. *Environ. Res.* **2022**, *214*, 114038. [CrossRef] [PubMed]
18. Josey, K.P.; Delaney, S.W.; Wu, X.; Nethery, R.C.; DeSouza, P.; Braun, D.; Dominici, F. Air Pollution and Mortality at the Intersection of Race and Social Class. *N. Engl. J. Med.* **2023**, *388*, 1396–1404. [CrossRef] [PubMed]
19. Helpman, L.; Pond, G.R.; Elit, L.; Anderson, L.N.; Seow, H. Endometrial cancer presentation is associated with social determinants of health in a public healthcare system: A population-based cohort study. *Gynecol. Oncol.* **2020**, *158*, 130–136. [CrossRef] [PubMed]
20. Popescu, I.; Duffy, E.; Mendelsohn, J.; Escarce, J.J. Racial residential segregation, socioeconomic disparities, and the White-Black survival gap. *PLoS ONE* **2018**, *13*, e0193222. [CrossRef]
21. Surveillance, Epidemiology and End Results Program (SEER). SEER*Stat Database: Incidence—SEER Research Plus Data, 18 Registries, Nov 2020 Sub (2000–2018)—Linked To County Attributes—Total U.S., 1969–2019 Counties, National Cancer Institute, DCCPS, Surveillance Research Program, Released April 2021 (November 2020 Submission). Available online: www.seer.cancer.gov (accessed on 8 April 2024).

22. Clarke, M.A.; Devesa, S.S.; Hammer, A.; Wentzensen, N. Racial and ethnic differences in hysterectomy-corrected uterine corpus cancer mortality by stage and histologic subtype. *JAMA Oncol.* **2022**, *8*, 895–903. [CrossRef] [PubMed]
23. Surveillance, Epidemiology and End Results Program (SEER). Race and Hispanic Ethnicity Changes (November 2021 Submission). Available online: https://seer.cancer.gov/seerstat/variables/seer/race_ethnicity/ (accessed on 25 August 2023).
24. Gomez, S.L.; Le, G.M.; West, D.W.; Satariano, W.A.; O'Connor, L. Hospital policy and practice regarding the collection of data on race, ethnicity, and birthplace. *Am. J. Public Health* **2003**, *93*, 1685–1688. [CrossRef] [PubMed]
25. Surveillance, Epidemiology and End Results Program (SEER). Localized/Regional/Distant Stage Adjustments 2023. Available online: https://seer.cancer.gov/seerstat/variables/seer/yr1975_2020/lrd_stage/index.html#footnotea (accessed on 8 April 2024).
26. Surveillance, Epidemiology and End Results Program (SEER). Summary Stage Manual—Female Genital System 2000. Available online: https://seer.cancer.gov/tools/ssm/ssm2000/breast_femgen.pdf (accessed on 7 February 2024).
27. Berkowitz, S.A.; Traore, C.Y.; Singer, D.E.; Atlas, S.J. Evaluating area-based socioeconomic status indicators for monitoring disparities within health care systems: Results from a primary care network. *Health Serv. Res.* **2015**, *50*, 398–417. [CrossRef] [PubMed]
28. Surveillance, Epidemiology and End Results Program (SEER). SEER*Stat Rate Exercise 5: Incidence Rates by County Attributes. Available online: https://seer.cancer.gov/seerstat/tutorials/rate5/webprint/ (accessed on 8 April 2023).
29. Mariotto, A.B.; Zou, Z.; Johnson, C.J.; Scoppa, S.; Weir, H.K.; Huang, B. Geographical, racial and socio-economic variation in life expectancy in the US and their impact on cancer relative survival. *PLoS ONE* **2018**, *13*, e0201034. [CrossRef] [PubMed]
30. Centers for Disease Control and Prevention (CDC), Division of Cancer Prevention and Control. Suppression of Rates and Counts (Published 2023). Available online: https://www.cdc.gov/cancer/uscs/technical_notes/stat_methods/suppression.htm (accessed on 23 May 2024).
31. U.S. Department of Commerce Economics and Statistics Administration U.S. Census Bureau. Census Regions and Divisions of the United States 2013. Available online: https://www2.census.gov/geo/pdfs/maps-data/maps/reference/us_regdiv.pdf (accessed on 1 July 2024).
32. Snider, N.G.; Hastert, T.A.; Nair, M.; Kc, M.; Ruterbusch, J.J.; Schwartz, A.G.; Peters, E.S.; Stoffel, E.M.; Rozek, L.S.; Purrington, K.S. Area-level Socioeconomic Disadvantage and Cancer Survival in Metropolitan Detroit. *Cancer Epidemiol. Biomark. Prev.* **2023**, *32*, 387–397. [CrossRef] [PubMed]
33. Madison, T.; Schottenfeld, D.; James, S.A.; Schwartz, A.G.; Gruber, S.B. Endometrial cancer: Socioeconomic status and racial/ethnic differences in stage at diagnosis, treatment, and survival. *Am. J. Public Health* **2004**, *94*, 2104–2111. [CrossRef] [PubMed]
34. Cheung, M.R. African American race and low income neighborhoods decrease cause specific survival of endometrial cancer: A SEER analysis. *Asian Pac. J. Cancer Prev.* **2013**, *14*, 2567–2570. [CrossRef] [PubMed]
35. Von Behren, J.; Abrahão, R.; Goldberg, D.; Gomez, S.L.; Setiawan, V.W.; Cheng, I. The influence of neighborhood socioeconomic status and ethnic enclave on endometrial cancer mortality among Hispanics and Asian Americans/Pacific Islanders in California. *Cancer Causes Control* **2018**, *29*, 875–881. [CrossRef] [PubMed]
36. Yang, T.C.; Zhao, Y.; Song, Q. Residential segregation and racial disparities in self-rated health: How do dimensions of residential segregation matter? *Soc. Sci. Res.* **2017**, *61*, 29–42. [CrossRef]
37. Shariff-Marco, S.; Gomez, S.L.; Canchola, A.J.; Fullington, H.; Hughes, A.E.; Zhu, H.; Pruitt, S.L. Nativity, ethnic enclave residence, and breast cancer survival among Latinas: Variations between California and Texas. *Cancer* **2020**, *126*, 2849–2858. [CrossRef]
38. Cha, J.; Bustamante, G.; Lê-Scherban, F.; Duprez, D.; Pankow, J.S.; Osypuk, T.L. Ethnic enclaves and incidence of cancer among US ethnic minorities in the multi-ethnic study of atherosclerosis. *J. Racial Ethn. Health Disparities* **2023**, 1–12, *preprint*. [CrossRef] [PubMed]
39. Karia, P.S.; Huang, Y.; Tehranifar, P.; Wright, J.D.; Genkinger, J.M. Racial and ethnic differences in type II endometrial cancer mortality outcomes: The contribution of sociodemographic, clinicopathologic, and treatment factors. *Gynecol. Oncol.* **2023**, *168*, 119–126. [CrossRef]
40. Doll, K.M. Investigating Black-White disparities in gynecologic oncology: Theories, conceptual models, and applications. *Gynecol. Oncol.* **2018**, *149*, 78–83. [CrossRef]
41. Williams, D.R.; Collins, C. Racial residential segregation: A fundamental cause of racial disparities in health. *Public Health Rep.* **2001**, *116*, 404–416. [CrossRef] [PubMed]
42. Doll, K.M.; Khor, S.; Odem-Davis, K.; He, H.; Wolff, E.M.; Flum, D.R.; Ramsey, S.D.; Goff, B.A. Role of bleeding recognition and evaluation in Black-White disparities in endometrial cancer. *Am. J. Obstet. Gynecol.* **2018**, *219*, e593. [CrossRef] [PubMed]
43. Doll, K.M.; Khor, S.; Odem-Davis, K.; He, H.; Wolff, E.M.; Flum, D.R.; Ramsey, S.D.; Goff, B.A. Assessment of Prediagnostic Experiences of Black Women with Endometrial Cancer in the United States. *JAMA Netw. Open* **2020**, *3*, e204954. [CrossRef]
44. Barrington, D.A.; Sinnott, J.A.; Nixon, D.; Padamsee, T.J.; Cohn, D.E.; Doll, K.M.; Donneyong, M.M.; Felix, A.S. More than treatment refusal: A National Cancer Database analysis of adjuvant treatment refusal and racial survival disparities among women with endometrial cancer. *Am. J. Obstet. Gynecol.* **2022**, *227*, e244. [CrossRef] [PubMed]
45. Moore, J.X.; Andrzejak, S.E.; Bevel, M.S.; Jones, S.R.; Tingen, M.S. Exploring racial disparities on the association between allostatic load and cancer mortality: A retrospective cohort analysis of NHANES, 1988 through 2019. *SSM Popul. Health* **2022**, *19*, 101185. [CrossRef]

46. Moore, J.X.; Bevel, M.S.; Aslibekyan, S.; Akinyemiju, T. Temporal changes in allostatic load patterns by age, race/ethnicity, and gender among the US adult population; 1988–2018. *Prev. Med.* **2021**, *147*, 106483. [CrossRef]
47. Hicks, M.L.; Hicks, M.M.; Mathews, R.P.; Khabele, D.; Clare, C.A.; Balogun, O.; Lawson, Y.R.; Tillman, R.H.; Butler, R.; Spann, C.O.; et al. Racial disparities in endometrial cancer: Where are we after 26 years? *Gynecol. Oncol.* **2024**, *184*, 236–242. [CrossRef]
48. Pinheiro, P.S.; Medina, H.; Callahan, K.E.; Kwon, D.; Ragin, C.; Sherman, R.; Kobetz, E.N.; Jemal, A. Cancer mortality among US blacks: Variability between African Americans, Afro-Caribbeans, and Africans. *Cancer Epidemiol.* **2020**, *66*, 101709. [CrossRef] [PubMed]
49. Rabe, M.; Exploring the Racial and Ethnic Diversity of Various Age Groups. 2023: United States Census Bureau. Available online: https://www.census.gov/newsroom/blogs/random-samplings/2023/09/exploring-diversity.html (accessed on 25 August 2023).
50. Wakkerman, F.C.; Wu, J.; Putter, H.; Jurgenliemk-Schulz, I.M.; Jobsen, J.J.; Lutgens, L.C.; Haverkort, M.A.; de Jong, M.; Mens, J.W.M.; Wortman, B.G.; et al. Prognostic impact and causality of age on oncological outcomes in women with endometrial cancer: A multimethod analysis of the randomised PORTEC-1, PORTEC-2, and PORTEC-3 trials. *Lancet Oncol.* **2024**, *24*, 779–789. [CrossRef] [PubMed]
51. Yu, B.J.; Gross, C.P.; Wilson, L.D.; Smith, B.D. NCI SEER public-use data: Applications and limitations in oncology research. *Oncology* **2009**, *23*, 288–295. [PubMed]

cancers

MDPI

Review

Early Detection and Prevention of Ovarian Cancer

Lauren Montemorano [1], Laura Huffman [2] and Lisa Barroilhet [2,*]

1 Department of Obstetrics and Gynecology, Kaiser Permanente, Oakland, CA 94611, USA; lmontemorano@gmail.com
2 Carbone Cancer Center, University of Wisconsin, Madison, WI 53792, USA; huffman3@wisc.edu
* Correspondence: barroilhet@wisc.edu

Simple Summary

Epithelial ovarian cancer (EOC) is the second most fatal gynecologic malignancy, with an estimated 21,410 new cases and 13,770 deaths reported in the United States in 2021. Throughout this paper, the term "ovarian cancer" will refer to malignancies originating in the ovary, fallopian tube, and peritoneal cavity, as these share similar histologic features and are managed using the same treatment strategies. While early-stage ovarian cancer is often curable, the majority of patients—approximately 65%—present with advanced disease. These cases typically require extensive surgery and chemotherapy and are associated with high recurrence and mortality rates. Consequently, improving early detection and implementing effective prevention strategies remain critical priorities. Currently, there is no screening test that has been shown to reduce mortality from ovarian cancer, even in high-risk populations. Prevention strategies are typically limited to surgical intervention. This review article summarizes the current state and potential future directions for the early detection and prevention of ovarian cancer.

Abstract

Background/Objectives: Epithelial ovarian cancer (EOC) encompasses ovarian, fallopian tube and peritoneal malignancies. It is a deadly disease and is rarely cured when diagnosed at advanced stages. Early-stage disease is often curable, but clinicians and researchers have been stymied in their attempts to reliably screen for this disease, even in high-risk populations. Effective prevention of ovarian cancer is usually limited to the use of combined oral contraceptives and removal of the ovaries and fallopian tubes. **Methods and Results**: We aim to review the current guidelines and the evidence reported for both the early detection and prevention of ovarian cancer. Novel imaging techniques, biomarkers, and surgical advances will be discussed. **Conclusions**: This review will offer (a) an understanding of the epidemiology of EOC (b) analysis and a discussion of relevant molecular markers that might be exploited for more accurate early detection (c) medical and surgical methods to prevent ovarian cancer.

Keywords: ovarian cancer; prevention; early detection

1. Introduction

Epithelial ovarian cancer (EOC) constitutes the second most lethal gynecologic malignancy, with a projected 21,410 new cases and 13,770 deaths recorded in the United States in 2021 [1]. For the purposes of this chapter, the term "ovarian cancer" will be used to describe cancers arising in the ovary, fallopian tube, and peritoneal cavity, all of which are histologically similar and treated identically. In ovarian cancer, early-stage disease is frequently curable. Unfortunately, the majority (upward of 65%) of patients present with advanced-stage disease, requiring invasive surgery and chemotherapy, with high rates of relapse and death [2]. Early detection and prevention are areas of critical need.

When we consider a patient's risk of ovarian cancer, the strongest patient-related risk factor is increasing age. The median age at diagnosis is 63 years old. The incidence of ovarian cancer increases with each additional year of life, climbing from 15.7 per 100,000 age 40 years to 54 per 100,000 at age 70 years [3]. Other reproductive factors can alter risk. Pregnancy and parity are associated with a decreased risk of ovarian cancer. Compared to nulliparous women, those who have had one live birth have approximately a 25% decrease in risk of ovarian cancer, while women with two or more live births have an approximate 42% risk reduction. Other protective factors include breastfeeding, bilateral tubal ligation, hysterectomy, and oral contraceptive pills, particularly with more than five years of continuous use [4].

Genetic factors, including family history and germline mutations, can drastically increase a woman's lifetime risk of developing ovarian cancer. Up to twenty percent of ovarian cancers are related to identified, inheritable mutations in certain genes. Hereditary ovarian cancer syndromes are most often linked to BRCA1 and BRCA2 genes, which are considered the highest risk genes, with lifetime risks as high as 46% and 20%, respectively [5]. Many other genes are emerging as being associated with higher risk, including BRIP1, PALB2, RAD51C, RAD51D, and SMARCA4, [6] with increased risks between 5–15%. Other patient-related factors that are associated with higher risk of ovarian cancer include race (white women are at highest risk), early menarche and later menopause [7].

2. Molecular Features, Pre-Cursor Lesions and Carcinogenesis

There are generally considered to be two groups of epithelial ovarian cancer. Type I cancer comprises low-grade serous, low-grade endometrioid, clear cell and mucinous carcinomas. The behavior of these cancers is typically slower-growing, and they often present with tumor confined to the ovary, rather than disseminated disease. Type I cancers are often characterized by specific somatic mutations, including *KRAS*, *BRAF*, *ERBB2*, *CTNNB1*, *PTEN*, *PIK3CA*, *ARID1A*, and *PPP2R1A*, which target cell signaling pathways. Type I tumors rarely harbor *TP53* mutations as opposed to Tyle II tumors. These cancers appear to develop in a stepwise fashion from precursor lesions, such as borderline tumors and endometriosis.

Type II cancers include high-grade serous, high-grade endometrioid, malignant mixed mesodermal tumors (carcinosarcomas), and undifferentiated carcinomas. They are aggressive, present in advanced stage, and have a very high frequency of *TP53* mutations (>95%). Serous tubal intraepithelial carcinomas (STIC lesions) are now known to be very common precursor lesions for high-grade serous carcinomas and are also *TP53* mutated. STIC lesions spread from the open fimbriated ends of the fallopian tubes throughout the peritoneal cavity [8–10]. A large group of studies included patients with and without known genetic risk and confirmed that STICs and early invasive tubal carcinomas occurred in both groups. STIC lesions have also been identified in women without ovarian cancer who have their ovaries and tubes removed for other indications. The presence of identical TP53 mutations in STICs and concomitant ovarian high-grade serous cancers supports a clonal relationship between the two.

Identifying a known precursor lesion is an important step as scientists and clinicians attempt to detect ovarian cancer at earlier stages. Ovarian cancer development is a multistage process, the details of which are still being elucidated. The identification and characterization of oncogenes has provided the structural link between the carcinogen/promoter model of malignancy and the biochemical pathways that first become altered as cells become cancerous. Researchers have shown that malignant cells must have the ability to generate their own mitogenic signals, resist growth-inhibitory controls, evade apoptosis and to proliferate without limits while acquiring vasculature [11]. Additionally, the process of STIC lesions undergoing malignant transformation may produce changes in cellular metabolism. The above pathways put into context and offer opportunities to explore different approaches to the prevention and early detection of ovarian cancer.

3. Traditional and Novel Detection Methods

3.1. Circulating Biomarkers and Cytology

3.1.1. Protein Biomarkers

Currently, Carbohydrate Antigen 125 (CA-125) and Human Epididymis protein 4 (HE4) are the only two markers that have been approved by the Federal Drug Administration (FDA) for monitoring treatment response and detecting disease recurrence in ovarian cancer patients. CA-125 is a mucin-type glycoprotein (MUC16) that is elevated in the serum of 83% of patients with ovarian cancer, but only in 50–60% of patients with stage I disease. Overexpression in other cancers, and benign diseases of the ovaries and reproductive tract (e.g., endometriosis), menstruation and pregnancy, give the test a lower specificity [12]. Due to low reliability (70% sensitivity and 87% specificity), CA-125 is not recommended for population screening [13].

Human epididymis protein (HE4) has been developed as a biomarker to improve the early detection of ovarian cancer. It is a glycoprotein belonging to the larger protein family called "WAP" for whey acidic proteins. It is secreted by epithelial ovarian cancer cells. HE4 levels are not elevated in endometriosis, which improves its specificity [14]. HE4 has proven the most useful in triaging patients with known pelvic masses to cancer specialists when indicated. Its sensitivity for detecting ovarian cancer is 65–83%, specificity is 78–99% [15].

Recently, OVERA® was approved as referral or triage test for patients presenting with any type of ovarian mass, whether it be symptomatic or asymptomatic. The test is a combination of CA-125, HE4, apolipoprotein A-1, follicle-stimulating hormone (FSH), and transferrin. Previous studies have shown a relatively high sensitivity (91%) and a lower specificity (69%) limiting its clinical usefulness with a false positive rate similar to other serum markers [14].

Other blood biomarkers have been evaluated for their ability to detect early-stage ovarian cancer. Many have been tested in combination with CA-125. HE4 and CA-125, when combined, have a sensitivity of 76.4% and specificity of 95% for predicting if a pelvic mass is cancerous [16]. Other biomarkers of interest were identified from a large cohort of ovarian cancer screening studies, and include transthyretin, CA72.4, and CA15.3 [17]. Transthyretin is a plasma protein synthesized by the liver and plays a role in cellular metabolism. CA72.4 and CA15.3 are glycoproteins most commonly associated with gastrointestinal and breast cancers, respectively. These biomarkers were tested in 810 invasive epithelial ovarian cancer cases and controls from the European Prospective Investigation into Cancer and Nutrition (EPIC) study. Receiver operator performance was highest for CA-125 (0.92), followed by HE4 (0.84), CA72.4 (0.77), and CA15.3 (0.73) [18].

3.1.2. Autoantibodies

Using proteomics, it is possible not only to identify new protein biomarkers, but also to find human immunoglobulin bound proteins and detect autoantibodies. Autoantibodies target tumor-associated antigens. Small volumes of cancer may not release adequate amounts of antigen to elevate serum protein levels, but could induce a human immune response and present an opportunity for identification of disease before widespread metastases or large primary tumors are present [19]. For example, the tumor suppressor gene *TP53* is mutated almost universally in high-grade serous ovarian cancers in addition to precursor STIC lesions [20]. TP53 autoantibody levels are increased in more than 20% of pre-treatment sera from patients with both early- and late-stage ovarian cancer. Previous studies showed that about 15% of ovarian cancer patients can be detected measuring autoantibodies against wild-type TP53 in sera [21]. Several other candidate autoantibodies have been identified, including homeobox gene A7 (HOXA7) interleukin 8 (IL-8), TRIM21, and NY-ESO01 [22–24], all of which are under investigation.

3.1.3. MicroRNAs

MicroRNAs (miRNAs) are short (~22 nucleotides), single-stranded non-coding RNAs that play a key role in regulating gene expression at the post-transcriptional level [25]. These molecules are found circulating in various body fluids, where they remain stable by associating with proteins or being enclosed in exosomes, microvesicles, or apoptotic bodies. Because their expression patterns shift in response to pathological conditions (like cancer), circulating miRNAs are being explored as promising biomarkers for cancer detection [26,27]. In ovarian cancer, miRNA expression is often dysregulated, which can be useful in distinguishing malignant cells from normal ovarian tissue [28].

Several serum miRNAs (Figure 1), including miR-182 and members of the miR-200 family (miR-200a, miR-200b, miR-200c), are elevated in patients with serous epithelial ovarian cancer (EOC). Notably, the combination of miR-200b and miR-200c demonstrated good diagnostic performance, with an AUC of 0.784 for distinguishing serous EOC from healthy controls [26]. Another combination—miR-205 and let-7f—showed even higher diagnostic accuracy (AUC = 0.831, 95% CI: 0.772–0.880), with 62.4% sensitivity and 92.9% specificity, particularly effective in detecting stage I disease [29]. Additionally, an eight-miRNA panel was able to differentiate early-stage EOC from benign ovarian tumors with 86% sensitivity and 83% specificity [30,31]. Ongoing prospective clinical trials, such as NCT05146505, are currently evaluating the utility of miRNAs for early ovarian cancer detection.

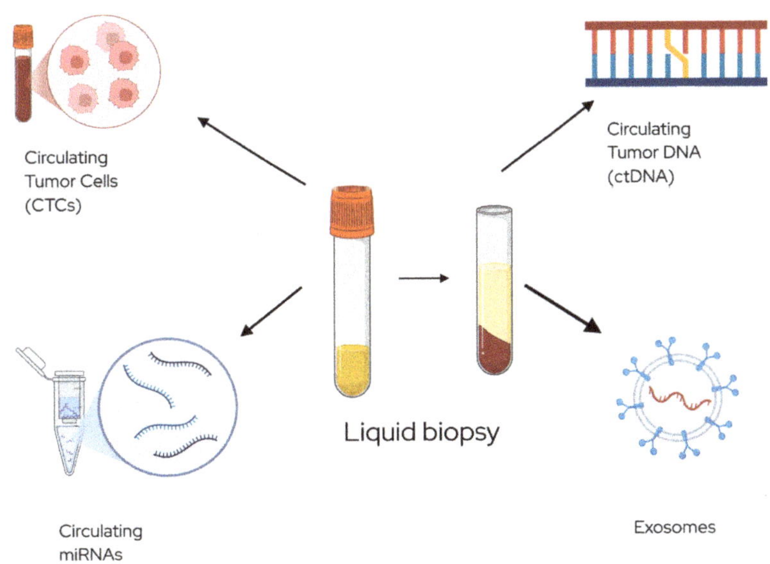

Figure 1. Examples of output from liquid biopsy. Created in BioRender. Barroilhet, L. (2025) https://BioRender.com/telkoq3 (accessed on 6 October 2025).

3.1.4. Urine Testing

The potential of urine as a source of ovarian cancer biomarkers is the subject of ongoing research [32]. Eosinophil protein X, a glycosylated form of eosinophil-derived neurotoxin and carboxylic acid-terminated fragments of osteopontin [33] have both been detected in the urine of ovarian cancer patients. Interestingly, eosinophil protein X has been shown to have antitumor and antiangiogenic activity via induction of apoptosis of endothelial cells and may also have therapeutic potential. Osteopontin is correlated with systemic inflammation. Potential urine biomarkers also include HE4 [34] and HMGA1 [35].

3.1.5. Circulating Tumor DNA

Circulating tumor DNA (ctDNA) refers to the fraction of cell-free DNA in the bloodstream that carries tumor-specific somatic mutations [36]. Assessing ctDNA in serum offers a promising, noninvasive approach for cancer detection and disease monitoring [37]. Advances in assay technologies have significantly enhanced sensitivity, enabling detection of cancers at early stages when ctDNA levels are very low. In a small cohort of seven ovarian cancer patients, Bettegowda et al. reported ctDNA detection in all cases (100%) [38]. Targeted deep sequencing has further improved detection, identifying mutations at allelic frequencies as low as 2% with greater than 97% sensitivity and specificity [39].

A novel multi-analyte blood test, CancerSEEK, combines analysis of eight protein biomarkers—including CA-125—with mutation screening across 1933 genomic sites, encompassing single-base substitutions, insertions, and deletions [40]. In a case–control study spanning eight cancer types, CancerSEEK demonstrated its highest accuracy in ovarian cancer, achieving 98% sensitivity and over 99% specificity, with only seven false positives among 812 cancer-free individuals. However, most participants had advanced-stage disease, which could have been detected using CA-125 alone. Further validation in early-stage ovarian cancer patients remains essential [40].

3.1.6. DNA Methylation

Aberrant DNA methylation, particularly in promoter regions, is recognized as an early molecular event in the development of cancer [31]. This epigenetic alteration can disrupt normal gene regulation, leading to silencing of tumor suppressor genes or activation of oncogenes [41]. In a study involving 250 patients with epithelial ovarian cancer (EOC) undergoing chemotherapy, along with a larger cohort from an ultrasound-based screening trial, a panel of three serum DNA methylation markers was evaluated. These markers successfully identified more than half of ovarian cancer cases in blood samples collected up to two years before clinical diagnosis [42]. While similar methylation-based approaches have been effectively translated into clinical diagnostics for colorectal cancer, the diverse histological subtypes of ovarian cancer present a significant obstacle to creating a biomarker panel that is both sensitive and specific [43].

3.1.7. Microbiome

The human body is colonized by diverse communities of microorganisms that reside within the mucosal surfaces, known as the microbiome. The microbiome has many necessary and important functions, including nutrient absorption, metabolism of hormones and vitamins, immune system modulation, and protection against pathogenic infections. Contemporary next-generation sequencing technologies have advanced our understanding that distinct bacterial populations also inhabit the uterus, fallopian tubes, and peritoneum [36,44].

In the majority of women of reproductive age, the microbiota of vagina and cervix microenvironment is dominated by Lactobacillus species. By contrast, the uterus, ovaries and fallopian tubes are sterile in healthy individual. Dysbiosis, or alterations to the bacterial communities and their diversity, within the microbiome has been associated with many diseases of the female reproductive tract, such as recurrent urinary tract infections, endometriosis, infertility, pre-invasive disease, and gynecologic cancers. When dysbiosis occurs, altered immune and metabolic signaling can cause chronic inflammation, epithelial barrier breach, changes in cellular proliferation and apoptosis, genome instability, angiogenesis and metabolic dysregulation. These pathophysiological changes could be a part of the complicated process of carcinogenesis in gynecologic organs [45]. Certain microbes can elicit DNA damage and apoptosis directly via the production of toxic metabolites or indirectly via reactive oxygen species. Furthermore, bacteria associated with malignant transformation, such as *Salmonella enterica* serovar *typhi*, *Helicobacter pylori*, and *Fusobacteria* spp., have been shown to upregulate oncogenic pathways, including Β-catenin signaling, and increase the production of pro-inflammatory cytokines [46]. Differences between the microbiome in benign and malignant disease processes are being studied (NCT 03388996).

3.1.8. Fallopian Tube Cytology and Tumor DNA Detection in Pap Smears

Recognition that many ovarian cancers arise in the fallopian tube has resulted in the development of fallopian tube tissue sampling devices. More recently, in a small feasibility study, Lum et al. attempted cytologic sampling of the fallopian tubes using hysteroscopic brush cytology without visualization [47]. While they were able to perform cellular collection laparoscopically, hysteroscopic methods were less successful [48]. Others have verified cytologic sampling from excised tubes [49], but again, this has not been duplicated consistently using a less invasive hysteroscopic approach. Several groups have started to characterize the cytology of the fallopian tube as well as endometrial cytology for early detection of ovarian cancer, although further analysis and standardization is needed [50,51].

Liquid-based cervical Pap smear testing, which combines cytological evaluation with DNA analysis, has recently been explored as a potential screening method for cancers beyond the cervix [52]. In a study by Kinde et al., DNA mutations were detected in liquid Pap smear samples from 100% of patients with endometrial cancer and 41% of those with ovarian cancer [53]. Additionally, in a small cohort of patients with advanced high-grade serous ovarian cancer, TP53 mutations were identified in three out of five individuals who had intact fallopian tubes and used a tampon prior to surgery. In contrast, none of the three patients with prior tubal ligation showed detectable mutations [20].

3.2. Imaging

3.2.1. Ultrasound with Contrast

Ultrasound is a safe, widely available, and relatively inexpensive imaging modality commonly used to visualize the female pelvic organs. However, even for screening high-risk populations for ovarian cancer, its utility has been limited. The introduction of molecularly targeted contrast microbubbles that can bind to specific molecules expressed in ovarian cancer has been the subject of recent research studies to improve early detection and potentially drug delivery [54–56].

Contrast microbubbles are micron-sized, intravascular contrast agents consisting of a gaseous core contained by a shell. The introduction of molecularly targeted contrast microbubbles that can bind to certain molecules expressed in cancer has made ultrasound a molecular imaging modality that allows improved detection, characterization, and monitoring. By functionalizing the shell with binding ligands to certain molecules, microbubble-enhanced ultrasound can visualize molecules such as kinase insert domain receptor (KDR), one of the key regulators of angiogenesis preferentially expressed in various cancers, including ovarian cancer [55].

Twenty-four women (age 48 to 79 years) with focal ovarian lesions were injected intravenously with MBKDR (0.03 to 0.08 mL/kg of body weight) ultrasound of the ovaries was performed starting 5 min after injection. Persistent focal MBKDR binding was assessed. Patients underwent surgical resection of the target lesions, and tissues were stained for CD31 and KDR by immunohistochemistry. Overall, the safety profile of MBKDR was good, with infrequent mild or moderate self-resolving adverse events [55]. Although microbubble enhancement holds promise for improving specificity, it may not improve detection of early-stage disease. In the case of precancerous lesions within the fallopian tubes, microbubbles may not add any diagnostic advantage.

3.2.2. Magnetic Resonance Imaging

Both magnetic resonance imaging (MRI) and magnetic relaxometry (MRX) can detect and locate magnetic nanoparticles without ionizing radiation. MRX is a highly sensitive technique that detects and quantifies magnetic nanoparticles in biological tissues by measuring their decay after removal of an external magnetic field. MRX is being studied for the early detection of many cancers, but is not ready for routine adaption into clinical practice [57,58]. For ovarian cancer, it holds the most promise in distinguishing benign endometriosis lesions from endometriosis-associated ovarian cancer [59,60].

4. Novel Prevention Methods

As discussed above, early detection methods for ovarian cancer are evolving, with no consistent recommendations for screening even high-risk patients. This informs the landscape (and underscores the importance) of ovarian cancer prevention [61].

4.1. Salpingectomy

Large clinical screening trials using transvaginal ultrasound and serum CA-125 have not demonstrated a reduction in EOC specific mortality. Accordingly, development of sensitive early detection methods that achieve high specificity represents a critical unmet need. Opportunistic salpingectomy performed as an alternative to tubal ligation at the time of hysterectomy can prevent cancers originating in the fallopian tubes [61]. Opportunistic salpingectomy at the time of cesarean delivery has replaced tubal ligation as the most common type of sterilization [62]. Postpartum opportunistic salpingectomy after vaginal delivery has also been shown to be a cost effective and safe option [63,64].

Removal of the ovaries and fallopian tubes is an effective approach to reducing ovarian cancer risk and is routinely recommended in high-risk populations. While this approach is very effective (>95% risk reduction), it often comes with undesirable side effects, such as a truncated fertility window and the negative effects of early menopause [65]. A proposed approach in high-risk women who are premenopausal is tubal removal (salpingectomy) with delayed oophorectomy closer to the age of natural menopause. Salpingectomy does not affect ovarian reserve in the short term and provides effective non-hormonal birth control. Concerns with salpingectomy for ovarian cancer reduction include the possibility of residual tubal tissue remaining on the ovarian surface. This concern was validated when tubes and ovaries were removed separately during surgery and examined histologically as two separate specimens; residual fimbrial tissue was found on the ovarian surface in 16% of cases [66]. This suggests that salpingectomy may not prevent all ovarian cancers of tubal origin. Ongoing, prospective, observational studies are following up participants who choose among different surgical prevention options. These studies include TUBA-WISP-II [67] (international), SOROCk (US) [68] and PROTECTOR (UK) [69].

Opportunistic salpingectomy is also being studied as a risk reduction technique in the general population. In this model, average-risk patients who are having abdominal surgery for another indication have their fallopian tubes removed at the same time. Retrospective population-based data on bilateral salpingectomy with ovarian conservation from Sweden and Denmark suggest that it is associated with a 42% and 65% ovarian cancer risk reduction, respectively [70,71]. Both studies compared salpingectomy with no surgery as opposed to hysterectomy without salpingectomy. We are awaiting mature survival data from a retrospective cohort studied entitled Hysterectomy and Opportunistic Salpingectomy [72]. Currently there is insufficient evidence to accurately estimate the magnitude of EOC risk reduction with opportunistic salpingectomy in average risk patients or salpingectomy alone in high-risk populations.

4.2. Chemoprevention

Combined oral contraceptives (COCs) have long been used to reduce ovarian cancer risk, although their use is more controversial in BRCA positive and other high-risk populations, as COC use has been linked to an increased risk of breast cancer [73]. A recently published meta-analysis found eight cohort studies that investigated oral contraceptive use in relation to ovarian cancer rate. Compared with non-users, all hormonal contraceptive users had a 36% lower rate ratio of ovarian cancer (0.65, 95% CI 0.60, 0.68) [74] without an increase in breast cancer risk. From a prevention standpoint, combined (estrogen/progesterone) oral contraceptives are the most effective non-surgical method available. Continuous COC used for more than 5 years can reduce the risk of ovarian cancer by >50% [75].

A recent meta-analysis based on retrospective data found that any intrauterine device (IUD) use, a common contraceptive option, was associated with a 32% reduction in the incidence of ovarian cancer [76].

Many analgesic drugs have been studied for the prevention of ovarian cancer, with mixed results. Meta-analyses and pooled analyses of cohort and case–control studies have found that aspirin may reduce ovarian cancer risk by 10–20%, particularly when used frequently (i.e., daily or almost daily) [77–80]. Aspirin is now being used as a cancer preventing agent (600 mg/day for at least two years) for Lynch syndrome patients, based on its ability to reduce the risk of colorectal, ovarian and endometrial cancers [81]. Results surrounding NSAID and acetaminophen use as it relates to ovarian cancer risk reduction are inconsistent [82–84].

Natural compounds with antiangiogenic properties, such as curcumin, resveratrol, and silibinin have shown promising preclinical effects in cancer prevention [85–89] as have retinoids [90]. Small molecular inhibitors that block oxidative phosphorylation and may prevent early changes in cellular metabolism related to glycolysis are also being studied [91].

5. Conclusions and Future Directions

In conclusion, multiple innovative methods for early detection and prevention are under investigation. The current state focuses on identification of high-risk patients and early adoption of chemoprevention or surgical tactics to reduce risk. Although successful early detection of ovarian cancer would undoubtedly save lives, no imaging, plasma or urine study is currently FDA-approved for this purpose. Stepwise or multi-modality screening may enhance sensitivity and specificity. Population-based screening is not currently recommended, as no screening test has been able to demonstrate reduced mortality from ovarian cancer, even when implemented in a high-risk population.

Author Contributions: Conceptualization, visualization, methodology, writing (original draft preparation, review and editing): L.M., L.H., L.B. Resources, data curation, supervision and project administration: L.B. All authors have read and agreed to the published version of the manuscript.

Funding: This research received no external funding.

Conflicts of Interest: The authors declare no conflicts of interest.

Abbreviations

The following abbreviations are used in this manuscript:

CA-125	Carbohydrate Antigen 125
DNA	Deoxyribonucleic acid
EOC	Epithelial ovarian cancer
HE4	Human epididymis protein 4
HGSOC	High-grade serous ovarian cancer
HMGA	High mobility group A
IUD	Intrauterine device
MRA	Magnetic relaxometry
MRI	Magnetic resonance imaging
NSAID	Non-steroidal anti-inflammatory drugs
STIC	Serous tubal intraepithelial carcinoma

References

1. Siegel, R.L.; Miller, K.D.; Fuchs, H.E.; Jemal, A. Cancer Statistics, 2021. *CA Cancer J. Clin.* **2021**, *71*, 7–33. [CrossRef]
2. Goff, B.A.; Mandel, L.; Muntz, H.G.; Melancon, C.H. Ovarian carcinoma diagnosis. *Cancer* **2000**, *89*, 2068–2075. [CrossRef]
3. Berek, J.S.; Renz, M.; Kehoe, S.; Kumar, L.; Friedlander, M. Cancer of the ovary, fallopian tube, and peritoneum: 2021 update. *Int. J. Gynaecol. Obstet.* **2021**, *155*, 61–85. [CrossRef] [PubMed]
4. Torre, L.A.; Trabert, B.; DeSantis, C.E.; Miller, K.D.; Samimi, G.; Runowicz, C.D.; Gaudet, M.M.; Jemal, A.; Siegel, R.L. Ovarian cancer statistics, 2018. *CA Cancer J. Clin.* **2018**, *68*, 284–296. [CrossRef]
5. Kotsopoulos, J.; Hathaway, C.A.; Narod, S.A.; Teras, L.R.; Patel, A.V.; Hu, C.; Yadav, S.; Couch, F.J.; Tworoger, S.S. Germline Mutations in 12 Genes and Risk of Ovarian Cancer in Three Population-Based Cohorts. *Cancer Epidemiol. Biomark. Prev.* **2023**, *32*, 1402–1410. [CrossRef]
6. Konstantinopoulos, P.A.; Norquist, B.; Lacchetti, C.; Armstrong, D.; Grisham, R.N.; Goodfellow, P.J.; Kohn, E.C.; Levine, D.A.; Liu, J.F.; Lu, K.H.; et al. Germline and Somatic Tumor Testing in Epithelial Ovarian Cancer: ASCO Guideline. *J. Clin. Oncol.* **2020**, *38*, 1222–1245. [CrossRef]
7. Webb, P.M.; Jordan, S.J. Epidemiology of epithelial ovarian cancer. *Best Pract. Res. Clin. Obstet. Gynaecol.* **2017**, *41*, 3–14. [CrossRef] [PubMed]
8. Roh, M.H.; Kindelberger, D.; Crum, C.P. Serous tubal intraepithelial carcinoma and the dominant ovarian mass: Clues to serous tumor origin? *Am. J. Surg. Pathol.* **2009**, *33*, 376–383. [CrossRef]
9. Crum, C.P.; Drapkin, R.; Miron, A.; Ince, T.A.; Muto, M.; Kindelberger, D.W.; Lee, Y. The distal fallopian tube: A new model for pelvic serous carcinogenesis. *Curr. Opin. Obstet. Gynecol.* **2007**, *19*, 3–9. [CrossRef] [PubMed]
10. Gockley, A.A.; Elias, K.M. Fallopian tube tumorigenesis and clinical implications for ovarian cancer risk-reduction. *Cancer Treat. Rev.* **2018**, *69*, 66–71. [CrossRef]
11. Hahn, W.C.; Weinberg, R.A. Rules for making human tumor cells. *N. Engl. J. Med.* **2002**, *347*, 1593–1603. [CrossRef] [PubMed]
12. Bast, R.C., Jr.; Xu, F.J.; Yu, Y.H.; Barnhill, S.; Zhang, Z.; Mills, G.B. CA 125: The past and the future. *Int. J. Biol. Markers* **1998**, *13*, 179–187. [CrossRef]
13. Moss, E.L.; Hollingworth, J.; Reynolds, T.M. The role of CA125 in clinical practice. *J. Clin. Pathol.* **2005**, *58*, 308–312. [CrossRef]
14. Dochez, V.; Caillon, H.; Vaucel, E.; Dimet, J.; Winer, N.; Ducarme, G. Biomarkers and algorithms for diagnosis of ovarian cancer: CA125, HE4, RMI and ROMA, a review. *J. Ovarian Res.* **2019**, *12*, 28. [CrossRef]
15. Caruso, G.; Weroha, S.J.; Cliby, W. Ovarian Cancer: A Review. *JAMA* **2025**, *334*, 1278–1291. [CrossRef]
16. Moore, R.G.; Brown, A.K.; Miller, M.C.; Skates, S.; Allard, W.J.; Verch, T.; Steinhoff, M.; Messerlian, G.; DiSilvestro, P.; Granai, C.O.; et al. The use of multiple novel tumor biomarkers for the detection of ovarian carcinoma in patients with a pelvic mass. *Gynecol. Oncol.* **2008**, *108*, 402–408. [CrossRef]

17. Cramer, D.W.; Bast, R.C., Jr.; Berg, C.D.; Diamandis, E.P.; Godwin, A.K.; Hartge, P.; Lokshin, A.E.; Lu, K.H.; McIntosh, M.W.; Mor, G.; et al. Ovarian cancer biomarker performance in prostate, lung, colorectal, and ovarian cancer screening trial specimens. *Cancer Prev. Res.* **2011**, *4*, 365–374. [CrossRef]
18. Terry, K.L.; Schock, H.; Fortner, R.T.; Hüsing, A.; Fichorova, R.N.; Yamamoto, H.S.; Vitonis, A.F.; Johnson, T.; Overvad, K.; Tjønneland, A.; et al. A Prospective Evaluation of Early Detection Biomarkers for Ovarian Cancer in the European EPIC Cohort. *Clin. Cancer Res.* **2016**, *22*, 4664–4675. [CrossRef] [PubMed]
19. Macdonald, I.K.; Parsy-Kowalska, C.B.; Chapman, C.J. Autoantibodies: Opportunities for Early Cancer Detection. *Trends Cancer* **2017**, *3*, 198–213. [CrossRef] [PubMed]
20. Erickson, B.K.; Kinde, I.; Dobbin, Z.C.; Wang, Y.; Martin, J.Y.; Alvarez, R.D.; Conner, M.G.; Huh, W.K.; Roden, R.B.S.; Kinzler, K.W.; et al. Detection of somatic TP53 mutations in tampons of patients with high-grade serous ovarian cancer. *Obstet. Gynecol.* **2014**, *124*, 881–885. [CrossRef]
21. Yang, W.L.; Gentry-Maharaj, A.; Simmons, A.; Ryan, A.; Fourkala, E.O.; Lu, Z.; Baggerly, K.A.; Zhao, Y.; Lu, K.H.; Bowtell, D.; et al. Elevation of TP53 Autoantibody Before CA125 in Preclinical Invasive Epithelial Ovarian Cancer. *Clin. Cancer Res.* **2017**, *23*, 5912–5922. [CrossRef] [PubMed]
22. Hurley, L.C.; Levin, N.K.; Chatterjee, M.; Coles, J.; Muszkat, S.; Howarth, Z.; Dyson, G.; Tainsky, M.A. Evaluation of paraneoplastic antigens reveals TRIM21 autoantibodies as biomarker for early detection of ovarian cancer in combination with autoantibodies to NY-ESO-1 and TP53. *Cancer Biomark.* **2020**, *27*, 407–421. [CrossRef] [PubMed]
23. Fortner, R.T.; Damms-Machado, A.; Kaaks, R. Systematic review: Tumor-associated antigen autoantibodies and ovarian cancer early detection. *Gynecol. Oncol.* **2017**, *147*, 465–480. [CrossRef]
24. Lokshin, A.E.; Winans, M.; Landsittel, D.; Marrangoni, A.M.; Velikokhatnaya, L.; Modugno, F.; Nolen, B.M.; Gorelik, E. Circulating IL-8 and anti-IL-8 autoantibody in patients with ovarian cancer. *Gynecol. Oncol.* **2006**, *102*, 244–251. [CrossRef]
25. Saliminejad, K.; Khorram Khorshid, H.R.; Soleymani Fard, S.; Ghaffari, S.H. An overview of microRNAs: Biology, functions, therapeutics, and analysis methods. *J. Cell. Physiol.* **2019**, *234*, 5451–5465. [CrossRef]
26. Kan, C.W.; Hahn, M.A.; Gard, G.B.; Maidens, J.; Huh, J.Y.; Marsh, D.J.; Howell, V.M. Elevated levels of circulating microRNA-200 family members correlate with serous epithelial ovarian cancer. *BMC Cancer* **2012**, *12*, 627. [CrossRef]
27. Ghafouri-Fard, S.; Shoorei, H.; Taheri, M. miRNA profile in ovarian cancer. *Exp. Mol. Pathol.* **2020**, *113*, 104381. [CrossRef]
28. Yokoi, A.; Yoshioka, Y.; Hirakawa, A.; Yamamoto, Y.; Ishikawa, M.; Ikeda, S.I.; Kato, T.; Niimi, K.; Kajiyama, H.; Kikkawa, F.; et al. A combination of circulating miRNAs for the early detection of ovarian cancer. *Oncotarget* **2017**, *8*, 89811–89823. [CrossRef]
29. Zheng, H.; Zhang, L.; Zhao, Y.; Yang, D.; Song, F.; Wen, Y.; Hao, Q.; Hu, Z.; Zhang, W.; Chen, K. Plasma miRNAs as diagnostic and prognostic biomarkers for ovarian cancer. *PLoS ONE* **2013**, *8*, e77853. [CrossRef]
30. Feng, S.; Pan, W.; Jin, Y.; Zheng, J. MiR-25 promotes ovarian cancer proliferation and motility by targeting LATS2. *Tumour Biol.* **2014**, *35*, 12339–12344. [CrossRef] [PubMed]
31. Wang, S.; Wu, W.; Claret, F.X. Mutual regulation of microRNAs and DNA methylation in human cancers. *Epigenetics* **2017**, *12*, 187–197. [CrossRef] [PubMed]
32. Chambers, A.F.; Vanderhyden, B.C. Ovarian cancer biomarkers in urine. *Clin. Cancer Res.* **2006**, *12*, 323–327. [CrossRef] [PubMed]
33. Rani, S.; Sehgal, A.; Kaur, J.; Pandher, D.K.; Punia, R.S. Osteopontin as a Tumor Marker in Ovarian Cancer. *J. Midlife Health* **2022**, *13*, 200–205. [CrossRef] [PubMed]
34. Jia, M.M.; Deng, J.; Cheng, X.L.; Yan, Z.; Li, Q.C.; Xing, Y.Y.; Fan, D.M.; Tian, X.Y. Diagnostic accuracy of urine HE4 in patients with ovarian cancer: A meta-analysis. *Oncotarget* **2017**, *8*, 9660–9671. [CrossRef]
35. Zhou, J.; Xie, M.; He, H.; Shi, Y.; Luo, B.; Gong, G.; Li, J.; Wang, J.; Wu, X.; Wen, J. Increases urinary HMGA1 in serous epithelial ovarian cancer patients. *Cancer Biomark.* **2015**, *15*, 325–331. [CrossRef]
36. Cheng, F.; Su, L.; Qian, C. Circulating tumor DNA: A promising biomarker in the liquid biopsy of cancer. *Oncotarget* **2016**, *7*, 48832–48841. [CrossRef]
37. Pantel, K.; Alix-Panabières, C. Liquid biopsy and minimal residual disease—Latest advances and implications for cure. *Nat. Rev. Clin. Oncol.* **2019**, *16*, 409–424. [CrossRef]
38. Bettegowda, C.; Sausen, M.; Leary, R.J.; Kinde, I.; Wang, Y.; Agrawal, N.; Bartlett, B.R.; Wang, H.; Luber, B.; Alani, R.M.; et al. Detection of circulating tumor DNA in early- and late-stage human malignancies. *Sci. Transl. Med.* **2014**, *6*, 224ra24. [CrossRef]
39. Forshew, T.; Murtaza, M.; Parkinson, C.; Gale, D.; Tsui, D.W.; Kaper, F.; Dawson, S.J.; Piskorz, A.M.; Jimenez-Linan, M.; Bentley, D.; et al. Noninvasive identification and monitoring of cancer mutations by targeted deep sequencing of plasma DNA. *Sci. Transl. Med.* **2012**, *4*, 136ra68. [CrossRef]
40. Cohen, J.D.; Li, L.; Wang, Y.; Thoburn, C.; Afsari, B.; Danilova, L.; Douville, C.; Javed, A.A.; Wong, F.; Mattox, A.; et al. Detection and localization of surgically resectable cancers with a multi-analyte blood test. *Science* **2018**, *359*, 926–930. [CrossRef]
41. Gao, Y.; Zhou, N.; Liu, J. Ovarian Cancer Diagnosis and Prognosis Based on Cell-Free DNA Methylation. *Cancer Control* **2024**, *31*, 10732748241255548. [CrossRef]

42. Widschwendter, M.; Zikan, M.; Wahl, B.; Lempiäinen, H.; Paprotka, T.; Evans, I.; Jones, A.; Ghazali, S.; Reisel, D.; Eichner, J.; et al. The potential of circulating tumor DNA methylation analysis for the early detection and management of ovarian cancer. *Genome Med.* **2017**, *9*, 116. [CrossRef] [PubMed]
43. Papakonstantinou, E.; Androutsopoulos, G.; Logotheti, S.; Adonakis, G.; Maroulis, I.; Tzelepi, V. DNA Methylation in Epithelial Ovarian Cancer: Current Data and Future Perspectives. *Curr. Mol. Pharmacol.* **2021**, *14*, 1013–1027. [CrossRef] [PubMed]
44. Łaniewski, P.; Ilhan, Z.E.; Herbst-Kralovetz, M.M. The microbiome and gynaecological cancer development, prevention and therapy. *Nat. Rev. Urol.* **2020**, *17*, 232–250. [CrossRef] [PubMed]
45. Siddiqui, R.; Makhlouf, Z.; Alharbi, A.M.; Alfahemi, H.; Khan, N.A. The Gut Microbiome and Female Health. *Biology* **2022**, *11*, 1683. [CrossRef]
46. Sipos, A.; Ujlaki, G.; Mikó, E.; Maka, E.; Szabó, J.; Uray, K.; Krasznai, Z.; Bai, P. The role of the microbiome in ovarian cancer: Mechanistic insights into oncobiosis and to bacterial metabolite signaling. *Mol. Med.* **2021**, *27*, 33. [CrossRef]
47. Lum, D.; Guido, R.; Rodriguez, E.; Lee, T.; Mansuria, S.; D'Ambrosio, L.; Austin, R.M. Brush cytology of the fallopian tube and implications in ovarian cancer screening. *J. Minim. Invasive Gynecol.* **2014**, *21*, 851–856. [CrossRef]
48. Maritschnegg, E.; Wang, Y.; Pecha, N.; Horvat, R.; Van Nieuwenhuysen, E.; Vergote, I.; Heitz, F.; Sehouli, J.; Kinde, I.; Diaz, L.A., Jr.; et al. Lavage of the Uterine Cavity for Molecular Detection of Müllerian Duct Carcinomas: A Proof-of-Concept Study. *J. Clin. Oncol.* **2015**, *33*, 4293–4300. [CrossRef]
49. Chen, H.; Klein, R.; Arnold, S.; Chambers, S.; Zheng, W. Cytologic studies of the fallopian tube in patients undergoing salpingo-oophorectomy. *Cancer Cell Int.* **2016**, *16*, 78. [CrossRef]
50. Otsuka, I.; Kameda, S.; Hoshi, K. Early detection of ovarian and fallopian tube cancer by examination of cytological samples from the endometrial cavity. *Br. J. Cancer* **2013**, *109*, 603–609. [CrossRef]
51. Rodriguez, E.F.; Lum, D.; Guido, R.; Austin, R.M. Cytologic findings in experimental in vivo fallopian tube brush specimens. *Acta Cytol.* **2013**, *57*, 611–618. [CrossRef]
52. Wang, Y.; Li, L.; Douville, C.; Cohen, J.D.; Yen, T.T.; Kinde, I.; Sundfelt, K.; Kjær, S.K.; Hruban, R.H.; Shih, I.M.; et al. Evaluation of liquid from the Papanicolaou test and other liquid biopsies for the detection of endometrial and ovarian cancers. *Sci. Transl. Med.* **2018**, *10*, eaap8793. [CrossRef]
53. Kinde, I.; Bettegowda, C.; Wang, Y.; Wu, J.; Agrawal, N.; Shih Ie, M.; Kurman, R.; Dao, F.; Levine, D.A.; Giuntoli, R.; et al. Evaluation of DNA from the Papanicolaou test to detect ovarian and endometrial cancers. *Sci. Transl. Med.* **2013**, *5*, 167ra4. [CrossRef]
54. Ba, S.; Yu, M. Ultrasound-stimulated microbubbles enhances radiosensitivity of ovarian cancer. *Acta Radiol.* **2022**, *63*, 1433–1440. [CrossRef]
55. Willmann, J.K.; Bonomo, L.; Testa, A.C.; Rinaldi, P.; Rindi, G.; Valluru, K.S.; Petrone, G.; Martini, M.; Lutz, A.M.; Gambhir, S.S. Ultrasound Molecular Imaging with BR55 in Patients with Breast and Ovarian Lesions: First-in-Human Results. *J. Clin. Oncol.* **2017**, *35*, 2133–2140. [CrossRef]
56. Luo, T.; Sun, J.; Zhu, S.; He, J.; Hao, L.; Xiao, L.; Zhu, Y.; Wang, Q.; Pan, X.; Wang, Z.; et al. Ultrasound-mediated destruction of oxygen and paclitaxel loaded dual-targeting microbubbles for intraperitoneal treatment of ovarian cancer xenografts. *Cancer Lett.* **2017**, *391*, 1–11. [CrossRef] [PubMed]
57. Nebgen, D.R.; Lu, K.H.; Bast, R.C., Jr. Novel Approaches to Ovarian Cancer Screening. *Curr. Oncol. Rep.* **2019**, *21*, 75. [CrossRef]
58. Vandsburger, M.H.; Radoul, M.; Addadi, Y.; Mpofu, S.; Cohen, B.; Eilam, R.; Neeman, M. Ovarian carcinoma: Quantitative biexponential MR imaging relaxometry reveals the dynamic recruitment of ferritin-expressing fibroblasts to the angiogenic rim of tumors. *Radiology* **2013**, *268*, 790–801. [CrossRef] [PubMed]
59. Matsubara, S.; Kawahara, N.; Horie, A.; Murakami, R.; Horikawa, N.; Sumida, D.; Wada, T.; Maehana, T.; Yamawaki, A.; Ichikawa, M.; et al. Magnetic resonance relaxometry improves the accuracy of conventional MRI in the diagnosis of endometriosis-associated ovarian cancer: A case report. *Mol. Clin. Oncol.* **2019**, *11*, 296–300. [CrossRef] [PubMed]
60. Kawahara, N.; Kobayashi, H.; Maehana, T.; Iwai, K.; Yamada, Y.; Kawaguchi, R.; Takahama, J.; Marugami, N.; Nishi, H.; Sakai, Y.; et al. MR Relaxometry for Discriminating Malignant Ovarian Cystic Tumors: A Prospective Multicenter Cohort Study. *Diagnostics* **2024**, *14*, 1069. [CrossRef]
61. Lawton, F.G.; Pavlik, E.J. Perspectives on Ovarian Cancer 1809 to 2022 and Beyond. *Diagnostics* **2022**, *12*, 791. [CrossRef]
62. Rufin, K.G.A.; do Valle, H.A.; McAlpine, J.N.; Elwood, C.; Hanley, G.E. Complications after opportunistic salpingectomy compared with tubal ligation at cesarean section: A retrospective cohort study. *Fertil. Steril.* **2024**, *121*, 531–539. [CrossRef]
63. Wagar, M.K.; Forlines, G.L.; Moellman, N.; Carlson, A.; Matthews, M.; Williams, M. Postpartum Opportunistic Salpingectomy Compared with Bilateral Tubal Ligation After Vaginal Delivery for Ovarian Cancer Risk Reduction: A Cost-Effectiveness Analysis. *Obstet. Gynecol.* **2023**, *141*, 819–827. [CrossRef]
64. Wagar, M.K.; Godecker, A.; Landeros, M.V.; Williams, M. Postpartum Salpingectomy Compared with Standard Tubal Ligation After Vaginal Delivery. *Obstet. Gynecol.* **2021**, *137*, 514–520. [CrossRef] [PubMed]

65. Collins, E.; Strandell, A.; Granåsen, G.; Idahl, A. Menopausal symptoms and surgical complications after opportunistic bilateral salpingectomy, a register-based cohort study. *Am. J. Obstet. Gynecol.* **2019**, *220*, 85.e1–85.e10. [CrossRef]
66. Gan, C.; Chenoy, R.; Chandrasekaran, D.; Brockbank, E.; Hollingworth, A.; Vimplis, S.; Lawrence, A.C.; Jeyarajah, A.R.; Oram, D.; Deo, N.; et al. Persistence of fimbrial tissue on the ovarian surface after salpingectomy. *Am. J. Obstet. Gynecol.* **2017**, *217*, 425.e416–425.e421. [CrossRef]
67. Steenbeek, M.P.; van Bommel, M.H.D.; intHout, J.; Peterson, C.B.; Simons, M.; Roes, K.C.B.; Kets, M.; Norquist, B.M.; Swisher, E.M.; Hermens, R.; et al. TUBectomy with delayed oophorectomy as an alternative to risk-reducing salpingo-oophorectomy in high-risk women to assess the safety of prevention: The TUBA-WISP II study protocol. *Int. J. Gynecol. Cancer* **2023**, *33*, 982–987. [CrossRef]
68. Huh, W.K.; Pugh, S.L.; Walker, J.L.; Pennington, K.; Jewell, E.L.; Havrilesky, L.J.; Carter, J.; Muller, C.; Drapkin, R.; Lankes, H.A. NRG-CC008: A nonrandomized prospective clinical trial comparing the non-inferiority of salpingectomy to salpingo-oophorectomy to reduce the risk of ovarian cancer among BRCA1 carriers [SOROCk]. *Am. Soc. Clin. Oncol.* **2022**, *40*, 16. [CrossRef]
69. Gaba, F.; Robbani, S.; Singh, N.; McCluggage, W.G.; Wilkinson, N.; Ganesan, R.; Bryson, G.; Rowlands, G.; Tyson, C.; Arora, R.; et al. Preventing Ovarian Cancer through early Excision of Tubes and late Ovarian Removal (PROTECTOR): Protocol for a prospective non-randomised multi-center trial. *Int. J. Gynecol. Cancer* **2021**, *31*, 286–291. [CrossRef] [PubMed]
70. Falconer, H.; Yin, L.; Grönberg, H.; Altman, D. Ovarian cancer risk after salpingectomy: A nationwide population-based study. *J. Natl. Cancer Inst.* **2015**, *107*, dju410. [CrossRef]
71. Guldberg, R.; Wehberg, S.; Skovlund, C.W.; Mogensen, O.; Lidegaard, O. Salpingectomy as standard at hysterectomy? A Danish cohort study, 1977–2010. *BMJ Open* **2013**, *3*, e002845. [CrossRef]
72. Idahl, A.; Darelius, A.; Sundfeldt, K.; Pålsson, M.; Strandell, A. Hysterectomy and opportunistic salpingectomy (HOPPSA): Study protocol for a register-based randomized controlled trial. *Trials* **2019**, *20*, 10. [CrossRef]
73. Heimdal, K.; Skovlund, E.; Møller, P. Oral contraceptives and risk of familial breast cancer. *Cancer Detect. Prev.* **2002**, *26*, 23–27. [CrossRef] [PubMed]
74. Jahanfar, S.; Mortazavi, J.; Lapidow, A.; Cu, C.; Al Abosy, J.; Morris, K.; Becerra-Mateus, J.C.; Steinfeldt, M.; Maurer, O.; Bohang, J.; et al. Assessing the impact of contraceptive use on reproductive cancer risk among women of reproductive age-a systematic review. *Front. Glob. Womens Health* **2024**, *5*, 1487820. [CrossRef]
75. Beral, V.; Doll, R.; Hermon, C.; Peto, R.; Reeves, G. Ovarian cancer and oral contraceptives: Collaborative reanalysis of data from 45 epidemiological studies including 23,257 women with ovarian cancer and 87,303 controls. *Lancet* **2008**, *371*, 303–314. [CrossRef] [PubMed]
76. Wheeler, L.J.; Desanto, K.; Teal, S.B.; Sheeder, J.; Guntupalli, S.R. Intrauterine Device Use and Ovarian Cancer Risk: A Systematic Review and Meta-analysis. *Obstet. Gynecol.* **2019**, *134*, 791–800. [CrossRef]
77. Zhang, D.; Bai, B.; Xi, Y.; Wang, T.; Zhao, Y. Is aspirin use associated with a decreased risk of ovarian cancer? A systematic review and meta-analysis of observational studies with dose-response analysis. *Gynecol. Oncol.* **2016**, *142*, 368–377. [CrossRef] [PubMed]
78. Trabert, B.; Poole, E.M.; White, E.; Visvanathan, K.; Adami, H.O.; Anderson, G.L.; Brasky, T.M.; Brinton, L.A.; Fortner, R.T.; Gaudet, M.; et al. Analgesic Use and Ovarian Cancer Risk: An Analysis in the Ovarian Cancer Cohort Consortium. *J. Natl. Cancer Inst.* **2019**, *111*, 137–145. [CrossRef]
79. Hurwitz, L.M.; Webb, P.M.; Jordan, S.J.; Doherty, J.A.; Harris, H.R.; Goodman, M.T.; Shvetsov, Y.B.; Modugno, F.; Moysich, K.B.; Schildkraut, J.M.; et al. Association of Frequent Aspirin Use with Ovarian Cancer Risk According to Genetic Susceptibility. *JAMA Netw. Open* **2023**, *6*, e230666. [CrossRef]
80. Trabert, B.; Ness, R.B.; Lo-Ciganic, W.H.; Murphy, M.A.; Goode, E.L.; Poole, E.M.; Brinton, L.A.; Webb, P.M.; Nagle, C.M.; Jordan, S.J.; et al. Aspirin, nonaspirin nonsteroidal anti-inflammatory drug, and acetaminophen use and risk of invasive epithelial ovarian cancer: A pooled analysis in the Ovarian Cancer Association Consortium. *J. Natl. Cancer Inst.* **2014**, *106*, djt431. [CrossRef]
81. Burn, J.; Sheth, H.; Elliott, F.; Reed, L.; Macrae, F.; Mecklin, J.P.; Möslein, G.; McRonald, F.E.; Bertario, L.; Evans, D.G.; et al. Cancer prevention with aspirin in hereditary colorectal cancer (Lynch syndrome), 10-year follow-up and registry-based 20-year data in the CAPP2 study: A double-blind, randomised, placebo-controlled trial. *Lancet* **2020**, *395*, 1855–1863. [CrossRef]
82. Moysich, K.B.; Mettlin, C.; Piver, M.S.; Natarajan, N.; Menezes, R.J.; Swede, H. Regular use of analgesic drugs and ovarian cancer risk. *Cancer Epidemiol. Biomark. Prev.* **2001**, *10*, 903–906.
83. Meier, C.R.; Schmitz, S.; Jick, H. Association between acetaminophen or nonsteroidal antiinflammatory drugs and risk of developing ovarian, breast, or colon cancer. *Pharmacotherapy* **2002**, *22*, 303–309. [CrossRef]
84. Sasamoto, N.; Babic, A.; Vitonis, A.F.; Titus, L.; Cramer, D.W.; Trabert, B.; Tworoger, S.S.; Terry, K.L. Common Analgesic Use for Menstrual Pain and Ovarian Cancer Risk. *Cancer Prev. Res.* **2021**, *14*, 795–802. [CrossRef]
85. Gor, R.; Ramachandran, I.; Ramalingam, S. Targeting the Cancer Stem Cells in Endocrine Cancers with Phytochemicals. *Curr. Top. Med. Chem.* **2022**, *22*, 2589–2597. [CrossRef]

86. Casey, S.C.; Amedei, A.; Aquilano, K.; Azmi, A.S.; Benencia, F.; Bhakta, D.; Bilsland, A.E.; Boosani, C.S.; Chen, S.; Ciriolo, M.R.; et al. Cancer prevention and therapy through the modulation of the tumor microenvironment. *Semin. Cancer Biol.* **2015**, *35*, S199–S223. [CrossRef] [PubMed]
87. Milani, A.; Basirnejad, M.; Shahbazi, S.; Bolhassani, A. Carotenoids: Biochemistry, pharmacology and treatment. *Br. J. Pharmacol.* **2017**, *174*, 1290–1324. [CrossRef] [PubMed]
88. Xu, X.L.; Deng, S.L.; Lian, Z.X.; Yu, K. Resveratrol Targets a Variety of Oncogenic and Oncosuppressive Signaling for Ovarian Cancer Prevention and Treatment. *Antioxidants* **2021**, *10*, 1718. [CrossRef] [PubMed]
89. Cho, H.J.; Suh, D.S.; Moon, S.H.; Song, Y.J.; Yoon, M.S.; Park, D.Y.; Choi, K.U.; Kim, Y.K.; Kim, K.H. Silibinin inhibits tumor growth through downregulation of extracellular signal-regulated kinase and Akt in vitro and in vivo in human ovarian cancer cells. *J. Agric. Food Chem.* **2013**, *61*, 4089–4096. [CrossRef]
90. Veronesi, U.; Decensi, A. Retinoids for ovarian cancer prevention: Laboratory data set the stage for thoughtful clinical trials. *J. Natl. Cancer Inst.* **2001**, *93*, 486–488. [CrossRef]
91. Kapur, A.; Mehta, P.; Simmons, A.D.; Ericksen, S.S.; Mehta, G.; Palecek, S.P.; Felder, M.; Stenerson, Z.; Nayak, A.; Dominguez, J.M.A.; et al. Atovaquone: An Inhibitor of Oxidative Phosphorylation as Studied in Gynecologic Cancers. *Cancers* **2022**, *14*, 2297. [CrossRef] [PubMed]

cancers

MDPI

Article

Medroxyprogesterone Acetate Inhibits Tumorigenesis in Mouse Models of Oviductal High-Grade Serous Carcinoma

Yali Zhai [1], Karan Bedi [2,3], Rong Wu [1], Ying Feng [4], Maranne E. Green [4], Celeste Leigh Pearce [5], Malcolm C. Pike [6], Eric R. Fearon [1,3,4,7] and Kathleen R. Cho [1,3,4,*]

1 Department of Pathology, University of Michigan Medical School, Ann Arbor, MI 48109, USA; yaliz@umich.edu (Y.Z.); rongwu@umich.edu (R.W.); fearon@umich.edu (E.R.F.)
2 Department of Biostatistics, University of Michigan School of Public Health, Ann Arbor, MI 48109, USA
3 Rogel Cancer Center, University of Michigan Medical School, Ann Arbor, MI 48109, USA
4 Department of Internal Medicine, University of Michigan Medical School, Ann Arbor, MI 48109, USA; yingfeng@umich.edu (Y.F.); megreen@umich.edu (M.E.G.)
5 Department of Epidemiology, University of Michigan School of Public Health, Ann Arbor, MI 48109, USA; lpearce@umich.edu
6 Department of Epidemiology & Biostatistics, Memorial Sloan Kettering Cancer Center, New York, NY 10065, USA; pikem@mskcc.org
7 Department of Human Genetics, University of Michigan Medical School, Ann Arbor, MI 48109, USA
* Correspondence: kathcho@umich.edu

Simple Summary

Oral contraceptive use is known to significantly reduce the risk of developing the most common and lethal type of ovarian cancer, known as high-grade serous carcinoma (HGSC). The risk reduction is likely attributable to synthetic progestins in oral contraceptives, but it is not known which of the many types of progestins are most effective, and whether combination with estrogen mitigates their protective effects. We used genetically engineered mouse models of tubo-ovarian HGSC to show that medroxyprogesterone acetate (MPA), a common constituent of contraceptive hormones, significantly inhibits tumor development and prolongs survival. In contrast, estradiol in combination with progesterone accelerates tumorigenesis and significantly shortens survival. The findings provide strong support for MPA as a chemo-preventive agent for HGSC, confirming observations based on large-scale retrospective studies in humans. The model system offers a platform for testing other progestins and dosing strategies to identify the most effective ones in preventing HGSC in high-risk populations.

Abstract

Background/Objectives: Tubo-ovarian high-grade serous carcinoma (HGSC) is a highly lethal malignancy, usually diagnosed at an advanced stage due to the lack of early symptoms and biomarkers. Contraceptive hormone use is associated with a reduced risk of HGSC, but the relative contributions of natural versus synthetic progestins, and their interaction with estrogens, are poorly understood. Methods: We evaluated the chemopreventive efficacy of a synthetic progestin medroxyprogesterone acetate (MPA), progesterone (P4), and combined 17β-estradiol-progesterone (E2 + P4) in a well-characterized genetically engineered mouse model (GEMM) of oviductal HGSC based on the conditional inactivation of one or both alleles of the *Brca1*, *Trp53*, *Rb1*, and *Nf1* tumor suppressor genes (*BPRN*-het and *BPRN*-homo mice, respectively). Mice received hormones or placebo via slow-release pellets implanted subcutaneously. After induction of tumor formation, the mice were monitored for tumor development, progression, and survival. Tumor incidence was assessed histologically, and hormone effects were further explored via RNA-seq analysis of oviductal tissues. Results: MPA significantly reduced HGSC incidence and delayed tumor progression compared to the placebo, P4, and P4 + E2 in both *BPRN*-homo and *BPRN*-het mice, with up to 78% tumor-free survival in the MPA-treated *BPRN*-het cohort. P4 monotherapy did not provide significant protection vs. the placebo, but the effects of P4 could have been impacted by a failure to achieve sustained release of the hormone beyond 4–8 weeks. The E2 + P4 combination accelerated tumorigenesis and reduced survival ($p < 0.0001$ in *BPRN*-homo and $p = 0.0004$ in *BPRN*-het mice). MPA did not affect tumorigenesis in a colon cancer GEMM, or the growth of mouse HGSC-derived cells in vivo, suggesting the role of MPA in the early stages of HGSC development. Gene expression analyses showed that P4 and MPA downregulated cholesterol homeostasis, early and late estrogen response, and epithelial–mesenchymal transition pathways, though only MPA afforded tumor protection. Conclusions: These findings demonstrate that a synthetic progestin, specifically MPA, confers robust protection against HGSC development, while a combination including E2 (E2 + P4) increases risk. This work also illustrates how HGSC GEMMs can be used to compare the chemo-preventive effects of various synthetic progestins on HGSC development in order to prioritize the most effective ones for use in preventing HGSC in both general and high-risk populations.

Keywords: high-grade serous carcinoma; medroxyprogesterone acetate; oral contraceptives; ovarian cancer prevention; genetically engineered mouse model; epithelial–mesenchymal transition

1. Introduction

Tubo-ovarian high-grade serous carcinoma (HGSC) is a highly aggressive and lethal type of cancer, often diagnosed at an advanced stage due to the absence of early symptoms and reliable biomarkers. Despite advancements in treatment, the prognosis for women with HGSC remains poor, with a five-year survival rate of approximately 40% [1]. Therefore, the development of effective preventive strategies is of high importance, particularly for women who are genetically predisposed to developing HGSC. Oral contraceptives are the most effective chemical drugs for preventing ovarian cancer known to date. Several epidemiological studies have reported that the use of oral contraceptives is inversely associated with ovarian cancer risk, with the protective effect increasing with duration of use [2–5]. The risk reduction does not appear to be specific to any particular formulation and is observed for all histologic types of ovarian cancer—though the risk reduction is lower for mucinous carcinoma in some studies [2,6]. A population-based study by Ness et al. found that after adjustment for current age, number of pregnancies, race, and family history of ovarian cancer, the use of low-estrogen/low-progestin oral contraceptives afforded an estimated risk reduction that was comparable to that for high-estrogen/high-progestin formulations [4]. Interestingly, a meta-analysis of 14 population-based studies found that hormone replacement therapy with unopposed estrogen increases the risk of ovarian cancer in a duration-dependent manner, while the addition of progestins at least partially blocks this effect [7]. Several questions regarding the protection afforded by oral contraceptive use remain unanswered. For example, though the protection appears to be independent of the estrogen and progestin doses, it is less clear whether progestin-only contraceptives provide a comparable level of protection to those containing both estrogen and progestin, or whether protective effects vary depending on the specific progestin used or their route of administration (oral, depot, or in intrauterine devices). Medroxyprogesterone acetate (MPA) is a synthetic progestin widely used in contraceptives and hormone replacement therapies. We have previously reported that depot-MPA use is associated with a decreased risk of ovarian cancer [8]. In contrast, natural progesterone (P4) has been proposed as a pro-tumorigenic endogenous factor in a mouse model of HGSC based on the inactivation of *Dicer1* and *Pten* [9].

In this study, we utilized well-characterized genetically engineered mouse models (GEMMs) of oviductal HGSC based on the conditional mono- or biallelic somatic inactivation of the *Brca1*, *Trp53*, *Rb1*, and *Nf1* tumor suppressor genes (TSGs) in the oviductal epithelium (*BPRN*-het and *BPRN*-homo mice, respectively) [10–13] to compare the efficacy of MPA, P4, and E2 + P4, delivered via slow-release pellets implanted subcutaneously, in preventing tumorigenesis.

2. Materials and Methods

2.1. Genetically Engineered Mice and Animal Care

The generation and characterization of *Ovgp1-iCreER*^{T2} transgenic mice with various combinations of floxed TSG alleles, including *BPRN*-homo and *BPRN*-het mice, have been described previously in detail [10,11]. *CDX2P-CreER*^{T2}; *Apc*$^{fl/+}$; *Kras*$^{LSL\text{-}G12D/+}$; *Trp53*$^{fl\text{-}R270H/fl}$ mice (referred to hereafter as *AKP* mice) have also been previously reported [14]. Genotypes of the individual mice included in this study were confirmed by polymerase chain reaction (PCR) analysis of tail-derived DNA. The mice were housed in standard housing and fed the same chow diet. All procedures involving mice for the research described herein have been approved by the University of Michigan's Institutional Animal Care and Use Committee (PRO00008343, PRO00010212, and PRO00012388).

2.2. In Vivo Induction of Oviductal Tumors and Implantation of Hormone Pellets

Female 6- to 8-week-old *Ovgp1-iCreER*^{T2}; *BPRN*-homo, and *BPRN*-het mice were implanted with slow-release (90 day) hormone pellets subcutaneously under the dorsal skin. The placebo (25 mg), MPA (25 mg), and E2 + P4 (0.05 mg 17β-estradiol + 25 mg P4) pellets were obtained from Innovative Research of America (Sarasota, FL, USA). Slow-release (90 day) P4 (25 mg) pellets were obtained from Belma Technologies (Liege, Belgium). Groups of *BPRN*-homo mice were implanted with placebo (n = 22), P4 (n = 21), MPA (n = 24), or E2 + P4 (n = 27) pellets and 72 h later were treated transiently with Tamoxifen (TAM, 0.2 g/kg body weight, injected intraperitoneally (i.p.) on day 1 and day 3) as described previously to induce Cre-mediated recombination of the floxed TSG alleles in the oviductal epithelium. Groups of *BPRN*-het mice were similarly implanted with placebo (n = 28), P4 (n = 24), MPA (n = 25), or E2 + P4 (n = 27) pellets, and tumors were then induced with TAM as described above. After 90 days, fresh pellets were implanted for a total of 180 days of hormone treatment. Similarly, 2- to –3-month-old *AKP* mice (5 males and 5 females per group) were implanted with either placebo or MPA pellets as described above. Subsequently, the *AKP* mice received tamoxifen (TAM) at 70 mg/kg daily for two consecutive days via i.p. injection. All mice were euthanized upon reaching predefined humane endpoints or at the end of the specified monitoring period.

2.3. Measurement of Hormone Levels

To determine whether the pellets released the hormones as expected, serum samples were collected from 3 mice in each group (placebo, P4, and E2 + P4) via tail vein at 1 week, 4 weeks, 8 weeks, and 12 weeks after pellet insertion. P4 and E2 levels were tested by the University of Virginia Center for Research in Reproduction (Charlottesville, VA, USA). To meet the larger sample size requirement for MPA assays, we collected plasma samples from 3 to 4 mice euthanized at weeks 2 and 8 in the MPA and placebo groups. At the end of the surveillance period, we also collected plasma samples from 3 euthanized mice. MPA levels were analyzed by the Pharmacokinetics Core in the College of Pharmacy at the University of Michigan.

2.4. Tissue Sampling, Histopathological and Immunohistochemical Analysis

Mice in the *BPRN*-homo and *BPRN*-het cohorts were euthanized at 52 weeks and 60 weeks post-TAM, respectively, or were sacrificed earlier if they reached humane endpoints. Necropsies were performed to determine the presence and extent of disease in hormone- vs. placebo-treated mice as previously described [11,13]. Briefly, mice were grossly inspected for oviductal tumors, metastatic disease, and ascites. Bilateral ovaries and oviducts, uterus, mesentery, omentum, pancreas, kidneys, liver, and lungs were harvested from each case. Representative formalin-fixed paraffin-embedded (FFPE) sections of each tissue were stained with hematoxylin and eosin (H&E) and examined by light microscopy. Oviducts without gross lesions were examined in their entirety by exhaustive sectioning and microscopic evaluation of alternating sections. The histology of each lesion was reviewed by two participants with expertise in gynecologic pathology. Lesions were classified as serous tubal intraepithelial carcinoma (STIC), early (invasive but confined to the oviduct) HGSC or carcinosarcoma (eHGSC or eCaSa), or as invasive HGSC or carcinosarcoma extending beyond the oviduct (HGSC or CaSa), including metastasis.

2.5. Cell Line Establishment and Cell Culture

2D tumor-derived cell cultures were generated following the protocols described by Shepherd and colleagues [15]. Briefly, the cell line *BPRN*-3867 was derived from an oviductal tumor arising in a *BPRN*-homo mouse and then transferred to monolayer cell culture; cells were maintained in DMEM/F12 (Gibco, Fisher Scientific, Waltham, MA, USA) supplemented with 10% heat-inactivated fetal bovine serum (FBS, HyClone, Logan, UT, USA) and penicillin–streptomycin (Gibco, Fisher Scientific).

2.6. Allografts and MPA Treatment

BPRN-3867 cells were used to generate tumor allografts to test the effects of MPA on established tumors. A total of six immunocompromised, 6-week-old Nod-SCID-IL2Rgamma (NSG) mice were divided into three groups. On day 0, two mice were implanted with placebo pellets (group 1) and two mice were implanted with MPA pellets (group 2). Two days later, all six mice were subcutaneously injected in the bilateral flanks with 2×10^6 cultured cells resuspended in 100 µL cold PBS and 100 µL Matrigel (Corning, New York, NY, USA). On day 9, the two mice not implanted earlier were implanted with MPA pellets (group 3). All mice were sacrificed and necropsied 60 days after cell injection. Tumor weights and volumes were obtained, with tumor volume estimated by the formula $L \times W \times H/2$ (cm^3). The excised tumors were then fixed in 10% buffered formalin, paraffin-embedded, and sectioned for histopathological examination.

2.7. Gene Expression

RNA sequencing (RNA-seq) was used to profile gene expression in the following mouse oviductal epithelial tissues: (1) no treatment (no pellets, no TAM, n = 4), (2) TAM only (no pellets, n = 4), (3) placebo pellets + TAM (n = 3), (4) P4 pellets + TAM (n = 3), (5) MPA pellets + TAM (n = 3), and (6) E2 + P4 pellets + TAM (n = 3). Slow-release pellets containing steroid hormones (MPA, P4, or E2 + P4) or placebo were implanted subcutaneously into 6- to 8-week-old female *BPRN*-het mice. Three days later, the mice received intraperitoneal (i.p) TAM injections to induce Cre expression. Mice were euthanized 14 days post-TAM, and the distal portion of the oviducts were collected for RNA isolation. RNA was extracted using the miRNeasy Mini Kit (Qiagen, Germantown, MD, USA). Library prep and RNA sequencing was carried out in the Advanced Genomics Core at the University of Michigan. Subsequently, 151 bp paired-end sequencing was performed according to the manufacturer's protocol (Illumina NovaSeq San Diego, CA, USA). The BCL Convert Conversion Software v4.0 (Illumina) was used to generate de-multiplexed Fastq files. Initial data analysis was performed by the University of Michigan Bioinformatics Core. Raw reads were trimmed using Cutadapt (v2.3) [16]. FastQC (v0.11.8) was used to ensure that only high-quality data were used for expression quantity and differential expression [17]. Fastq Screen v was used to screen for various types of contamination [18]. Reads were mapped to the reference genome GRCm38 (ENSEMBL), using STAR v2.7.8a [19] and assigned count estimates to genes with RSEM v1.3.3 [20]. Alignment options followed ENCODE standards for RNA-seq (https://github.com/alexdobin/STAR/blob/master/doc/STARmanual.pdf accessed on 23 September 2023). Multiqc v1.7 compiled the results from several of these tools and provided a detailed and comprehensive quality control report [21].

Differential gene expressions across groups were analyzed with DESeq2 in R. For enrichment tests, we mapped mouse genes to human homologs using biomaRt ensembl version 100. The obtained differentially expressed genes (DEGs, defined as fold change (FC) $\geq$1.5 and FDR-adjusted p value $\leq$ 0.05) were used for gene set enrichment analysis with a web-based functional enrichment analysis tool called WebGestalt (WEB-based GEne SeT AnaLysis Toolkit) available at http://www.webgestalt.org/. Gene counts produced by RSEM and the raw sequencing data are available in the NCBI's Gene Expression Omnibus (GEO) database under the accession number GSE309487.

2.8. Statistical Analysis

The survival rates of tumor-bearing placebo- and hormone-treated mice were estimated by Kaplan–Meier analysis using the GraphPad Prism software version 10. The Cochran–Mantel–Haenszel Chi-square test of association was performed using SPSS 9.0 to evaluate the effects of each type of hormone on tumor incidence and progression compared to the placebo. Differences at $p < 0.05$ were considered statistically significant.

3. Results

3.1. MPA Inhibits HGSC Tumorigenesis in BPRN-homo and BPRN-het Mice

A schematic diagram of the overall experimental design is provided in Figure 1 and representative photomicrographs of oviductal tumors that arose in our GEMMs are shown in Supplemental Figure S1.

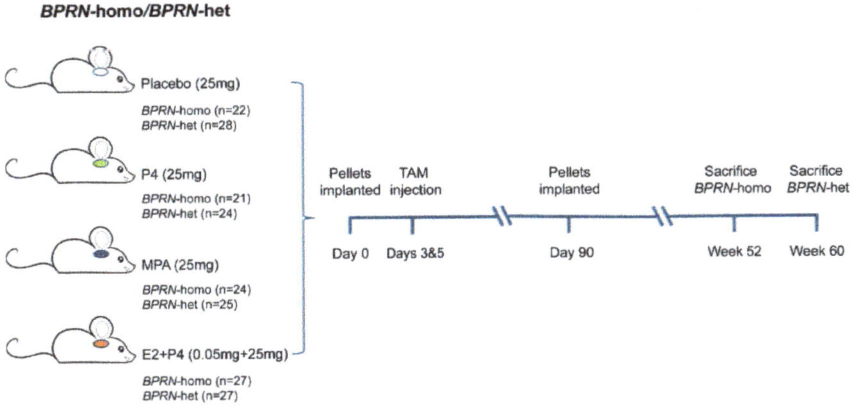

Figure 1. Schematic diagram of experimental design. Groups of *BPRN*-homo and *BPRN*-het mice (21–28 mice per group) were implanted with hormone or placebo pellets and monitored for at least one year after the induction of tumor formation with TAM injection.

Tumor development in cohorts of TAM-treated *BPRN* mice after the implantation of placebo, P4, MPA, and E2 + P4 pellets in *BPRN*-homo and *BPRN*-het mice are summarized in the upper and lower panels of Figure 2, respectively.

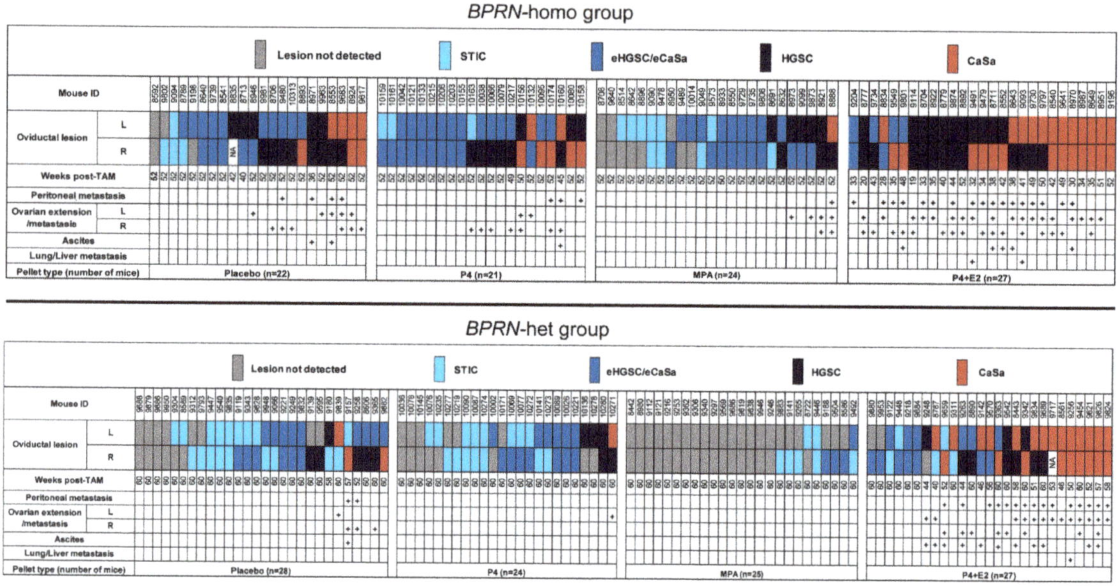

Figure 2. Summary of tumor development in placebo-, P4-, MPA-, and E2 + P4-treated *Ovgp1-iCreERT2; BPRN*-homo (**top**) and *Ovgp1-iCreERT2; BPRN*-het (**bottom**) mice. Weeks post-TAM indicates the time point at which the mice were euthanized. Oviductal lesions are indicated by colored boxes as indicated: serous tubal intraepithelial carcinoma (STIC); early invasive HGSC or carcinosarcoma (CaSa) confined to the oviduct (eHGSC/eCaSa); and invasive HGSC or CaSa extending beyond the oviduct.

Compared with mice treated with placebo, P4 alone, or E2 + P4, we found that both *BPRN*-homo and *BPRN*-het mice treated with MPA showed significantly fewer and less advanced oviductal tumors. Specifically, in *BPRN*-homo mice 52 weeks after TAM injection, 10 of 48 (21%) oviducts in MPA-treated vs. 4 of 43 (9%) oviducts of placebo-treated mice were free of detectable tumors (Table 1). The preventive effect of MPA was even more pronounced in the *BPRN*-het mice 60 weeks after TAM injection. In these mice, 39 of 50 (78%) MPA-treated vs. 14 of 56 (25%) placebo-treated oviducts were tumor-free (Table 2). In addition, tumor progression was significantly slowed in the MPA-treated groups, with no advanced HGSCs or carcinosarcomas (CaSas) observed in *BPRN*-het mice (0% MPA vs. 16% placebo), and fewer HGSCs and CaSas observed in the *BPRN*-homo mice (19% MPA vs. 47% placebo). Statistical analysis by the Cochran–Mantel–Haenszel Chi-square association test showed significant differences in tumor incidence and progression between the MPA and placebo-treated groups in both *BPRN*-het and *BPRN*-homo mice, with significantly fewer and less advanced tumors in both MPA-treated groups compared to the placebo-treated groups (p values of <0.001 and $p = 0.004$, respectively). Our data shows that the synthetic progestin MPA is highly effective at inhibiting or delaying oviductal HGSC development in our GEMMs.

Table 1. Comparison of oviductal HGSC development in hormone-treated *BPRN*-homo mice.

	Case/Total (%)				
Cohorts	**No Lesion**	**STIC**	**eHGSC/eCaSa**	**HGSC/CaSa**	***p* Value**
Placebo	4/43 (9.3)	4/43 (9.3)	15/43 (34.88)	20/43 (46.51)	
P4	0/42 (0.0)	3/42 (7.14)	23/42 (54.8)	16/42 (38.1)	=0.48
MPA	10/48 (20.8)	10/48 (20.8)	19/48 (39.58)	9/48 (18.75)	=0.004 *
E2 + P4	0/54 (0.0)	0/54 (0.0)	7/54 (12.96	47/54 (87.04)	<0.001 ***

*, $p < 0.05$ and ***, $p < 0.001$ versus control; Cochrane–Mantel–Haenszel Chi-square test of association. Numbers shown reflect two oviducts (right and left) for each mouse included in the analysis.

Table 2. Comparison of oviductal HGSC development in the hormone-treated *BPRN*-het mice.

	Case/Total (%)				
Cohorts	**No Lesion**	**STIC**	**eHGSC/eCaSa**	**HGSC/CaSa**	***p* Value**
Placebo	14/56 (25.0)	18/56 (32.14)	15/56 (26.79)	9/56 (16.07)	
P4	14/48 (29.17)	15/48 (31.30)	13/48 (27.1)	6/48 (12.5)	=0.58
MPA	39/50 (78.0)	7/50 (14.0)	4/50 (8.00)	0/50 (0.00)	<0.001 ***
E2 + P4	3/53 (5.7)	5/53 (9.43)	13/53 (24.52)	32/53 (60.38)	<0.001 ***

***, $p < 0.001$ versus control; Cochrane–Mantel–Haenszel Chi-square test of association. Numbers shown reflect two oviducts (right and left) for each mouse included in the analysis.

3.2. *Tumor Development in BPRN-homo and BPRN-het Mice Treated with P4 Alone Is Comparable to Placebo*

While MPA markedly inhibited the development and/or progression of HGSC in our GEMMs, P4 did not show a similar protective effect. As shown in Figure 2 and Tables 1 and 2, the incidence of oviductal lesions in P4-treated *BPRN*-homo and *BPRN*-het mice was similar to that in the placebo-treated control groups ($p = 0.48$ and $p = 0.56$, respectively).

3.3. *Progesterone (P4) in Combination with Estradiol (E2) Accelerates BPRN Tumor Development*

Interestingly, the combination of P4 with E2 appeared to accelerate the initiation and progression of HGSC in both *BPRN*-homo and *BPRN*-het mice (Figure 2). Among *BPRN*-homo mice receiving the E2 + P4 combination, 87% developed advanced HGSC or CaSa, compared to only 47% of the placebo-treated mice (Table 1). Similarly, 60% of E2 + P4-treated *BPRN*-het mice developed advanced HGSCs or CaSas, compared to only 16% of the placebo-treated mice. Differences in the presence and extent of tumors between the E2 + P4- and placebo-treated mice in both the *BPRN*-homo and *BPRN*-het groups were highly significant ($p < 0.001$ and $p < 0.001$, respectively). In addition, the Kaplan–Meier survival analysis showed that after E2 + P4 treatment, the survival of mice in the *BPRN*-homo and *BPRN*-het groups was significantly shorter than that of mice in the placebo control groups (median survival, *BPRN*-homo: E2 + P4 38.8 weeks vs. placebo 50.4 weeks, $p < 0.0001$, log-rank Mantel–Cox test (Figure 3A); and *BPRN*-het: E2 + P4 55 weeks vs. placebo 59.5 weeks, $p = 0.0004$ (Figure 3B)). These findings suggest that E2 has tumor-promoting effects in these GEMMs that are not ameliorated by the addition of P4.

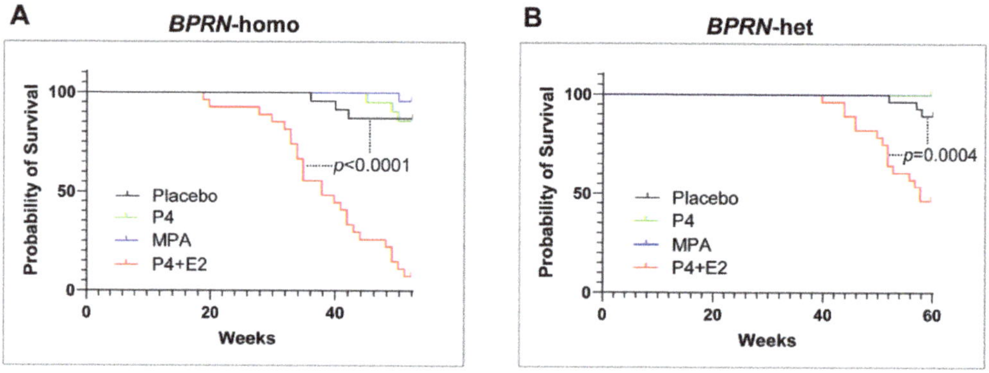

Figure 3. Kaplan–Meier survival curves of placebo- (black), P4- (green), MPA- (blue), and E2 + P4- (red) treated Ovgp1-iCreERT2; BPRN-homo (**A**) and Ovgp1-iCreERT2; BPRN-het (**B**) mice after TAM injection. $p < 0.0001$ when comparing the BPRN-homo E2 + P4 mice with the placebo mice (median survival, 38.8 weeks vs. 50.4 weeks) and $p = 0.0004$ when comparing the BPRN-het E2 + P4 mice with the placebo mice (median survival, 55 weeks vs. 59.5 weeks) using the log-rank (Mantel–Cox) test.

3.4. Some Commercially Available Slow-Release Pellets Have Suboptimal Performance

To monitor the effectiveness of the slow-release pellets used in this study, hormone levels were measured at a series of time points after the implantation of pellets intended to consistently release P4, MPA, or E2 + P4 hormones over a period of approximately 90 days after subcutaneous implantation in the mice. As shown in Supplemental Figure S2 (left panel), the estradiol (E2) levels in mice implanted with E2 + P4 pellets peaked at week 1, had declined by week 4, and were comparable to baseline levels at week 8. Serum progesterone levels in the P4 and E2 + P4 treatment groups also peaked at week 1, then decreased significantly by week 4, and further decreased to the same level as in the placebo group by week 8 (Supplemental Figure S2, center panel). Based on these findings, it is possible that more sustained delivery of P4 could have resulted in either tumor-promoting or tumor-inhibiting effects, and our failure to observe an impact of P4 alone on tumorigenesis should be interpreted with caution. Interestingly, even short (4–8 week) exposure to E2 in combination with P4 had striking tumor-promoting effects. Plasma levels of MPA were quite consistent between mice, persisted longer, and remained elevated, even at the study endpoint (52–60 weeks, Supplemental Figure S2, right panel).

3.5. MPA Has No Effect on Tumorigenesis in a Colon Cancer GEMM

To evaluate the specificity of the effects of MPA on oviductal tumorigenesis, we tested the effects of MPA in a GEMM of colon cancer based on the conditional somatic inactivation of the *Apc* and *Trp53* tumor suppressor genes and activation of *Kras* (*AKP* mice) in the colonic epithelium [14]. Slow-release (90 day) MPA or placebo pellets were subcutaneously implanted into cohorts of male and female *AKP* mice and three days after pellet implantation, the mice were injected with TAM on two consecutive days to induce colonic tumor formation. In this mouse colon carcinoma model, we did not observe any difference in survival between the MPA- and placebo-treated groups of male or female mice (Supplemental Figure S3). This finding supports the conclusion that the effects of MPA are specific to certain tumor types rather than influencing tumorigenesis more broadly.

3.6. MPA Has No Effect on the Growth of Established HGSCs

Based on the data shown above, *BPRN*-homo and *BPRN*-het mice treated with MPA during the period when oviductal tumors are developing show marked reduction in their numbers and extents, but it is unclear if MPA inhibits tumor initiation, progression, or both. To test MPA's effect on established HGSCs, MPA or placebo pellets were subcutaneously implanted into NSG mice before or after injection of a tumorigenic cell line (*BPRN*-3867) derived from an oviductal tumor that arose in a TAM-treated *BPRN* mouse. MPA failed to exhibit any in vivo anti-tumor activity, regardless of whether MPA treatment was started before or after tumor cell injection (Figure 4A–D). These results indicate that MPA does not affect the growth of fully transformed cells or inhibit their ability to establish tumor allografts. Furthermore, the findings suggest that MPA is more likely to exert its effects in the early stages of tumorigenesis, either on the oviductal epithelium itself, through effects on non-neoplastic cells in the oviductal microenvironment, or both.

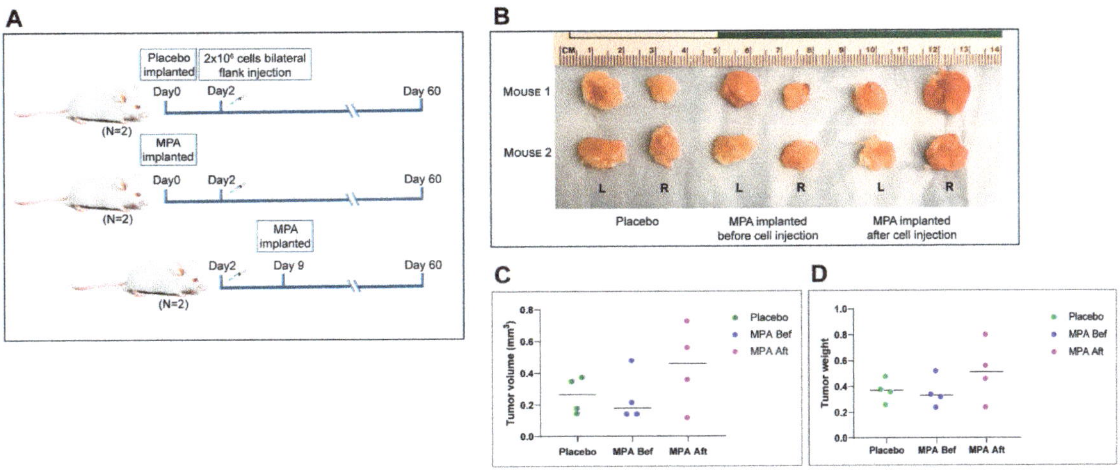

Figure 4. In vivo treatment of allografted tumors using MPA pellets. (**A**) Schematic diagram of experimental design. Three groups of mice (2 mice per group) were injected with *BPRN*-3867 tumor-derived cells before (Bef)/after (Aft) MPA or placebo pellet implantation and analyzed. (**B**) Images of tumors resected from each group at the study endpoint. L: left flank tumor and R: right flank tumor. (**C**) Tumor volumes in each group, measured using a vernier caliper. (**D**) Tumor weights in each group.

3.7. Identification of Differentially Expressed Genes

To further examine the effects of each type of hormone on mouse oviductal tissue in the earliest phases of tumor development, RNA-Seq analysis was used to compare gene expression in the oviducts of *BPRN*-homo mice treated with placebo, MPA, P4, or E2 + P4 (n = 3 for each group) or no pellets (n = 4) 14 days after the induction of tumor formation via TAM injection. A group of mice (n = 4) with no TAM injection and no pellet implantation was also included. Principal components analysis showed that the MPA and P4 groups had similar global gene expression, which was distinct from the E2 + P4 and placebo groups (Figure 5A). Based on differential gene expression analysis (FDR-adjusted $p \leq 0.05$, fold change ≥ 1.5 in either direction), gene set enrichment testing was performed using differentially expressed genes (DEGs) against curated Hallmark gene sets from MSigDB. Compared with the placebo group, there were three gene sets in the gene list of downregulated genes in the MPA group (Figure 5B), nine gene sets in the gene list of downregulated genes in the P4 group (Figure 5C), and two gene sets in the gene list of upregulated genes in the E2 + P4 group (Figure 5D).

Supplemental Table S1 lists additional upregulated and downregulated Hallmark gene sets enriched for differentially expressed genes, providing more gene sets with trend changes. Notably, consistent with the aggressiveness of the E2 + P4 group, the most enriched upregulated gene set was the epithelial–mesenchymal transition (EMT) set, with 13 of 200 EMT genes, including *COL11A1*, *ECM1*, *TGFBI*, *BGN*, *TNFRSF11B*, *TNFRSF12A*, and *TIMP1*. Furthermore, several matrix metalloproteinases were also upregulated in the E2 + P4 group, including *MMP7*, *MMP13*, and *MMP19*. In contrast, we observed downregulation of the EMT gene set in the P4 group (17 of 200 EMT genes) including *CCN1*, *CDH11*, *SERPINE1*, and some types of collagens (*COL1A1*, *COL1A2*, *COL3A1*, *COL4A2*, and *COL6A2*). A trend for downregulation of the EMT gene set was also noted in the MPA group ($p = 0.0079925$, FDR = 0.067), including several downregulated EMT genes (*CCN1*, *COL1A1*, *COL1A2*, and *COL3A1*). In addition, both the P4 and MPA groups had downregulation of Hallmark gene sets for estrogen response and cholesterol homeostasis, indicating that MPA and P4 share some similarities in regulating gene expression. KEGG pathway analysis of differentially upregulated and downregulated genes identified some differences between the MPA and P4 groups vs. the placebo group that may be related to the tumor-inhibiting effects of MPA but not P4 (Figure 6). Specifically, regulation of lipolysis in adipocytes, cholesterol metabolism, and PPAR signaling were upregulated by MPA treatment, but not by P4.

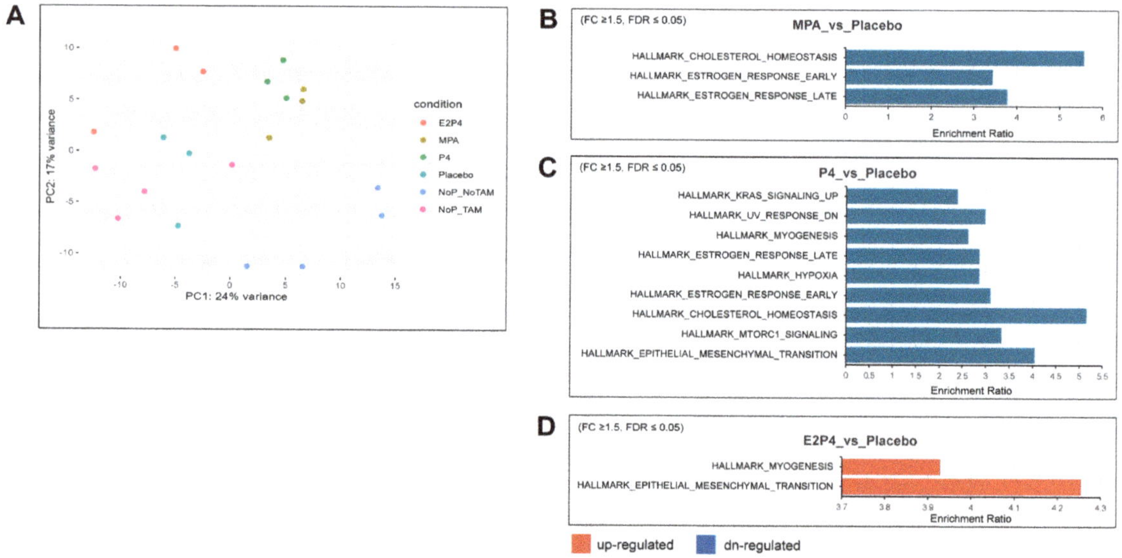

Figure 5. Hallmark gene sets enriched for differentially expressed genes in the MPA, P4, and E2 + P4 groups compared with the placebo group. (**A**) Global gene expression analyses were performed on oviductal tissue RNA from E2P4, MPA, P4, placebo (n = 3 for each group), and no pellets (NoP, n = 4) mice treated with TAM for 14 days and on oviductal tissue RNA from no TAM/no pellet mice (n = 4). Principal components analysis showed that the MPA group and P4 group had close global gene expression pattens, whereas the no pellet/no TAM group had a global gene expression pattern different from that of the other groups. (**B**,**C**) Downregulated (blue) Hallmark gene sets overrepresented in MPA and P4 compared with the placebo group (fold change ≥ 1.5, FDR ≤ 0.05). (**D**) Upregulated (red) Hallmark gene sets overrepresented in E2 + P4 compared with the placebo group (fold change ≥ 1.5, FDR ≤ 0.05).

Lists of the 50 most upregulated and downregulated genes in the placebo-treated mice vs. mice treated with P4, MPA, or E2 + P4 are provided in Supplemental Tables S2, S3, and S4, respectively.

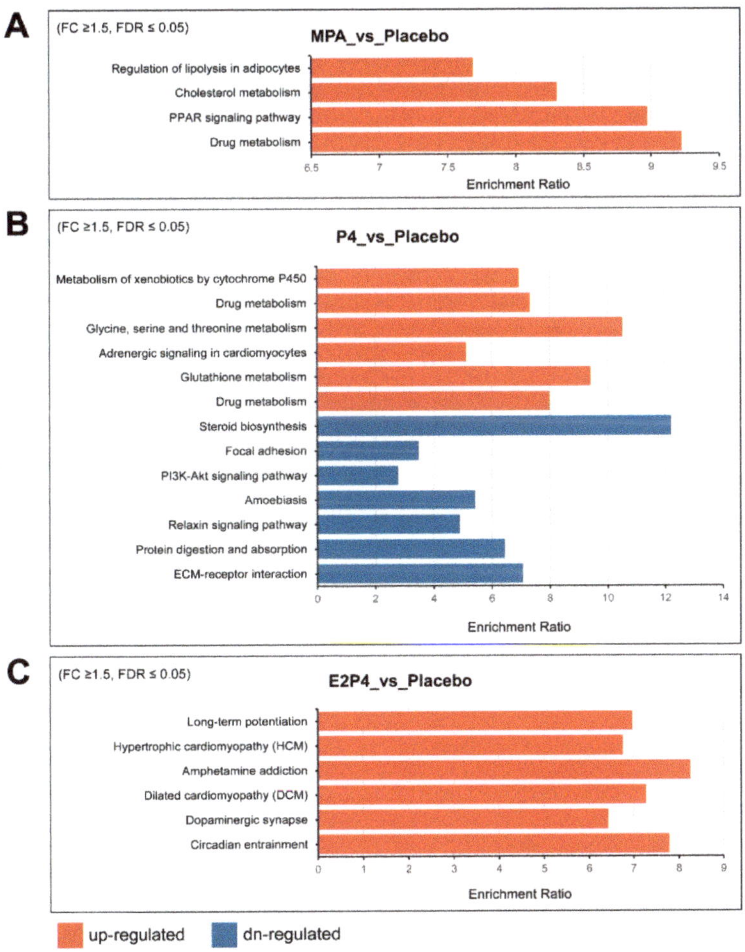

Figure 6. KEGG pathway analysis of differentially expressed upregulated and downregulated genes in oviductal tissues of BPRN-conditional knockout mice. (**A**) KEGG pathways overrepresented in the lists of genes upregulated (red) in mouse oviductal tissues of MPA-treated mice compared with placebo mice after 14 days of TAM treatment. (**B**) KEGG pathways overrepresented in the lists of genes up- (red) and downregulated (blue) in mouse oviductal tissues of P4-treated mice compared with placebo mice after 14 days of TAM treatment. (**C**) KEGG pathways overrepresented in the lists of genes upregulated (red) in mouse oviductal tissues of E2P4-treated mice compared with placebo mice after 14 days of TAM treatment. The enriched pathways with FDR $\leq$ 0.05 are shown.

4. Discussion

In this work, we showed that medroxyprogesterone acetate significantly reduced tumor incidence and delayed tumor progression in mouse models of oviductal high-grade serous carcinoma. The preventive effect of MPA was more pronounced in *BPRN*-het than in *BPRN*-homo mice. This latter finding is consistent with prior work using these model systems, which suggested the impact of interventions during the long (several month) latency period between tumor induction and emergence of early oviductal lesions is more apparent in mice with fewer initially targeted gene mutations. For example, we found that a preventive effect of multiparity was seen in *BPRN*$^{fl/+}$ mice (carrying two floxed alleles of *Brca1*, *Trp53*, and *Rb1*, but only one floxed *Nf1* allele), but not in *BPRN*-homo mice [12]. In both *BPRN*-het and *BPRN*-homo mice, an anti-tumor effect was not observed with P4 alone, and addition of 17β-estradiol to P4 significantly enhanced tumorigenesis and reduced survival in both model systems. While the protective versus deleterious impact of MPA and E2, respectively, in our GEMMs are concordant with large-scale epidemiological studies in humans [7,8,22], some limitations of our study are acknowledged. First, our GEMMs are based on simultaneous mono- or biallelic inactivation of four TSGs that are not broadly representative of human HGSCs, including those that are HRD-proficient. While all four of these TSGs are on the list of genes with statistically recurrent somatic alterations in the HGSC Cancer Genome Atlas [23], human tumors do not arise via simultaneous inactivation of all four TSGs. Second, given the length of time and expense associated with testing the effects of hormones as chemo-preventive agents in our GEMMs, only a limited number of chemo-preventive regimens could be tested in this initial study. Third, mice are not humans and may not have the same pharmacodynamics with respect to hormone dosing and tissue response. Finally, our selection of hormones and hormone combinations was impacted by what is commercially available (alone and in combination) for use in mice and, as noted above, some of the pellets did not consistently release hormones over the designated 90-day period. As a consequence, the effects of P4 and E2 + P4 may have been significantly influenced by shorter and lower dosing than we anticipated. Nonetheless, we found that MPA alone had a robust preventive effect on oviductal tumorigenesis and even intermittent exposure to E2 + P4 accelerated tumorigenesis in our model systems. Notably, a large-scale clinical trial assessing the long-term influence of estrogen alone or in combination with MPA for menopausal hormone therapy found that estrogen alone increased ovarian cancer incidence and mortality, while estrogen plus MPA did not [24]. Similarly, in pooled data from case–control studies, Lee and colleagues found that estrogen in combination with a progestin for menopausal therapy was not associated with an increased risk of ovarian cancer, but information on the progestin dose and type was not available for analysis [25]. Unfortunately, we were not able to test the effects of E2 + MPA in our model system because pellets with that hormone combination are not readily available.

The mechanisms by which MPA confers protection from oviductal HGSC development remain unclear. MPA blocks follicular maturation and ovulation, thickens the cervical mucus to impede sperm motility, and renders the endometrium less favorable for implantation [26]. MPA may also have direct effects on the oviductal/fallopian tube epithelium itself, but the protective effect of MPA does not appear to be due to nonspecific repression of cell proliferation, as MPA did not suppress the growth of established *BPRN* HGSC allografts or alter the proliferative index of oviductal epithelium based on Ki-67 expression (Cho lab, unpublished data), nor did it inhibit tumor development or progression in a mouse colon cancer GEMM. These findings, along with the absence of any detectable lesions in the majority of the MPA-treated *BPRN*-het mice, suggest that MPA acts on specific hormone-responsive cell types primarily at the initiation or pre-malignant transformation stages of tumorigenesis, perhaps by altering the fate of at-risk oviductal epithelial cells or by modifying the tissue microenvironment in a manner that is unfavorable for malignant transformation. Interestingly, Nelson and colleagues investigated the effects of depo-medroxyprogesterone acetate (DMPA) on the development of fallopian tube/oviductal carcinomas in a different transgenic mouse model of oviductal HGSC (Mogp-Tag mice) [27]. In this model, mice treated with DMPA accumulated fewer p53 positive cells, p53 signature lesions, serous tubal intraepithelial carcinomas (STICs), and adenocarcinomas in their oviducts. DMPA was found to induce cleaved caspase-3 in the fallopian tube/oviductal epithelium, suggesting the involvement of apoptosis in DMPA-related clearance of genetically damaged cells in the oviduct. Like us, these investigators suggest that progestins target early rather than late events in the development of HGSC. While our gene expression analysis of hormone-exposed oviductal epithelium from *BPRN* mice did not identify apoptosis pathway genes differentially expressed in the MPA- vs. placebo-treated oviductal epithelium, our data suggest possible roles for MPA in cholesterol metabolism and the downregulation of EMT genes, either in the oviductal epithelium, other cells in the oviductal microenvironment during tumor development, or both. While MPA and P4 have similar effects on EMT signaling, cholesterol homeostasis, and estrogen-responsive gene sets, their different effects on tumor development and progression suggest that there are important distinctions between synthetic and natural progestins in terms of receptor activity or downstream signaling, which need further investigation. Interestingly, simvastatin, a drug that lowers cholesterol levels by inhibiting a key enzyme in cholesterol synthesis (HMG-CoA reductase), has been shown to interfere with cancer stem cell plasticity and reduce metastasis in ovarian cancer [28], and statin use has been associated with a lower risk of both serous and non-serous types of ovarian carcinoma [29].

Our findings regarding the E2 + P4 combination are quite striking, as the combination exhibited a strong effect in accelerating HGSC development in *BPRN* mice, which is consistent with epidemiological data showing that estrogen therapy alone or in combination is a risk factor for ovarian cancer [7,24,25]. Unfortunately, we were unable to include an E2-only cohort of mice in this study due to the failure of the E2 pellets we tested to release the hormone in a suitable fashion. Notably, our gene expression analysis revealed that EMT was the most highly enriched Hallmark gene set upregulated by E2 + P4 compared to placebo, while EMT pathway genes were downregulated by P4 and MPA. Previous studies have shown that estrogen can promote EMT through the estrogen receptor, downregulating E-cadherin and upregulating Snail and vimentin in breast and prostate cancer cells [30–32]. Wang and colleagues also demonstrated that estrogen can promote EMT in a subset of *Brca1*-deficient breast cancer cells through an ER-independent signaling pathway [33]. Estrogen's interactions with its receptor are also known to regulate the expression and activity of various MMPs, some of which are known to be associated with EMT and promote tumor growth and metastasis [34–37]. Our results are concordant with those of Laviolette et al., who employed yet another transgenic model of ovarian cancer (tgCAG-LS-Tag mice) to show that estrogen promoted tumor development and progression while P4 had no impact [38]. In contrast, another study reported that in the *Dicer1* $^{flox/flox}$ *Pten* $^{flox/flox}$ *Amhr2* $^{cre/+}$ ovarian cancer GEMM, treatment with progesterone (P4) alone significantly accelerated tumor development and progression in all ovariectomized *Dicer1-Pten* double-knockout (DKO) mice, whereas the addition of combined estrogen and progesterone treatment inhibited this rapid progression [9]. These observations highlight the importance of hormone balance, timing, and receptor context in mediating cancer risks. The selected engineered alleles, specificity of the induced genetic alterations to the oviductal epithelium, and genetic background in different mouse models are undoubtedly influencing factors.

The relationship between hormones and ovarian cancer is complex and depends on many factors, including the type and duration of hormone exposure, individual genetics and medical history, and the specific subtype of ovarian cancer. Our ability to test and compare the chemo-preventive effects of specific natural and synthetic hormones and hormone combinations in a controlled genetic context and well-characterized model system represents a significant advance. Our work supports the use of synthetic progestins in chemo-preventive regimens aimed at reducing the risk of tubo-ovarian HGSC and perhaps other types of ovarian carcinomas. What is less clear is whether some progestins are more effective than others at reducing risk and whether the magnitude of risk reduction may be related to the type of progestin, dose, and/or dosing method. Importantly, as noted by Stanczyk over two decades ago, all progestins are not created equal [39]. Some are structurally related to progesterone (including MPA), while others are structurally related to testosterone (such as norethindrone and levonorgestrel). Moreover, the various progestins used in oral contraceptives and hormone replacement therapies have differences in metabolism, pharmacokinetics, receptor interactions, intracellular actions, and biological effects that may impact their efficacy as chemo-preventive agents for HGSC [40]. Given the popularity of hormone-releasing intrauterine devices for contraception, our model system could be used to test whether the progestins in these devices (e.g., levonorgestrel) confers comparable anti-tumor protection to MPA. Similar studies could be performed with norethindrone and other common constituents of contraceptive medications that may be equally efficacious for contraception but have different impacts on reducing the risk of developing one of the most lethal cancers affecting women.

5. Conclusions

The current study illustrates the utility of high-fidelity mouse models of HGSC in testing various chemoprevention strategies that could reduce the risk of HGSC in both the general population and genetically predisposed women. Our findings provide experimental evidence confirming a robust protective effect of medroxyprogesterone acetate and supporting a pro-tumorigenic effect of estradiol. The GEMMs offer a suitable platform for testing other progestins and dosing strategies to identify the most effective ones in preventing HGSC.

Supplementary Materials: The following supporting information can be downloaded at https://www.mdpi.com/article/10.3390/cancers17213456/s1. Figure S1: Representative photomicrographs of H&E-stained sections showing progression of oviductal lesions in *BPRN*-homo mice. Figure S2: Serum/plasma hormone levels (E2, P4, and MPA) in *BPRN*-homo mice at different time points with P4, MPA, and E2 + P4 treatment. Figure S3: Kaplan–Meier survival curves of TAM-treated *AKP* mice implanted with pellets containing MPA vs. placebo. Table S1: Hallmark gene sets enriched with differentially expressed genes in the E2 + P4 and MPA groups compared with the placebo group. Table S2: Top 50 upregulated and 50 downregulated genes in the oviducts of P4- vs. placebo-treated mice. Table S3: Top 50 upregulated and 50 downregulated genes in the oviducts of MPA- vs. placebo-treated mice. Table S4: Top 50 upregulated and 50 downregulated genes in the oviducts of E2 + P4- vs. placebo-treated mice.

Author Contributions: Y.Z., E.R.F., and K.R.C. conceived the study and designed the experiments. Y.Z. performed all experiments and analyses. K.B. performed the RNA-Seq data analysis. Y.Z., R.W., and K.R.C. conducted the histopathological evaluation of mouse tumors. Y.F. and M.E.G. performed the colon cancer mouse model study. Y.Z., K.B., M.C.P., C.L.P., E.R.F., and K.R.C. analyzed the data and wrote the manuscript. All authors have read and agreed to the published version of the manuscript.

Funding: The research reported in this paper was supported by the National Cancer Institute of the National Institutes of Health under award numbers R01CA226756 (KRC, YZ, ERF) and P30CA046592 (KB, KRC, ERF) and included use of the Rogel Cancer Center's Pharmacokinetics and Cancer Data Science Shared Resources.

Institutional Review Board Statement: All procedures involving mice for the research described herein have been approved by the Institutional Review Board of University of Michigan's Institutional Animal Care and Use Committee (protocol code PRO00008343, PRO00010212, and PRO00012388, and the dates of approval are 11 June 2018, 17 May 2021, and 25 February 2025, respectively).

Informed Consent Statement: Not applicable.

Data Availability Statement: The original RNA-seq data presented in the study are openly available in the NCBI's Gene Expression Omnibus (GEO) database under the accession number GSE309487. The remaining original contributions presented in this study are included in the article/Supplementary Material. Further inquiries can be directed to the corresponding author.

Conflicts of Interest: The authors declare no potential conflicts of interest.

Abbreviations

The following abbreviations are used in this manuscript:

HGSC	High-grade serous carcinoma
OCs	Oral contraceptives
MPA	Medroxyprogesterone acetate
LD	Progesterone
P4	17β-estradiol
E2	Genetically engineered mouse model
GEMM	Brca1, Trp53, Rb1, and Nf1
BPRN	Tumor suppressor genes
TSGs	Serous tubal intraepithelial carcinomas
STICs	Early high-grade serous carcinoma
eHGSC	Carcinosarcoma
CaSa	Fallopian tube
FT	High-grade serous carcinoma

References

1. Hollis, R.L. Molecular characteristics and clinical behaviour of epithelial ovarian cancers. *Cancer Lett.* **2023**, *555*, 216057. [CrossRef] [PubMed]
2. Collaborative Group on Epidemiological Studies of Ovarian Cancer; Beral, V.; Doll, R.; Hermon, C.; Peto, R.; Reeves, G. Ovarian cancer and oral contraceptives: Collaborative reanalysis of data from 45 epidemiological studies including 23,257 women with ovarian cancer and 87,303 controls. *Lancet* **2008**, *371*, 303–314. [CrossRef]
3. Havrilesky, L.J.; Moorman, P.G.; Lowery, W.J.; Gierisch, J.M.; Coeytaux, R.R.; Urrutia, R.P.; Dinan, M.; McBroom, A.J.; Hasselblad, V.; Sanders, G.D.; et al. Oral contraceptive pills as primary prevention for ovarian cancer: A systematic review and meta-analysis. *Obs. Gynecol.* **2013**, *122*, 139–147. [CrossRef]
4. Ness, R.B.; Grisso, J.A.; Klapper, J.; Schlesselman, J.J.; Silberzweig, S.; Vergona, R.; Morgan, M.; Wheeler, J.E. Risk of ovarian cancer in relation to estrogen and progestin dose and use characteristics of oral contraceptives. SHARE Study Group. Steroid Hormones and Reproductions. *Am. J. Epidemiol.* **2000**, *152*, 233–241. [CrossRef]
5. Reid, B.M.; Permuth, J.B.; Sellers, T.A. Epidemiology of ovarian cancer: A review. *Cancer Biol. Med.* **2017**, *14*, 9–32. [CrossRef]
6. Wentzensen, N.; Poole, E.M.; Trabert, B.; White, E.; Arslan, A.A.; Patel, A.V.; Setiawan, V.W.; Visvanathan, K.; Weiderpass, E.; Adami, H.O.; et al. Ovarian Cancer Risk Factors by Histologic Subtype: An Analysis From the Ovarian Cancer Cohort Consortium. *J. Clin. Oncol.* **2016**, *34*, 2888–2898. [CrossRef]
7. Pearce, C.L.; Chung, K.; Pike, M.C.; Wu, A.H. Increased ovarian cancer risk associated with menopausal estrogen therapy is reduced by adding a progestin. *Cancer* **2009**, *115*, 531–539. [CrossRef]
8. Phung, M.T.; Lee, A.W.; Wu, A.H.; Berchuck, A.; Cho, K.R.; Cramer, D.W.; Doherty, J.A.; Goodman, M.T.; Hanley, G.E.; Harris, H.R.; et al. Depot-Medroxyprogesterone Acetate Use Is Associated with Decreased Risk of Ovarian Cancer: The Mounting Evidence of a Protective Role of Progestins. *Cancer Epidemiol. Biomark. Prev.* **2021**, *30*, 927–935. [CrossRef]
9. Kim, O.; Park, E.Y.; Kwon, S.Y.; Shin, S.; Emerson, R.E.; Shin, Y.H.; DeMayo, F.J.; Lydon, J.P.; Coffey, D.M.; Hawkins, S.M.; et al. Targeting progesterone signaling prevents metastatic ovarian cancer. *Proc. Natl. Acad. Sci. USA* **2020**, *117*, 31993–32004. [CrossRef]
10. Zhai, Y.; Wu, R.; Kuick, R.; Sessine, M.S.; Schulman, S.; Green, M.; Fearon, E.R.; Cho, K.R. High-grade serous carcinomas arise in the mouse oviduct via defects linked to the human disease. *J. Pathol.* **2017**, *243*, 16–25. [CrossRef]
11. McCool, K.W.; Freeman, Z.T.; Zhai, Y.; Wu, R.; Hu, K.; Liu, C.J.; Tomlins, S.A.; Fearon, E.R.; Magnuson, B.; Kuick, R.; et al. Murine Oviductal High-Grade Serous Carcinomas Mirror the Genomic Alterations, Gene Expression Profiles, and Immune Microenvironment of Their Human Counterparts. *Cancer Res.* **2020**, *80*, 877–889. [CrossRef] [PubMed]
12. Hou, X.; Zhai, Y.; Hu, K.; Liu, C.J.; Udager, A.; Pearce, C.L.; Fearon, E.R.; Cho, K.R. Aging accelerates while multiparity delays tumorigenesis in mouse models of high-grade serous carcinoma. *Gynecol. Oncol.* **2022**, *165*, 552–559. [CrossRef] [PubMed]

13. Chen, L.; Zhai, Y.; Wang, Y.; Fearon, E.R.; Nunez, G.; Inohara, N.; Cho, K.R. Altering the Microbiome Inhibits Tumorigenesis in a Mouse Model of Oviductal High-Grade Serous Carcinoma. *Cancer Res.* **2021**, *81*, 3309–3318. [CrossRef] [PubMed]
14. Feng, Y.; Zhu, N.; Bedi, K.; Li, J.; Perera, C.; Green, M.; Assarzadegan, N.; Zhai, Y.; Liu, Q.; Baladandayuthapani, V.; et al. SOX9 suppresses colon cancer via inhibiting epithelial-mesenchymal transition and SOX2 induction. *J. Clin. Invest.* **2025**, *135*, e184115. [CrossRef]
15. Shepherd, T.G.; Theriault, B.L.; Campbell, E.J.; Nachtigal, M.W. Primary culture of ovarian surface epithelial cells and ascites-derived ovarian cancer cells from patients. *Nat. Protoc.* **2006**, *1*, 2643–2649. [CrossRef]
16. Martin, M. Cutadapt removes adapter sequences from high-throughput sequencing reads. *EMBnet J.* **2011**, *17*, 10–12. [CrossRef]
17. Andrews, S. FastQC: A Quality Control Tool for High Throughput Sequence Data. 2010. Available online: https://www.bioinformatics.babraham.ac.uk/projects/fastqc/ (accessed on 23 September 2023).
18. Wingett, S.W.; Andrews, S. FastQ Screen: A tool for multi-genome mapping and quality control. *F1000Research* **2018**, *7*, 1338. [CrossRef]
19. Dobin, A.; Davis, C.A.; Schlesinger, F.; Drenkow, J.; Zaleski, C.; Jha, S.; Batut, P.; Chaisson, M.; Gingeras, T.R. STAR: Ultrafast universal RNA-seq aligner. *Bioinformatics* **2013**, *29*, 15–21. [CrossRef]
20. Li, B.; Dewey, C.N. RSEM: Accurate transcript quantification from RNA-Seq data with or without a reference genome. *BMC Bioinform.* **2011**, *12*, 323. [CrossRef]
21. Ewels, P.; Magnusson, M.; Lundin, S.; Kaller, M. MultiQC: Summarize analysis results for multiple tools and samples in a single report. *Bioinformatics* **2016**, *32*, 3047–3048. [CrossRef]
22. Johansson, A.; Schmitz, D.; Hoglund, J.; Hadizadeh, F.; Karlsson, T.; Ek, W.E. Investigating the Effect of Estradiol Levels on the Risk of Breast, Endometrial, and Ovarian Cancer. *J. Endocr. Soc.* **2022**, *6*, bvac100. [CrossRef]
23. The Cancer Genome Atlas Research Network. Integrated genomic analyses of ovarian carcinoma. *Nature* **2011**, *474*, 609–615. [CrossRef]
24. Chlebowski, R.T.; Aragaki, A.K.; Pan, K.; Haque, R.; Rohan, T.E.; Song, M.; Wactawski-Wende, J.; Lane, D.S.; Harris, H.R.; Strickler, H.; et al. Menopausal Hormone Therapy and Ovarian and Endometrial Cancers: Long-Term Follow-Up of the Women's Health Initiative Randomized Trials. *J. Clin. Oncol.* **2024**, *42*, 3537–3549. [CrossRef] [PubMed]
25. Lee, A.W.; Wu, A.H.; Wiensch, A.; Mukherjee, B.; Terry, K.L.; Harris, H.R.; Carney, M.E.; Jensen, A.; Cramer, D.W.; Berchuck, A.; et al. Estrogen Plus Progestin Hormone Therapy and Ovarian Cancer: A Complicated Relationship Explored. *Epidemiology* **2020**, *31*, 402–408. [CrossRef] [PubMed]
26. Sathe, A.; Patel, P.; Gerriets, V. Medroxyprogesterone. In *StatPearls*; StatPearls Publishing: Treasure Island, FL, USA, 2025.
27. Nelson, O.L.; Rosales, R.; Turbov, J.M.; Thaete, L.G.; Cline, J.M.; Rodriguez, G.C. Progestin Significantly Inhibits Carcinogenesis in the Mogp-TAg Transgenic Mouse Model of Fallopian Tube Cancer. *Cancer Prev. Res.* **2022**, *15*, 75–86. [CrossRef]
28. Kato, S.; Liberona, M.F.; Cerda-Infante, J.; Sanchez, M.; Henriquez, J.; Bizama, C.; Bravo, M.L.; Gonzalez, P.; Gejman, R.; Branes, J.; et al. Simvastatin interferes with cancer 'stem-cell' plasticity reducing metastasis in ovarian cancer. *Endocr. Relat. Cancer* **2018**, *25*, 821–836. [CrossRef]
29. Akinwunmi, B.; Vitonis, A.F.; Titus, L.; Terry, K.L.; Cramer, D.W. Statin Therapy and Association with Ovarian Cancer risk in the New England Case Control (NEC) Study. *Int. J. Cancer* **2018**, *144*, 991–1000. [CrossRef]
30. Di Zazzo, E.; Galasso, G.; Giovannelli, P.; Di Donato, M.; Bilancio, A.; Perillo, B.; Sinisi, A.A.; Migliaccio, A.; Castoria, G. Estrogen Receptors in Epithelial-Mesenchymal Transition of Prostate Cancer. *Cancers* **2019**, *11*, 1418. [CrossRef]
31. Planas-Silva, M.D.; Waltz, P.K. Estrogen promotes reversible epithelial-to-mesenchymal-like transition and collective motility in MCF-7 breast cancer cells. *J. Steroid Biochem. Mol. Biol.* **2007**, *104*, 11–21. [CrossRef]
32. Jimenez-Salazar, J.E.; Rivera-Escobar, R.M.; Damian-Ferrara, R.; Maldonado-Cubas, J.; Rincon-Perez, C.; Tarrago-Castellanos, R.; Damian-Matsumura, P. Estradiol-Induced Epithelial to Mesenchymal Transition and Migration Are Inhibited by Blocking c-Src Kinase in Breast Cancer Cell Lines. *J. Breast Cancer* **2023**, *26*, 446–460. [CrossRef]
33. Wang, C.; Bai, F.; Zhang, L.H.; Scott, A.; Li, E.; Pei, X.H. Estrogen promotes estrogen receptor negative BRCA1-deficient tumor initiation and progression. *Breast Cancer Res.* **2018**, *20*, 74. [CrossRef]
34. Gorodeski, G.I. Estrogen decrease in tight junctional resistance involves matrix-metalloproteinase-7-mediated remodeling of occludin. *Endocrinology* **2007**, *148*, 218–231. [CrossRef] [PubMed]
35. Nilsson, U.W.; Garvin, S.; Dabrosin, C. MMP-2 and MMP-9 activity is regulated by estradiol and tamoxifen in cultured human breast cancer cells. *Breast Cancer Res. Treat.* **2007**, *102*, 253–261. [CrossRef] [PubMed]
36. Khalili-Tanha, G.; Radisky, E.S.; Radisky, D.C.; Shoari, A. Matrix metalloproteinase-driven epithelial-mesenchymal transition: Implications in health and disease. *J. Transl. Med.* **2025**, *23*, 436. [CrossRef] [PubMed]
37. Oh, S.; Kwon, H.J.; Jung, J. Estrogen exposure causes the progressive growth of SK-Hep1-derived tumor in ovariectomized mice. *Toxicol. Res.* **2022**, *38*, 1–7. [CrossRef]

38. Laviolette, L.A.; Garson, K.; Macdonald, E.A.; Senterman, M.K.; Courville, K.; Crane, C.A.; Vanderhyden, B.C. 17beta-estradiol accelerates tumor onset and decreases survival in a transgenic mouse model of ovarian cancer. *Endocrinology* **2010**, *151*, 929–938. [CrossRef]
39. Stanczyk, F.Z. All progestins are not created equal. *Steroids* **2003**, *68*, 879–890. [CrossRef]
40. Stanczyk, F.Z.; Hapgood, J.P.; Winer, S.; Mishell, D.R., Jr. Progestogens used in postmenopausal hormone therapy: Differences in their pharmacological properties, intracellular actions, and clinical effects. *Endocr. Rev.* **2013**, *34*, 171–208. [CrossRef]

cancers

Article

Bilateral Salpingo-Oophorectomy Is Superior to Salpingectomy Alone in Preventing Non-Tubal Tumor Development in a Mouse Model of High-Grade Serous Carcinoma

Yali Zhai [1], Eric R. Fearon [1,2,3,4] and Kathleen R. Cho [1,2,4,*]

1 Department of Pathology, University of Michigan Medical School, Ann Arbor, MI 48109, USA; yaliz@med.umich.edu (Y.Z.); fearon@med.umich.edu (E.R.F.)
2 Department of Internal Medicine, University of Michigan Medical School, Ann Arbor, MI 48109, USA
3 Department of Human Genetics, University of Michigan Medical School, Ann Arbor, MI 48109, USA
4 Rogel Cancer Center, University of Michigan, Ann Arbor, MI 48109, USA
* Correspondence: kathcho@med.umich.edu

Simple Summary

The most common type of ovarian cancer, high-grade serous carcinoma (HGSC), arises more often from the fallopian tube than the ovary. In women genetically predisposed to developing HGSC, prophylactic removal of their fallopian tubes and ovaries is known to greatly reduce cancer risk, but it is unclear how much benefit removal of the ovaries provides beyond removal of the fallopian tubes alone. In this study, mice genetically engineered to develop HGSC in their oviducts (mouse fallopian tubes) were used to compare preventive effects of removing the oviducts alone (bilateral salpingectomy, BS) with removal of oviducts and ovaries (bilateral salpingo-oophorectomy, BSO). Both procedures significantly reduced HGSC incidence, but only BSO completely prevented tumor development. In mice that developed ovarian cancer after BS, tumors appeared to originate from tubal-type epithelium residing in the ovary (endosalpingiosis), suggesting endosalpingiosis as a potential cell of origin for HGSCs arising outside of the fallopian tubes.

Abstract

Background/Objectives: High-grade serous carcinoma (HGSC) is the leading cause of ovarian cancer-related mortality. It usually arises from fallopian tube epithelium, with a smaller subset arising in non-tubal sites including the ovary or peritoneum. The origin of HGSCs without evidence of tubal involvement remains unclear. Moreover, in women with genetic predisposition to developing HGSC, the additional protection afforded by prophylactic removal of the ovaries in addition to the fallopian tubes has not yet been established. Methods: We used a well-characterized genetically engineered mouse model (GEMM) of oviductal HGSC based on conditional, somatic inactivation of the *Brca1*, *Trp53*, *Rb1*, and *Nf1* tumor suppressor genes (*BPRN* mice) to compare preventive effects for HGSC via bilateral salpingectomy versus bilateral salpingo-oophorectomy. We also explored the origins of non-tubal HGSCs in ectopic tubal-type epithelium (endosalpingiosis) present in the mouse ovaries and peritoneum. Results: While bilateral salpingectomy significantly reduced the incidence of HGSCs in the GEMM model, bilateral salpingo-oophorectomy completely prevented tumor development. We identified an example of HGSC with apparent origin in endosalpingiosis, implicating endosalpingiosis as a likely precursor for non-tubal HGSC. Conclusions: Our findings confirm the superiority of bilateral salpingo-oophorectomy over salpingectomy alone in reducing HGSC risk and affirm the rationale for surgical strategies to reduce HGSC risk in women carrying pathogenic variants of *BRCA1/2*

https://doi.org/10.3390/cancers17172759

and other genes associated with homologous recombination deficiency. Our findings also illustrate how work with GEMMs can advance new insights into HGSC pathogenesis.

Keywords: endosalpingiosis; risk-reducing surgery; genetically engineered mouse model; high-grade serous carcinoma; primary peritoneal

1. Introduction

It is now widely accepted that high-grade serous carcinoma (HGSC), the most common and most lethal type of "ovarian" cancer, usually arises from fallopian tube epithelium (FTE). However, up to a third of HGSCs lack evidence of a tubal origin, even after microscopic examination of all resected tubal tissue [1]. There are uncertainties regarding the cells from which non-tubal HGSCs arise, but it has been proposed that ovarian and/or primary peritoneal HGSCs might develop either from occult tubal precursors, from ovarian surface epithelium or benign serous ovarian tumors, or from normal or neoplastic FTE that becomes denuded and subsequently implanted in non-tubal sites, including the ovary and peritoneal surfaces [2–5]. It has also been proposed that latent cells present in ectopic locations outside of the tubes retain the capacity for multi-lineage differentiation and can reconstitute glands lined by tubal-type epithelium (Mullerian metaplasia hypothesis) [6,7]. Tubal-type tissue outside of the fallopian tube is referred to as endosalpingiosis and is comparable to its better-known endometrial counterpart—endometriosis. Endosalpingiosis is characterized by single or small groups of simple glands that are sometimes cystically dilated. The glands are lined by the same cell types present in normal FTE, including ciliated, secretory and intercalated (peg) cells. Endosalpingiosis is most frequently found on the pelvic peritoneum covering the uterus, fallopian tubes, ovaries, and cul de sac, and in pelvic lymph nodes and the omentum [8]. Endosalpingiosis is usually an incidental finding identified intraoperatively or more often microscopically during histopathologic review of surgical specimens. Because endosalpingiosis has uncertain clinical significance, its presence is often not mentioned in diagnostic pathology reports and hence its prevalence is unclear. One prospective study identified endosalpingiosis in roughly a quarter of surgically resected omental specimens [8]. In the human ovary, endosalpingiosis manifests as benign glands with tubal-type differentiation in the ovarian cortex. A subset of ovarian "inclusion" cysts are lined by tubal-type epithelium and can be considered cystic endosalpingiosis.

We have developed genetically engineered mouse models (GEMMs) of HGSC in which the *Ovgp1* promoter is used in transgenic mice to drive expression of a Tamoxifen (TAM)-regulated Cre recombinase protein. The Cre recombinase can be used to inactivate concurrently the *Brca1*, *Trp53*, *Rb1*, and *Nf1* (*BPRN*) tumor suppressor genes (TSGs) in the oviductal epithelium (equivalent to human FTE). Mice with the *Ovgp1-CreER*^{T2} transgene and homozygous for floxed (fl) alleles of *Brca1*, *Trp53*, *Rb1*, and *Nf1* are referred to here as "*BPRN*" mice. *BPRN* mice treated transiently with TAM dosing at 6–8 weeks of age develop oviductal tumors that closely mimic their human HGSC counterparts with respect to associated precursor lesions, tumor morphology, and various biological features [9,10]. The murine oviductal HGSCs mirror the genomic alterations, gene expression profiles, and immune microenvironment of human HGSCs, based on an integrated molecular analysis of the mouse tumors with comparison to human HGSCs [11]. In the *BPRN* GEMM model, the earliest detectable oviductal lesions, equivalent to serous tubal intraepithelial carcinomas in women, are usually first observed 6–8 months after tumor induction, which is around the time of mouse menopause [10,12]. Because ovarian endosalpingiosis-like lesions are known to occur in mice [13–15], *BRPN* mice provide a unique model to test whether HGSC can arise from non-neoplastic tubal-type epithelium outside of the mouse oviduct, and to compare the protective effects of removing the oviducts alone versus the oviducts and ovaries from mice following tumor induction. We report our studies to determine if mice whose oviducts are surgically removed shortly before tumor induction with TAM acquire HGSCs outside of the oviduct, for example, in association with ovarian endosalpingiosis-like lesions. We also report whether removal of both ovaries and oviducts from the mice after tumor induction conferred additional protection from development of HGSC beyond removing the oviducts alone. These latter studies model the effects of risk-reducing salpingectomy (RRS) versus risk-reducing salpingo-oophorectomy (RRSO) in women with known genetic predisposition to HGSC, such as carriers of *BRCA1/2* pathogenic variants.

2. Materials and Methods

2.1. Genetically Modified Mice and Animal Care

Ovgp1-iCreER^{T2} mice and those also carrying combined floxed *Brca1*, *Trp53*, *Rb1*, *Nf1* TSG alleles (*BPRN* mice) and tumor induction with TAM have been described previously in detail [9,11,16]. Genotypes of the mice included in this study were confirmed by PCR analysis of tail-derived DNA using PCR primers reported previously [7]. The same primers were used to detect recombination of floxed TSG alleles in mouse tumors that developed in TAM-treated *BPRN* mice. Genomic DNA was isolated from formalin-fixed paraffin-embedded tumor tissues using the FFPE Tissue Kit (Qiagen, Germantown, MD, USA), following the manufacturer's instructions, and PCR amplified using GoTaq® Master Mixes (Promega, Madison, WI, USA). All procedures involving mice described in this study were approved by the University of Michigan's Institutional Animal Care and Use Committee (PRO00008343 and PRO00010212).

2.2. Bilateral Salpingectomy and Salpingo-Oophorectomy

Survival surgeries were performed on mice anesthetized by isoflurane, as previously described [9,11,16]. To determine effects of bilateral salpingectomy (BS) on the frequency of developing ovarian or primary peritoneal HGSC, removal of both oviducts was performed on a cohort (n = 25) of 8 week old *BPRN* mice. The uterus and ovaries were left intact. Another cohort of *BPRN* mice (n = 27) underwent removal of oviducts and ovaries (bilateral salpingo-oophorectomy, BSO) at 8 weeks of age. In this cohort, the uterus was left intact. A smaller cohort (n = 7) of age-matched *BPRN* mice underwent sham surgery in which the peritoneal cavity was opened, the ovaries and oviducts were exteriorized and then returned, and surgical sites were closed. After two weeks to allow recovery from the surgical procedure, mice were transiently treated with TAM (0.2 g/kg, intraperitoneally on day 1 and day 3) to inactivate the conditional TSG alleles in oviductal-type (OVGP1-expressing) epithelium present in the ovaries, peritoneum, or other sites. Mice were followed for 70 weeks post-TAM or euthanized earlier if humane endpoints were reached. Mice were necropsied and grossly evaluated for tumor presence and extent. In each case, the ovaries and oviducts (if present), uteri, lungs, multiple abdominal organs, and omentum/abdominal fat were examined microscopically. The presence or absence of other neoplasms arising spontaneously in the aged animals was also noted.

2.3. Histopathology and Immunohistochemistry

Formalin-fixed and paraffin-embedded (FFPE) tissue sections were stained with hematoxylin and eosin (H&E) for evaluation by light microscopy. To identify microscopic lesions in grossly normal oviducts and ovaries, each oviduct and/or ovary was serially sectioned (5 μm intervals) in its entirety and alternate sections were either stained with

H&E or retained for immunohistochemical (IHC) staining according to standard methods as described previously [9,11,16]. At least 12 H&E-stained sections representing serial levels of omentum were examined per mouse, while other organs were microscopically assessed by representative sections only. The following primary antibodies were used: rabbit anti-OVGP1 polyclonal antibody (Abcam, Cambridge, UK), rabbit anti-PAX8 (Proteintech, Chicago, IL, USA), rabbit anti-WT1 antibody (Santa Cruz, Dallas, TX, Thermo Fisher, Waltham, MA, USA), rabbit anti-acetyl tubulin (Cell Signaling, Danvers, MA, USA) and rat anti-cytokeratin 8 (CK8) (Developmental Studies Hybridoma Bank, Iowa City, IA, USA), rat anti-Ki-67 (Invitrogen, Thermo Fisher). Immunostaining of mouse tissues was scored based on the percentage and intensity of positive cells: "−", absent in all cells; "+/−", <10% positive cells with barely perceptible staining; "+", 10–25% positive cells with weak staining; "++", 26–50% positive cells with moderate staining; "+++", >50% positive cells with strong staining.

2.4. Statistical Analysis

The Mantel-Haenszel Chi-square test was performed using SPSS 29 software to compare the effects of the two surgical interventions (BS or BSO) vs. sham surgery. Differences with $p < 0.05$ were considered statistically significant. Survival rates of mice in the three cohorts (sham, BS, and BSO) were estimated by Kaplan–Meier analysis using GraphPad Prism software version 10.

3. Results

3.1. Endosalpingiosis Is Present in a Subset of Mice Following Bilateral Salpingectomy

In prior studies, we used *OVGP1-iCreER*^{T2} mice also expressing a Cre-inducible eYFP reporter allele (*R26R*$^{LSL\text{-}eYFP}$) to detect oviductal epithelium located outside of the oviduct, and we identified ovarian endosalpingiosis-like lesions in 14% of TAM-treated adult *OVGP1-iCreER*^{T2};*R26R*$^{LSL\text{-}eYFP}$ mice [17]. The endosalpingiotic inclusion glands and cysts were lined by secretory and ciliated cells and expressed PAX8, tubulin and OVGP1. We have also incidentally observed ovarian endosalpingiosis in the course of other work using our HGSC GEMMs (unpublished data). In this study, we identified endosalpingiosis in 6 of 50 (12%) ovaries from the 25 TAM-treated *BPRN* mice that underwent BS. A representative example is shown in Figure 1, in which a group of glands, some cystically dilated, are present in the ovarian stroma and are lined by cytologically bland epithelium (Figure 1A,B) that expresses markers of oviductal differentiation, including tubulin, OVGP1, PAX8 and WT1 (Figure 1C–F). Detailed information on the histopathologic features of endosalpingiosis and quantitative data of immunohistochemical staining are provided in Supplemental Table S1.

In addition, we identified focal endosalpingiosis with similar morphological and immunohistochemical features in the omentum of one *BPRN* mouse after BS (Figure 2A–F).

In mice, endosalpingiosis involving non-ovarian tissues is likely less common than ovarian endosalpingiosis, perhaps because the mouse ovarian bursa limits movement or dispersion of detached tubal epithelium to non-ovarian tissues. Also, the incidence of non-ovarian endosalpingiosis may be underestimated due to our inability to examine the omentum, abdomino-pelvic lymph nodes, and peritoneal surfaces for these microscopic lesions in an exhaustive fashion.

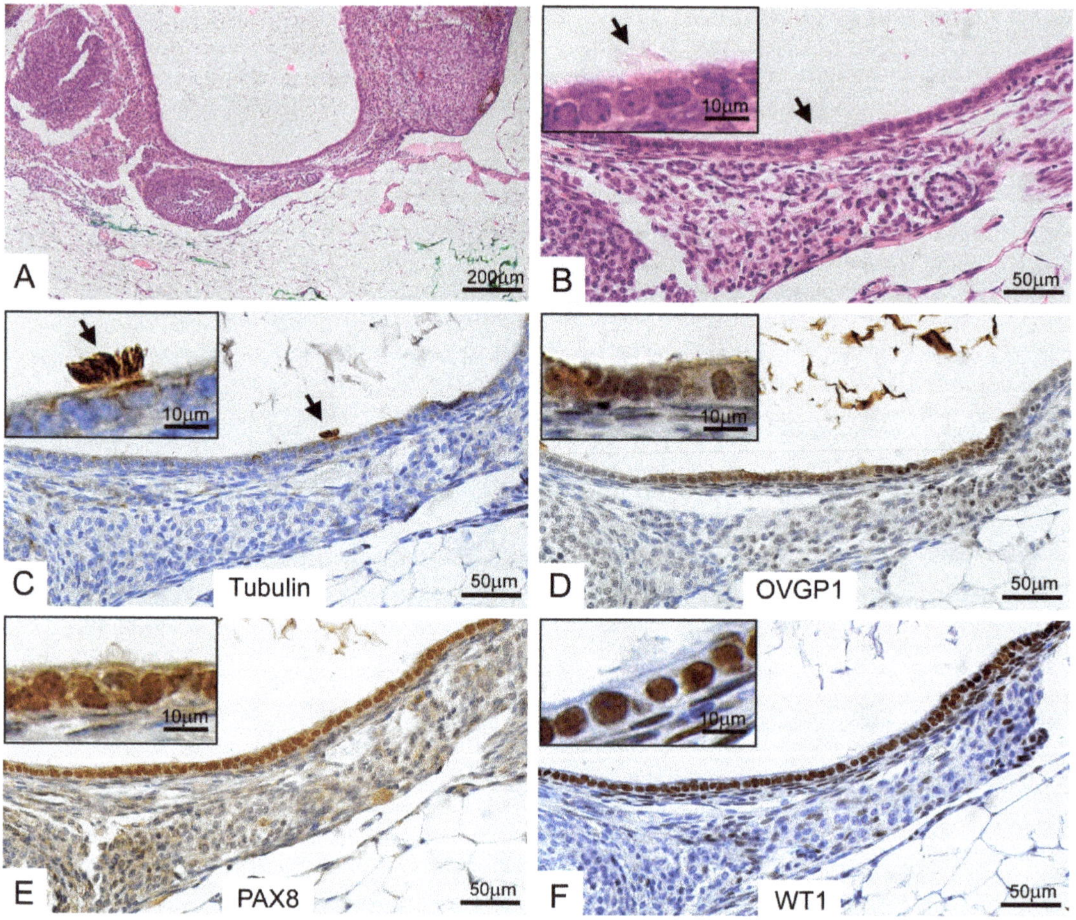

Figure 1. Murine ovarian endosalpingiosis expresses markers of tubal-type differentiation. H&E-stained sections of an ovary with cystic endosalpingiosis in a *BPRN* mouse treated with bilateral salpingectomy prior to TAM-induced gene targeting in oviductal epithelium (**A**,**B**). Immunohistochemical staining of sections showing expression of tubulin (**C**), OVGP1 (**D**), PAX8 (**E**) and WT1 (**F**). Black arrows indicate ciliated cells. Insets in selected panels show higher magnifications. Scale bars represent 200 μm, 50 μm and 10 μm as indicated on each panel.

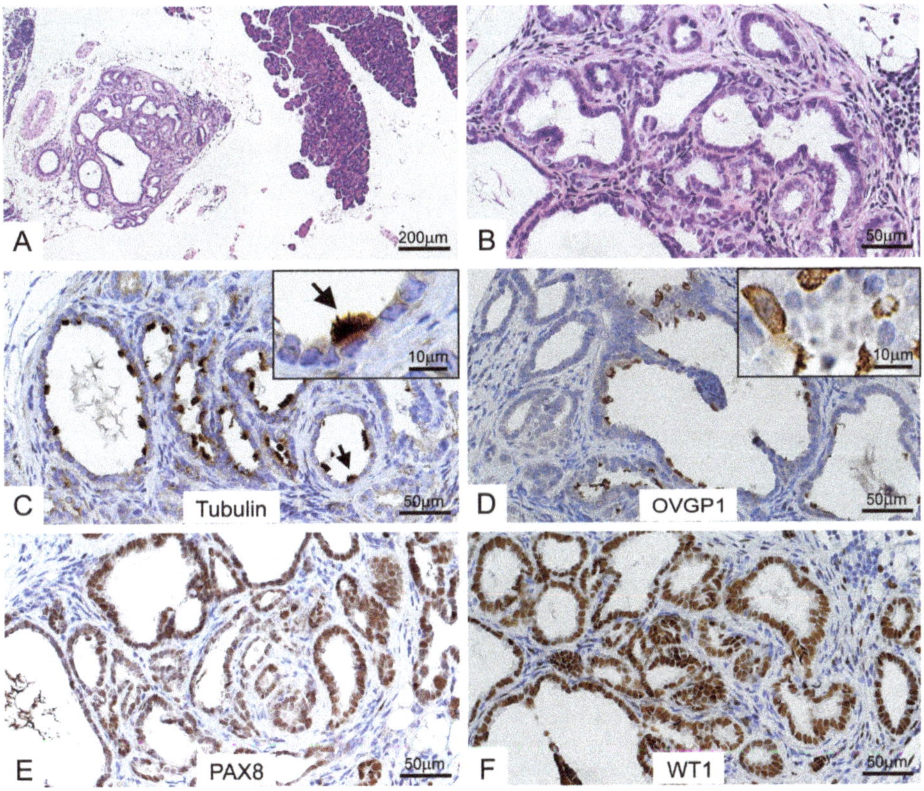

Figure 2. Murine peritoneal endosalpingiosis expresses markers of tubal-type differentiation. H&E-stained sections of peritoneal endosalpingiosis in a *BPRN* mouse that underwent bilateral salpingectomy prior to TAM-induced gene targeting in oviductal epithelium (**A**,**B**). Immunohistochemical staining of sections showing expression of tubulin (**C**), OVGP1 (**D**), PAX8 (**E**) and WT1 (**F**). Black arrows indicate ciliated cells. Insets in selected panels show tubulin and OVGP1 at higher magnification. Scale bars represent 200 µm, 50 µm and 10 µm as indicated on each panel.

3.2. Bilateral Salpingectomy Significantly Reduces the Incidence of Non-Tubal HGSC Development in TAM-Treated BPRN Mice, but Not as Completely as Bilateral Salpingo-Oophorectomy

Tumor development in cohorts of TAM-treated *BPRN* mice after undergoing sham surgery, BS, or BSO are summarized in Figure 3.

In agreement with our prior work, all *BPRN* mice that underwent sham surgery developed HGSC confined to the oviduct (eHGSC) or oviductal HGSC or carcinosarcoma (CaSa, a HGSC variant that has undergone epithelial-to-mesenchymal transition) extending beyond the oviduct, with more than half of the mice reaching humane endpoints before the end of the surveillance period (Figure 3, left section). Most of these mice had metastatic disease by study endpoint. In the cohort of 25 TAM-treated *BPRN* mice that underwent BS, 9 mice developed ovarian primary tumors over the 70-week surveillance period (Figure 3, center section). Three tumors were early HGSC (eHGSC, confined to the ovary) and six were more advanced tumors (HGSC or CaSa) that had spread beyond the ovary directly into adjacent soft tissue and/or metastasized to peritoneal surfaces. Representative gross photographs of advanced tumors arising in the ovaries of mice after BS are shown in Supplemental Figure S1. Two of the mice that developed advanced tumors reached humane endpoints earlier than 70 weeks post-TAM. None of the mice that underwent BS developed tumors with features of primary peritoneal HGSC, i.e., peritoneal disease in the absence of an ovarian tumor, although we cannot exclude the possibility of an occult peritoneal tumor given our inability to exhaustively examine the peritoneal surfaces in the mice.

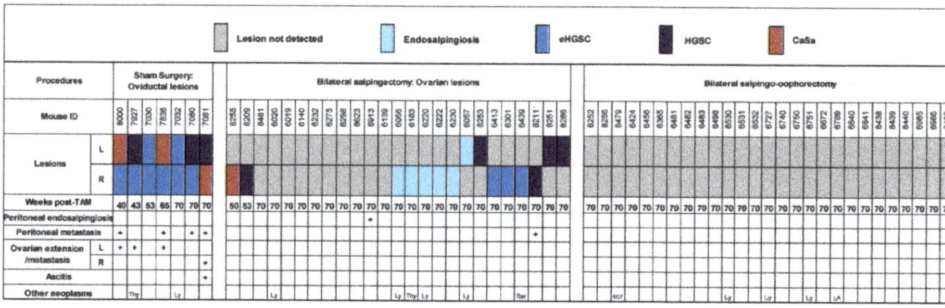

Figure 3. Summary of findings in TAM-treated *BPRN* mice following sham surgery, bilateral salpingectomy, or bilateral salpingo-oophorectomy. Weeks post-TAM indicates the time point at which mice were sacrificed. L, left; R, right oviduct (left section) or ovary (center section). Colored boxes indicate type of lesion identified (eHGSC: early HGSC confined to oviduct or ovary; HGSC; CaSa: carcinosarcoma). Additional tumors observed included Thy: Thyroid tumor; Ly: Lymphoma; Sar: Sarcoma; SGT: salivary gland tumor; LA: Lung adenocarcinoma.

As expected, tumors that developed in the ovaries after BS (Figure 4 upper panels) are morphologically comparable to the oviductal tumors that arose in the sham surgery cohort (Figure 4, lower panels), express markers characteristic of HGSCs—including CK8, PAX8, and WT1, and display recombination of the targeted TSGs (Supplemental Figure S2). In contrast, no peritoneal or other tumors were observed in any of the 27 mice that underwent removal of both oviducts and ovaries (Figure 3, right section). Statistical analysis showed that compared with the sham surgery group, tumor development was significantly reduced by both BS and BSO ($p < 0.001$ for both groups by the Mantel-Haenszel Chi-square test), although only the latter afforded complete protection from tumor development. Kaplan–Meier survival analysis of the mice in this study is shown in Supplemental Figure S3. Our findings on the effects of BS versus BSO in our *BRPN* GEMM provide unique insights regarding the limited protection against cancer development afforded by RRS.

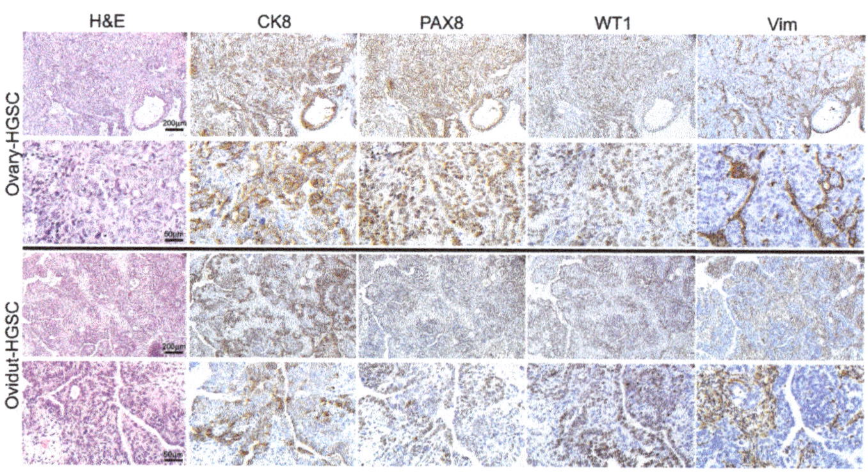

Figure 4. Tumors arising in the *BRPN* mice after sham surgery or bilateral salpingectomy, have comparable morphology and immunophenotype. The upper panels show H&E and immunohistochemical

stains of a representative HGSC arising in the ovary of a TAM-treated *BPRN* mouse after bilateral salpingectomy. Lower panels show H&E and immunohistochemical stains of a representative HGSC arising in the oviduct of a TAM-treated *BPRN* mouse after sham surgery. Sections were immunostained using antibodies for detection of cytokeratin 8 (CK8), PAX8, WT1 and vimentin (Vim). Scale bars represent 200 μm and 50 μm as indicated.

3.3. HGSC Is Associated with Endosalpingiosis in Mice

As noted above, some HGSCs in women likely arise outside of the fallopian tubes. The cells of origin for non-tubal HGSC lesions remain unclear. In our studies, an eHGSC arising in the ovary of a salpingectomized mouse appeared to arise from associated cystic endosalpingiosis. Figure 5 shows representative photomicrographs of the ovary in the mouse with an eHGSC lesion.

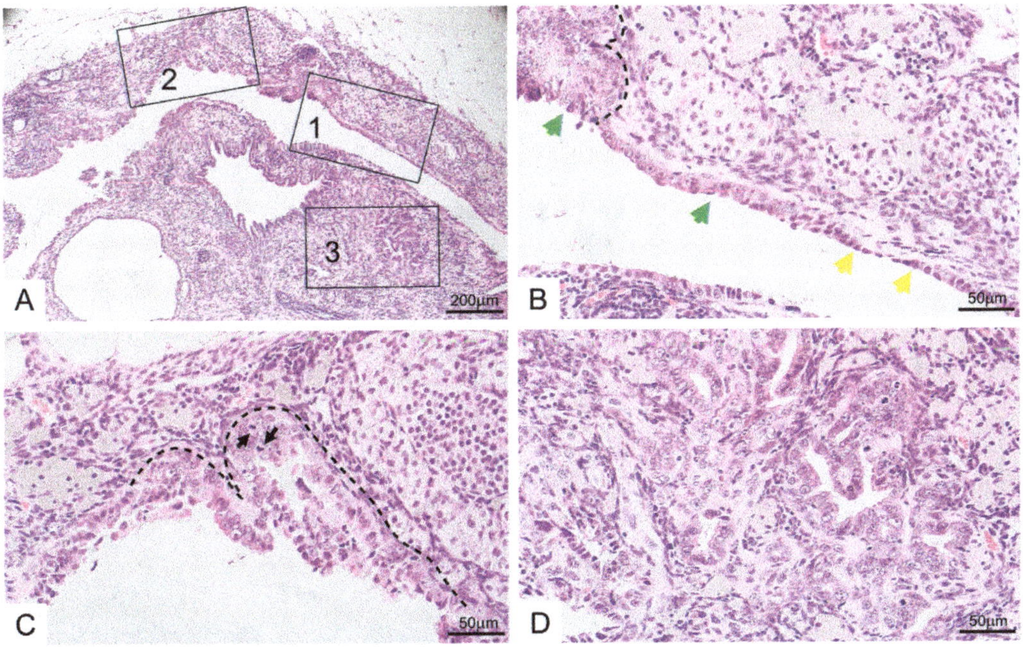

Figure 5. HGSC is associated with and appears to arise from ovarian endosalpingiosis in a TAM-treated *BPRN* mouse after bilateral salpingectomy. (**A**) Low-magnification photomicrograph of H&E-stained section of ovary showing eHGSC in continuity with endosalpingiosis. Boxed areas 1–3 are shown at higher magnification in (**B–D**) panels, respectively. (**B**) Area in Box 1 showing transition of benign-appearing oviductal-type epithelium (yellow arrows) to highly atypical epithelium (green arrows) consistent with serous intraepithelial carcinoma (SIC). (**C**) Area in Box 2 showing SIC lesion. Black arrows indicate mitotic figures. (**D**) Area in Box 3 showing invasive HGSC, still confined to ovary (eHGSC). Scale bars represent 200 μm and 50 μm as indicated.

The ovary shows endosalpingiosis in which benign-appearing epithelium is in continuity with epithelium displaying marked cytologic atypia (in situ carcinoma, referred to here as serous intraepithelial carcinoma, SIC), and then transitions to locally invasive carcinoma (eHGSC). Some markers of tubal differentiation present in the endosalpingiosis such as tubulin and OVGP1 are diminished in the invasive carcinoma, which retains expression of WT1, PAX8. Increased Ki-67 positive cells were observed in the SIC and eHGSC lesions (Supplemental Figure S4). Our observation of HGSC arising from endosalpingiosis provides a plausible explanation for why even salpingo-oophorectomy does not afford complete protection in women at high-risk of HGSC development [18].

4. Discussion

Most ovarian cancer-associated deaths among women in the United States are attributable to HGSC and there is general consensus that most HGSCs arise in the fallopian tubes—specifically from precursor lesions known as serous tubal intraepithelial carcinomas (STICs). Based on comprehensive DNA, RNA, and miRNA profiling, Ducie and colleagues showed that the molecular features of HGSCs with and without associated STIC lesions are mostly shared, supporting a common biologic origin they favored to be the distal fallopian tube [19]. Notably, however, when correlating RNA expression data between HGSCs and normal tissues, they found nearly 12% of HGSCs were better correlated to normal ovarian surface epithelium (OSE) than to normal FTE, supporting potential ovarian origin. Using transgenic mice, Zhang and colleagues showed that HGSCs can arise from either FTE or OSE and comparison of FTE- and OSE-derived mouse tumors to human HGSCs suggested that HGSCs can arise from both cell types [2]. Because the subset of HGSCs apparently arising outside of the fallopian tube (e.g., ovary or peritoneum) usually present at advanced stage [20], only a few STIC-like lesions or early HGSCs arising outside of the tubes have been reported, for example, in ovarian serous cystadenoma, serous cystadenofibroma, or benign serous cysts [3,4,21]. Collectively, these data support the conclusion that while most HGSCs indeed arise from the fallopian tube, a subset does not. However, it is still unclear from which cells these non-tubal HGSCs arise. Dubeau has suggested the "secondary Müllerian" system including endometriosis, endosalpingiosis, and paraovarian/paratubal cysts, may be playing roles in the pathogenesis of Mullerian carcinomas that appear to arise outside of the primary Mullerian tract [22,23]. Primary peritoneal HGSCs have been reported in women decades after BSO. Because the time needed for progression from STIC to HGSC is much shorter (approximately 6 years) [24], these tumors are more likely arising from endosalpingiosis [25–27]. A recent systematic review of peritoneal HGSCs arising after risk-reducing surgery concluded that most available evidence supports the dissemination of precursor cells originating in the fallopian tube, with ovarian origin considered less likely [28]. The dissemination of precursor cells could include not only benign tubal epithelium (i.e., endosalpingiosis), but also neoplastic epithelium from STICs and other early serous proliferations in the tube including p53 signature lesions through "precursor escape" [5,29].

HGSC GEMMs, in which the timing of tumor development is known and under experimental control, provide an ideal system with which to test if HGSCs can arise outside of the oviduct, and from what types of precursors. We found that removal of both oviducts prior to tumor induction markedly reduced but did not completely eliminate the development of HGSC, and the tumors are likely arising from ovarian endosalpingiosis lesions already present at the time of transient treatment with TAM to induce tumor formation. In contrast, removal of both oviducts and ovaries was completely protective, perhaps because endosalpingiosis in mice is predominantly found in the ovaries, which were removed prior to TAM treatment. We also observed a case of ovarian endsalpingiosis contiguous with SIC and invasive HGSC, providing histopathologic evidence for endosalpingiosis as a precursor from which non-tubal HGSCs can arise. While our findings provide support for the superiority of BSO over BS alone in preventing the development of HGSC, a limitation is that findings based solely on the *BPRN* model system may not be generalizable to other models not tested here. Moreover, anatomical differences between mice and humans may have impacted our findings compared to those expected in humans. Because dissemination of tubal epithelium (normal or neoplastic) to the peritoneal cavity but not the ovary is prevented by the ovarian bursa, it is not surprising that we have never identified presumptive primary peritoneal HGSC in *BPRN* mice. While BS and BSO are expected to disrupt the bursa, the timing of surgical intervention and TAM treatment could also impact the frequency and distribution of sites at which tumors arise.

In the general population, opportunistic salpingectomy is associated with reduced ovarian cancer risk [30]. For women with known genetic predisposition for developing HGSC (e.g., *BRCA1/2* mutation carriers), primary prevention strategies include RRSO, RRS, and salpingectomy with delayed oophorectomy (SDO) [31–33]. In high-risk populations RRSO has been shown to be very effective in reducing the risk of ovarian cancer [27,34], and in a recent study of more than 2500 women with germline pathogenic variant *BRCA1/2* alleles who underwent RRSO, the 20-year cumulative incidence rate of HGSC was 1.5% for *BRCA1* and 0.2% for *BRCA2* [18]. RRS, while less studied, is being investigated as a potential interim measure for younger at-risk women who wish to delay the onset of menopause but still want to reduce cancer risk. Our findings support the conclusion that RRSO affords more protection from developing HGSC than RRS alone. Importantly, our work does not address whether RRS with delayed oophorectomy might offer substantial HGSC risk-reduction while avoiding the consequences of inducing premature menopause. Experiments could be done in which cohorts of mice undergo BS, BSO or SDO at defined intervals after TAM exposure to determine whether and how the type and timing of surgery might influence protective effects and/or tumor distribution, and the results could inform studies in humans. Several ongoing clinical trials (NCT04294927, NCT04251052, and ISRCTN25173360) are evaluating the long-term outcomes and effectiveness of RRS or SDO versus RRSO, particularly in relation to quality of life, cancer risk reduction, and overall survival [35–37].

5. Conclusions

The current study illustrates the impact of GEMMs in advancing a more complete understanding of HGSC pathogenesis and how this knowledge can potentially be used to optimize prevention strategies for both the at-risk and general populations. Our findings provide experimental evidence for the superiority of RRSO relative to other strategies for reducing the risk of HGSC, while also providing a basis for further investigation of alternative preventive measures, including salpingectomy with delayed oophorectomy.

Supplementary Materials: The following supporting information can be downloaded at: https://www.mdpi.com/article/10.3390/cancers17172759/s1, Table S1: Summary of Endosalpingiosis, eHGSC and HGSC/CaSa Histopathology and Immunohistochemical Staining; Figure S1: Gross photographs of representative ovarian (Ov) tumors arising in TAM-treated *BPRN* mice after bilateral salpingectomy; Figure S2: Genotyping of mouse HGSCs arising in the ovaries of *BPRN* mice after bilateral salpingectomy; Figure S3: Kaplan-Meier survival analysis of mouse cohorts following sham surgery (red: $n = 7$), BS (green: $n = 25$), or BSO (blue; $n = 27$); Figure S4: Representative photomicrographs of endosalpingiosis and contiguous eHGSC in the ovary of mouse #6413 immunostained for markers of tubal-type differentiation.

Author Contributions: Y.Z., E.R.F. and K.R.C. conceived the study and designed the experiments. Y.Z. performed all experiments and analyses. Y.Z., E.R.F. and K.R.C. wrote the manuscript. All authors have read and agreed to the published version of the manuscript.

Funding: Research reported in this paper was supported by the National Cancer Institute of the National Institutes of Health under award numbers R01CA226756 (KRC, YZ, ERF) and P30CA046592 (KRC, ERF).

Data Availability Statement: Data is contained within the article or Supplementary Material—The original contributions presented in this study are included in the article/Supplementary Material. Further inquiries can be directed to the corresponding author.

Conflicts of Interest: The authors declare no potential conflicts of interest.

Abbreviations

The following abbreviations are used in this manuscript:

HGSC	High-grade serous carcinoma
GEMM	Genetically engineered mouse model
BPRN	Brca1, Trp53, Rb1, and Nf1
TSGs	Tumor suppressor genes
FTE	Fallopian tube epithelium
RRS	Risk-reducing salpingectomy
RRSO	Risk-reducing salpingo-oophorectomy
STICs	Serous tubal intraepithelial carcinomas
eHGSC	Early High-grade serous carcinoma
SIC	Serous intraepithelial carcinoma
CaSa	Carcinosarcoma
SDO	Salpingectomy with delayed oophorectomy

References

1. Przybycin, C.G.; Kurman, R.J.; Ronnett, B.M.; Shih Ie, M.; Vang, R. Are all pelvic (nonuterine) serous carcinomas of tubal origin? *Am. J. Surg. Pathol.* **2010**, *34*, 1407–1416. [CrossRef]
2. Zhang, S.; Dolgalev, I.; Zhang, T.; Ran, H.; Levine, D.A.; Neel, B.G. Both fallopian tube and ovarian surface epithelium are cells-of-origin for high-grade serous ovarian carcinoma. *Nat. Commun.* **2019**, *10*, 5367. [CrossRef] [PubMed]
3. Craig, E.; Clarke, R.; Rushton, G.; McCluggage, W.G. Serous Tubal Intraepithelial Carcinoma (STIC)-like Lesions Arising in Ovarian Serous Cystadenofibroma: Report of 2 Cases. *Int. J. Gynecol. Pathol.* **2015**, *34*, 535–540. [CrossRef] [PubMed]
4. Ramanayake, N.; Russell, P.; Yang, V. High grade serous intraepithelial carcinoma arising in a benign ovarian serous cyst—A bridge too far? *Pathology* **2018**, *50*, 485–489. [CrossRef] [PubMed]
5. Soong, T.R.; Howitt, B.E.; Miron, A.; Horowitz, N.S.; Campbell, F.; Feltmate, C.M.; Muto, M.G.; Berkowitz, R.S.; Nucci, M.R.; Xian, W.; et al. Evidence for lineage continuity between early serous proliferations (ESPs) in the Fallopian tube and disseminated high-grade serous carcinomas. *J. Pathol.* **2018**, *246*, 344–351. [CrossRef]
6. Prentice, L.; Stewart, A.; Mohiuddin, S.; Johnson, N.P. What is endosalpingiosis? *Fertil. Steril.* **2012**, *98*, 942–947. [CrossRef]
7. Kurman, R.J.; Shih Ie, M. The origin and pathogenesis of epithelial ovarian cancer: A proposed unifying theory. *Am. J. Surg. Pathol.* **2010**, *34*, 433–443. [CrossRef]
8. Zinsser, K.R.; Wheeler, J.E. Endosalpingiosis in the omentum: A study of autopsy and surgical material. *Am. J. Surg. Pathol.* **1982**, *6*, 109–117. [CrossRef]
9. Zhai, Y.; Wu, R.; Kuick, R.; Sessine, M.S.; Schulman, S.; Green, M.; Fearon, E.R.; Cho, K.R. High-grade serous carcinomas arise in the mouse oviduct via defects linked to the human disease. *J. Pathol.* **2017**, *243*, 16–25. [CrossRef]
10. Hou, X.; Zhai, Y.; Hu, K.; Liu, C.J.; Udager, A.; Pearce, C.L.; Fearon, E.R.; Cho, K.R. Aging accelerates while multiparity delays tumorigenesis in mouse models of high-grade serous carcinoma. *Gynecol. Oncol.* **2022**, *165*, 552–559. [CrossRef]
11. McCool, K.W.; Freeman, Z.T.; Zhai, Y.; Wu, R.; Hu, K.; Liu, C.J.; Tomlins, S.A.; Fearon, E.R.; Magnuson, B.; Kuick, R.; et al. Murine Oviductal High-Grade Serous Carcinomas Mirror the Genomic Alterations, Gene Expression Profiles, and Immune Microenvironment of Their Human Counterparts. *Cancer Res.* **2020**, *80*, 877–889. [CrossRef]
12. Diaz Brinton, R. Minireview: Translational animal models of human menopause: Challenges and emerging opportunities. *Endocrinology* **2012**, *153*, 3571–3578. [CrossRef]
13. Tan, O.L.; Hurst, P.R.; Fleming, J.S. Location of inclusion cysts in mouse ovaries in relation to age, pregnancy, and total ovulation number: Implications for ovarian cancer? *J. Pathol.* **2005**, *205*, 483–490. [CrossRef]
14. Burdette, J.E.; Oliver, R.M.; Ulyanov, V.; Kilen, S.M.; Mayo, K.E.; Woodruff, T.K. Ovarian epithelial inclusion cysts in chronically superovulated CD1 and Smad2 dominant-negative mice. *Endocrinology* **2007**, *148*, 3595–3604. [CrossRef]
15. Bristol-Gould, S.K.; Hutten, C.G.; Sturgis, C.; Kilen, S.M.; Mayo, K.E.; Woodruff, T.K. The development of a mouse model of ovarian endosalpingiosis. *Endocrinology* **2005**, *146*, 5228–5236. [CrossRef] [PubMed]

16. Wu, R.; Zhai, Y.; Kuick, R.; Karnezis, A.N.; Garcia, P.; Naseem, A.; Hu, T.C.; Fearon, E.R.; Cho, K.R. Impact of oviductal versus ovarian epithelial cell of origin on ovarian endometrioid carcinoma phenotype in the mouse. *J. Pathol.* **2016**, *240*, 341–351. [CrossRef]
17. Wang, Y.; Sessine, M.S.; Zhai, Y.; Tipton, C.; McCool, K.; Kuick, R.; Connolly, D.C.; Fearon, E.R.; Cho, K.R. Lineage tracing suggests that ovarian endosalpingiosis does not result from escape of oviductal epithelium. *J. Pathol.* **2019**, *249*, 206–214. [CrossRef] [PubMed]
18. Stroot, I.A.S.; Bart, J.; Hollema, H.; Wagner, M.M.; Yigit, R.; van Doorn, H.C.; de Hullu, J.A.; Gaarenstroom, K.N.; van Beurden, M.; van Lonkhuijzen, L.; et al. High-grade serous carcinoma occurring after risk-reducing salpingo-oophorectomy in BRCA1/2 germline pathogenic variant carriers. *J. Natl. Cancer Inst.* **2025**, *117*, 719–727. [CrossRef] [PubMed]
19. Ducie, J.; Dao, F.; Considine, M.; Olvera, N.; Shaw, P.A.; Kurman, R.J.; Shih, I.M.; Soslow, R.A.; Cope, L.; Levine, D.A. Molecular analysis of high-grade serous ovarian carcinoma with and without associated serous tubal intra-epithelial carcinoma. *Nat. Commun.* **2017**, *8*, 990. [CrossRef]
20. Hou, T.; Liang, D.; He, J.; Chen, X.; Zhang, Y. Primary peritoneal serous carcinoma: A clinicopathological and immunohistochemical study of six cases. *Int. J. Clin. Exp. Pathol.* **2012**, *5*, 762–769.
21. Novikov, F.V.; Anufriev, A.G.; Efremov, G.D. High-grade Serous Carcinoma Occurring in a Serous Cystadenoma on the Background of a Serous Tubal Intraepithelial Carcinoma (STIC)-like Lesion: A Case Report With Literature Review. *Int. J. Gynecol. Pathol.* **2024**, *43*, 626–630. [CrossRef]
22. Dubeau, L. The cell of origin of ovarian epithelial tumors and the ovarian surface epithelium dogma: Does the emperor have no clothes? *Gynecol. Oncol.* **1999**, *72*, 437–442. [CrossRef]
23. Dubeau, L.; Drapkin, R. Coming into focus: The nonovarian origins of ovarian cancer. *Ann. Oncol.* **2013**, *24* (Suppl. 8), viii28–viii35. [CrossRef]
24. Wu, N.Y.; Fang, C.; Huang, H.S.; Wang, J.; Chu, T.Y. Natural history of ovarian high-grade serous carcinoma from time effects of ovulation inhibition and progesterone clearance of p53-defective lesions. *Mod. Pathol.* **2020**, *33*, 29–37. [CrossRef]
25. Rodowa, M.; Salehi, A.H.; McGee, J. Primary Peritoneal Cancer Two Decades after a Bilateral Salpingo-Oophorectomy. *Case Rep. Obstet. Gynecol.* **2019**, *2019*, 1870834. [CrossRef] [PubMed]
26. Casey, M.J.; Synder, C.; Bewtra, C.; Narod, S.A.; Watson, P.; Lynch, H.T. Intra-abdominal carcinomatosis after prophylactic oophorectomy in women of hereditary breast ovarian cancer syndrome kindreds associated with BRCA1 and BRCA2 mutations. *Gynecol. Oncol.* **2005**, *97*, 457–467. [CrossRef] [PubMed]
27. Piver, M.S. Prophylactic Oophorectomy: Reducing the U.S. Death Rate from Epithelial Ovarian Cancer. A Continuing Debate. *Oncologist* **1996**, *1*, 326–330. [CrossRef]
28. Gootzen, T.A.; Bouwmeester, A.B.; de Hullu, J.A.; Piek, J.M.; van der Laak, J.A.; Simons, M.; Steenbeek, M.P. Pathogenesis of peritoneal high-grade serous carcinoma after risk-reducing surgery: A systematic review. *J. Pathol. Clin. Res.* **2025**, *11*, e70037. [CrossRef] [PubMed]
29. Soong, T.R.; Howitt, B.E.; Horowitz, N.; Nucci, M.R.; Crum, C.P. The fallopian tube, "precursor escape" and narrowing the knowledge gap to the origins of high-grade serous carcinoma. *Gynecol. Oncol.* **2019**, *152*, 426–433. [CrossRef]
30. Hanley, G.E.; Pearce, C.L.; Talhouk, A.; Kwon, J.S.; Finlayson, S.J.; McAlpine, J.N.; Huntsman, D.G.; Miller, D. Outcomes From Opportunistic Salpingectomy for Ovarian Cancer Prevention. *JAMA Netw. Open* **2022**, *5*, e2147343. [CrossRef]
31. Kauff, N.D.; Satagopan, J.M.; Robson, M.E.; Scheuer, L.; Hensley, M.; Hudis, C.A.; Ellis, N.A.; Boyd, J.; Borgen, P.I.; Barakat, R.R.; et al. Risk-reducing salpingo-oophorectomy in women with a BRCA1 or BRCA2 mutation. *N. Engl. J. Med.* **2002**, *346*, 1609–1615. [CrossRef]
32. Daly, M.B.; Dresher, C.W.; Yates, M.S.; Jeter, J.M.; Karlan, B.Y.; Alberts, D.S.; Lu, K.H. Salpingectomy as a means to reduce ovarian cancer risk. *Cancer Prev. Res.* **2015**, *8*, 342–348. [CrossRef]
33. Nebgen, D.R.; Hurteau, J.; Holman, L.L.; Bradford, A.; Munsell, M.F.; Soletsky, B.R.; Sun, C.C.; Chisholm, G.B.; Lu, K.H. Bilateral salpingectomy with delayed oophorectomy for ovarian cancer risk reduction: A pilot study in women with BRCA1/2 mutations. *Gynecol. Oncol.* **2018**, *150*, 79–84. [CrossRef]
34. Palaiologos, K.; Ellaboudy, A.; Abdullah, M.; Karan, S.; Saha, A. Prophylactic Bilateral Salpingo-oophorectomy in BRCA2 Mutation with Incidental Finding of Serous Tubal Intraepithelial Carcinoma (STIC) and Subsequent Diagnosis of Primary Peritoneal Carcinoma (PPC): A Case Report and Review of Current Literature. *Cureus* **2020**, *12*, e9301. [CrossRef] [PubMed]
35. Boerner, T.; Long Roche, K. Salpingectomy for the Risk Reduction of Ovarian Cancer: Is It Time for a Salpingectomy-alone Approach? *J. Minim. Invasive Gynecol.* **2021**, *28*, 403–408. [CrossRef] [PubMed]

36. Steenbeek, M.P.; van Bommel, M.H.D.; intHout, J.; Peterson, C.B.; Simons, M.; Roes, K.C.B.; Kets, M.; Norquist, B.M.; Swisher, E.M.; Hermens, R.; et al. TUBectomy with delayed oophorectomy as an alternative to risk-reducing salpingo-oophorectomy in high-risk women to assess the safety of prevention: The TUBA-WISP II study protocol. *Int. J. Gynecol. Cancer* **2023**, *33*, 982–987. [CrossRef] [PubMed]
37. Steenbeek, M.P.; Harmsen, M.G.; Hoogerbrugge, N.; de Jong, M.A.; Maas, A.; Prins, J.B.; Bulten, J.; Teerenstra, S.; van Bommel, M.H.D.; van Doorn, H.C.; et al. Association of Salpingectomy With Delayed Oophorectomy Versus Salpingo-oophorectomy With Quality of Life in BRCA1/2 Pathogenic Variant Carriers: A Nonrandomized Controlled Trial. *JAMA Oncol.* **2021**, *7*, 1203–1212. [CrossRef]

cancers

MDPI

Article

Implementation of a Traceback Testing Program for Ovarian Cancer: Findings from the FACTS Study

Nora B. Henrikson [1,*], M. Cabell Jonas [2], Paula R. Blasi [1], Adam H. Buchanan [3], Pim Suwannarat [2], Kathleen Leppig [4], Aaron Scrol [1], Tracey Leitzel [3], Adrienne N. Deneal [2], Daniela Canedo [2], Arvind Ramaprasan [1], Sundeep S. Basra [2], Jennifer Brown [2], Marilyn Odums [2], Yirui Hu [3], Katrina M. Romagnoli [3], Estella Khieu [5], Elsa Balton [4], Saumya Patel [3], Muki Kunnmann [3], Dina Hassen [3], Jing Hao [3], Meredith Lewis [3], Rachel Schwiter [3], Jessica Goehringer [3], Heather M. Ramey [3], Shanshan Gustafson [2], Katrina Hsieh [2], Ilene Ladd [3] and Alanna K. Rahm [3]

1 Kaiser Permanente Washington Health Research Institute, 1730 Minor Ave, Suite 1360, Seattle, WA 98101, USA
2 Mid-Atlantic Permanente Research Institute, 700-B 2nd Street NE, Washington, DC 20002, USA
3 Geisinger Department of Genomic Health, 658 Center Street, Danville, PA 17822, USA
4 Division of Medical Genetics, Department of Medicine, University of Washington, 1705 NE Pacific St., Seattle, WA 98195, USA
5 Swedish Health Services Maternal and Fetal Specialty Center, Seattle, WA 98104, USA
* Correspondence: nora.b.henrikson@kp.org; Tel.: +1-206-287-4675

Simple Summary: Genetic testing can help people with cancer and their families. Traceback testing finds people with cancer who missed genetic testing at diagnosis and offers it to them. We studied a Traceback program in three U.S. health systems. We identified 597 eligible patients and reached 59% of them. Of those reached, 133 (38%) completed genetic testing. Ten people (8% of those tested) had results that indicated increased cancer risks. However, no at-risk relatives completed free testing, despite reminders and support. Participants and staff thought that the program was important, but the extra work to accurately identify people who are eligible for Traceback testing could limit long-term success. Overall, the program found and reached people who were eligible for Traceback testing, but no at-risk relatives received testing during the study. Future studies can explore the program's potential.

Abstract: Background: Traceback testing—identifying and offering testing to people with previous cancer diagnoses who have not received current standard genetic testing—could benefit patients and their at-risk relatives. Methods: We conducted a multisite, nonrandomized pilot implementation study of a Traceback program at three integrated United States health systems. We assessed the reach, fidelity, effectiveness, and acceptability of the program using quantitative and qualitative methods. Results: We identified 597 eligible individuals using administrative data and manual chart review. We attempted to reach everyone identified (100% fidelity). We successfully contacted 354 people, for a reach of 59% of confirmed eligible individuals. In total, 133 people completed Traceback genetic testing. Ten of these (8%) received pathogenic or likely pathogenic results;. Nine of these ten people received positive results for which cascade testing of at-risk relatives would be indicated. None of their relatives underwent cascade testing during the study period. Thirty-six received variants of uncertain significance (VUS). Traceback programs were acceptable to participants and implementers and thought to be applicable to other genetic screening conditions. The time and resources required to accurately identify Traceback-eligible individuals are likely determinants of future sustainability. Conclusions: Education about free cascade testing, reminder calls to probands, and offers to directly contact at-risk relatives did not result in cascade testing in this pilot study. However, participant and

 https://doi.org/10.3390/cancers17071154

implementer discussions suggest that the potential benefits of Traceback programs and high participant acceptability are worthy of further study.

Keywords: ovarian cancer; cancer prevention; gynecologic cancer risk reduction

1. Introduction

Ovarian cancer is the sixth leading cause of cancer death in women. When detected early, the relative survival rate of ovarian cancer is over 90% [1], presenting an opportunity for primary prevention and early detection that could reduce the human suffering associated with ovarian cancer. Pathogenic variants in *BRCA1* and *BRCA2* and several other genes are risk factors for ovarian cancer [2], conferring a 17–44% lifetime risk of ovarian cancer for females with pathogenic variants in the *BRCA1* or *BRCA2* genes, as well as increased risks for breast (72%) and other cancers [3]. Current guidelines recommend genetic testing of all individuals with newly diagnosed ovarian cancer [4]. However, genetic testing at diagnosis has been limited, leaving many families without knowledge of potential pathogenic variants in the family. Furthermore, guidelines have evolved over time, leaving many people without current standard genetic testing. At-risk relatives of patients with a known pathogenic variant are recommended to be referred for familial variant testing, known as cascade testing. Identified carriers can choose surveillance for early cancer detection and/or risk-reducing surgery, both of which are associated with substantially improved health outcomes [5]. However, cascade testing uptake among eligible relatives is low—only about 20–30% [6]—even when there is no charge billed to family members for cascade testing [7]. This represents a missed opportunity for implementation of precision medicine and improved health outcomes in families.

The National Cancer Institute's (NCI) Traceback framework for family-based outreach consists of three phases: (1) identification of individuals with ovarian cancer without prior genetic testing (i.e., probands); (2) genetic testing of these individuals; and (3) cascade testing of at-risk family members [6,8]. Following this, the Feasibility and Acceptability of Cascade Traceback Screening (FACTS) study implemented Traceback testing for living ovarian cancer survivors without documentation of current standard genetic testing at diagnosis in three United States health systems [9]. Here, we describe the results of our implementation and mixed methods evaluation.

2. Materials and Methods

We conducted a multisite, nonrandomized pilot implementation study. Our project was informed by the Conceptual Model for Implementation Research, which provides guidance for measuring outcomes across multiple levels [10]. The model also both distinguishes between and links implementation processes and outcomes and informed data collection and interpretation of results across all aims. Study activities were approved by the Institutional Review Board of Kaiser Permanente Interregional IRB and by Geisinger IRB.

We implemented and evaluated the Traceback program at three integrated health systems: Kaiser Permanente Washington (KPWA), Kaiser Permanente Mid-Atlantic States (KPMAS), and Geisinger (GE). These three systems represent geographic diversity, racial and ethnic diversity, diversity of insurance models [11], and the opportunity to offer genetic testing to individuals and their at-risk relatives.

2.1. Population and Eligibility

Inclusion criteria included living and currently receiving care in one of the three health systems with a prior diagnosis of ovarian, peritoneal, or fallopian tube cancer and no history of current standard panel germline genetic testing for hereditary cancer risk at diagnosis. Patients diagnosed since 1980 and still living were eligible, as were individuals who were tested and received a negative result prior to 2014, when current panel testing became standard. Exclusion criteria included being in hospice care, having an ovarian tumor with borderline histology, and inability to provide verbal consent.

2.2. Intervention

The Traceback framework includes three phases to facilitate the identification of and genetic testing for inherited pathogenic variants: (1) identifying eligible probands, (2) proband testing, and (3) cascade testing of at-risk relatives [6]. Each of the three study sites implemented these three phases. First, we identified potentially eligible participants using administrative and tumor registry data. All sites used the same programming specifications to identify eligible participants in their tumor registries. KPMAS also used diagnosis codes to identify eligible patients within two additional internal patient databases. Unexpectedly, manual chart review by a clinician was necessary to confirm eligibility, in particular prior diagnosis (e.g., for tumors with borderline histology) and presence or absence of previous germline genetic testing that was not reliably accessible within the administrative or tumor registry databases (e.g., in scanned PDF reports). Therefore, we conducted manual chart reviews to confirm eligibility before contacting individuals.

To implement testing of eligible probands, we designed intervention materials and workflows based on our preliminary work [12]. To design the Traceback programs, we conducted interactive qualitative interviews using human-centered design methods with patients at each site (total $n = 70$) with ovarian, fallopian tube, or peritoneal cancer (probands) and people with a family history of ovarian cancer (relatives). Participants preferred to be approached using personalized communication, with access to a physician discussion if needed for follow-up questions. They preferred messaging focused on clinical benefits to relatives, rationale for testing, and ease of access to testing [12].

Each site designed their specific workflows according to local context and workflow needs and adapted the same recruitment materials. Descriptions of each site's intervention processes are shown in Figure 1. Invitations to participate in Traceback testing were sent via mail or patient portal, with follow-up reminder calls to schedule a pre-test genetic counseling visit for interested participants (KPWA and GE), with subsequent laboratory-based sample collection. At KPMAS, the language in outreach materials differed slightly based on whether the patient had out-of-date testing (prior to 2014) or no evidence of testing. At KPMAS, in lieu of pre-test genetic counseling, a study team nurse conducted outreach reminder calls and ordered a genetic testing kit to be sent directly to interested participants' homes. All participants received testing via the Invitae Common Hereditary Cancer Panel (San Francisco, CA, USA), a primary panel of 48 genes (*APC, ATM, AXIN2, BAP1, BARD1, BMPR1A, BRCA1, BRCA2, BRIP1, CDH1, CDK4, CDKN2A, CHEK2, CTNNA1, DICER1, EPCAM, FH, GREM1, HOXB13, KIT, MBD4, MEN1, MLH1, MSH2, MSH3, MSH6, MUTYH, NF1, NTHL1, PALB2, PDGFRA, PMS2, POLD1, POLE, PTEN, RAD51C, RAD51D, SDHA, SDHB, SDHC, SDHD, SMAD4, SMARCA4, STK11, TP53, TSC1, TSC2, VHL*).

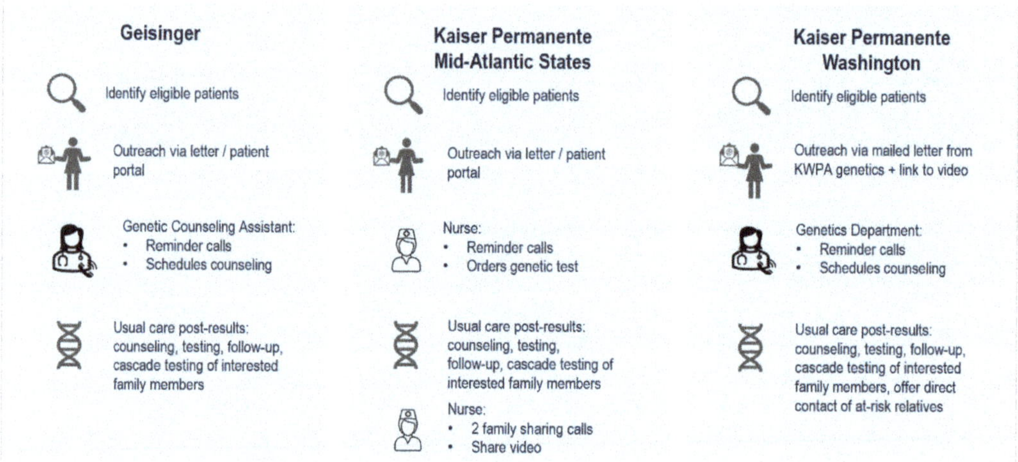

Figure 1. Overview of Traceback testing programs at each site.

All 3 sites used their usual care procedures for return of results and clinical follow-up. Individuals with positive results were recommended to inform their at-risk relatives of their result and to inform relatives of the availability of free cascade testing from the testing laboratory. Individuals with VUS were informed of the result, that their family members should not have genetic testing for the VUS as the presence or absence is not sufficient to inform risk, and that variant classification may be updated in the future as additional knowledge is gained. At KPMAS, after the results were returned, the nurse additionally offered reminder calls and shared a video made by the study team describing Traceback testing for participants to share with their at-risk relatives. At KPWA, the genetic counselor offered to directly contact at-risk relatives to recommend cascade testing, consistent with ethical guidance that suggests contacting relatives directly is acceptable with the proband's consent [13,14]. As part of other ongoing clinical efforts, Geisinger was also able to offer direct contact to at-risk relatives with proband consent later in the program implementation. Enrollment was completed between March 2022 and March 2023. Results were returned between August 2022 and July 2023.

2.3. Data Collection and Analysis

To assess reach, fidelity, and effectiveness, all sites monitored and recorded the numbers of patients who were eligible, invited, and tested. We assessed testing uptake and results of proband genetic testing and cascade testing in relatives across the three sites. To assess representativeness, we also collected demographic information, including age at diagnosis and at the time of our eligibility assessment. Information regarding stage at diagnosis was not reliably available, so was not included. We collected the numbers of pathogenic results and variant of uncertain significance (VUS) results returned to patients. To evaluate cascade testing uptake, we recorded the number of first or second degree (if FDR was deceased) at-risk relatives for each participating proband. We worked with Invitae to receive the numbers of at-risk relatives completing cascade testing during Invitae's free 150-day testing period offered to at-risk relatives.

To assess acceptability, we conducted qualitative interviews with program participants. We invited all individuals invited to Traceback testing to participate in an interview, regardless of whether they accepted genetic testing or not. About four months after participants were invited to receive genetic testing, we contacted participants again via letter, secure message, or phone call to invite them to participate in a telephone interview about their experience with the Traceback program. We followed a purposeful consecutive sampling approach by scheduling interviews with all those interested until we reached our goal of interviewing at least 45 probands, or about 15 per study site. Participants received USD 25 honoraria for their time. All interviews were conducted between August 2022 and July 2023.

Trained interviewers (P.R.B., A.D., C.J., D.C., J.G., T.L., A.M.K., S.P., R.S.) conducted one-on-one interviews with consenting participants; in some instances, a second interviewer joined the call to assist with taking notes. Interviewers used a secure Microsoft Teams conference line to conduct and record the audio-only interviews, which lasted 20–60 min. Interviewers followed a semi-structured interview protocol. Questions related to the reasons for participating, reaction to and experiences with the Traceback program, impacts of the program, and suggestions for improvement (the interview guide is available in the Supplementary Materials). Interviewers wrote brief field notes following each interview. Audio recordings were transcribed verbatim using professional transcription services. All transcripts, field notes, and memos identified participants by study identification number only.

We used Atlas.ti (Berlin, Germany; KPWA and Geisinger) and NVivo (Burlington, MA; KPMAS) to store, organize, and code anonymized interview transcripts. We conducted thematic analysis using a template analysis approach that relied on both a priori and emergent codes [15–17]. We developed a priori codes based on our research questions and interview field notes and we identified additional codes from the data. After piloting and finalizing our coding approach and definitions, we independently coded the remaining transcripts and drafted coding memos describing preliminary themes and supporting quotations. We examined similarities and differences in findings across study sites. The study team engaged in iterative discussions to further identify connections in the data, interpret findings, and focus the scope of the analysis. Investigators were not known to participants prior to this study. The authors practiced reflexivity by documenting and discussing the potential influence of their personal characteristics and perspectives on their interviews and analyses.

To assess stakeholder perspectives on Traceback programs, we conducted virtual, semi-structured discussions with program implementers at each FACTS clinical site between July and September 2023. Implementers included medical geneticists, genetic counselors, genetic counseling assistants, and nurses who were involved in the Traceback program at each clinical site. Discussions took place after the Traceback program patient outreach and genetic testing had concluded at each site. The study investigators invited implementers via phone or email. Discussion questions were guided by the Conceptual Model for Implementation Research and focused on implementation, service, and client outcomes. Topics included overall experience with the program, perceived benefits of the program, factors influencing sustainability, and improvements needed. The Acceptability of Intervention Measure (AIM) and Intervention Appropriateness Measures (IAM [18]) were administered at the close of each interview, with response options of Agree or Disagree (adapted from original 5-point scale).

Discussions were led and attended by study investigators (A.K.R., C.J., N.B.H.). Discussions were audio-recorded on a secure Microsoft Teams conference line. Investigators took notes on interview findings from the recordings, and prepared written summaries of each site's interviews. Through a series of analysis meetings, the team summarized the findings across discussions.

3. Results

Across the three sites, we identified 6386 people through administrative data who were potentially eligible for Traceback testing. Of these, 597 were confirmed eligible after manual chart review. Exclusions included individuals with tumors not known to be associated with hereditary cancer predisposition, individuals with a non-ovarian primary cancer that had spread to the ovary, and those with benign ovarian tumors. Individuals with serous or endometrioid ovarian carcinomas were included on a case-by-case basis during manual chart review by clinicians at KPMAS and KPWA, and included at Geisinger. We attempted to reach everyone identified (100% fidelity). Of these, 173 (29%) passively refused, defined as no response to outreach. We successfully contacted 354 people, for a reach of 59% of confirmed eligible individuals (range 48–68%).

In total, 133 people (22% of those eligible, 38% of reached individuals) completed Traceback genetic testing (Table 1). Participants were representative of non-participants for age at diagnosis, race/ethnicity, and ovarian vs. other cancer type (fallopian or peritoneal). Age at eligibility assessment was different at two of the three sites. At KPMAS, participants were older at eligibility assessment than nonparticipants (68.0 vs. 65.2, $p = 0.024$), as were KPWA (73.8 vs. 70.7, $p = 0.042$). Age at diagnosis was similar between sites (mean age: 57.4 years). Between sites, KPWA had significantly older participants at enrollment/eligibility assessment compared to KPMAS and Geisinger (73.8 years vs. 68.0 and 69.8; $p = 0.029$ for differences across sites). KPMAS had the most diverse participants, with 35% reporting Black and 12% Asian race, compared to 5% Black and 5% Asian at KPWA and 0% Black and 6% Asian at Geisinger ($p < 0.001$), reflecting their patient population and diversity-focused sampling strategy. Most (92%) participants had ovarian cancer; 8% had peritoneal or fallopian cancer. Rates of completing testing varied across sites, from 8% to 33%.

Table 1. Characteristics of Traceback program participants.

	Total, *n* (%)	KPWA *n* (%)	KPMAS *n* (%)	Geisinger *n* (%)	*p* Value, Difference Between Sites
Eligible	597	149	224	224	
Participated in testing	133	37	74	18	
Age at eligibility assessment, mean (SD)	69.8 (10.9)	73.8 (9.1)	68.0 (12.0)	69.8 (10.9)	0.029
Age at diagnosis, mean years (SD)	57.4 (10.3)	58.1 (10.5)	56.6 (9)	58.3 (11.8)	0.821
Years between eligibility assessment and age at diagnosis	13.2 (8.2)	13.2 (8.2)	13.2 (8.2)	13.2 (8.2)	13.2 (8.2)
Female	133 (100)	41(100)	74 (100)	18 (100)	NA
Race					<0.001
Asian	12 (9.0)	2 (4.9)	9 (12.2)	1 (5.6)	
Black	28 (21.1)	2 (4.9)	26 (35.1)	0 (0.0)	
Other race	7 (5.3)	2 (4.9)	5 (6.8)	0 (0.0)	
White	86 (64.7)	35 (85.4)	34 (45.9)	17 (94.4)	
Hispanic ethnicity	7 (5)	1 (2)	6 (8)	0	0.24
Ovarian cancer diagnosis	122 (92)	34 (83)	72 (99)	16 (89)	0.008
Peritoneal or fallopian cancer diagnosis	9 (8)	7 (17)	2 (1)	2 (11)	

Table 2 shows the genetic testing and cascade testing results. In total, 10 Traceback testing participants received pathogenic or likely pathogenic results, representing a positive rate of 8% of people completing testing and 3% of all confirmed eligible for Traceback testing, including seven with a pathogenic variant in a gene with strong or definitive evidence of association with ovarian cancer (*ATM* [2], *BRCA1* [1], *BRCA2* [2], *RAD51D* [2]; range of pathogenic findings across sites was 5–9% of people completing testing; Table 3). Thirty-six participants received variants of uncertain significance per each site's normal clinical practice; VUS rates were similar across sites. Nine of ten people received positive results for which cascade testing of at-risk relatives would be indicated. Genetic counselors identified 58 eligible at-risk relatives among these 9 participants. No probands provided consent for the study team to contact their relatives directly. Over the study period, no at-risk relatives underwent cascade variant testing, which was provided at no charge by the testing laboratory.

Table 2. Results of Traceback genetic testing program.

	Total *n*	**%**	**GE**	**%**	**KPWA**	**%**	**KPMAS**	**%**
Potentially eligible individuals identified through administrative data, *n*	6386		1173		2274		2939	
Potentially eligible, alive and age 18+ at data pull, chart reviewed [6]	1568		450		240		878	
Confirmed eligible for Traceback testing and attempted to reach, *n*	597	20% [1]	224	50%	149	62%	224	10%
Eligible individuals reached, *n* [7]	354	59% [2]	129	58%	72	48%	153	68%
Interested in genetic counseling/testing, *n* (KPWA and Geisinger only)	74		33		41		NA	
Attended genetic counseling appointment	73		30		41		2	
Consented to genetic testing (no pretest counseling) (KPMAS only)	123		NA		NA		123	
Completed Traceback genetic testing, *n*	133	22% [2]	18	8%	37	25%	74	33%
		38% [3]		14%		51%		48%
Number of participants receiving positive result	10	2% [2]	1	0.4%	2	1%	7	3%
		8% [4]		5.6%		5%		9%
Variant of unknown significance (VUS) result	36	6% [2]	5	2%	10	7%	21	9%
		28% [4]		28%				28%
Total number of at-risk relatives identified	58		NA [5]		21		37	
At-risk relatives completing cascade genetic testing, *n*	0				0		0	

GE = Geisinger; KPWA = Kaiser Permanente Washington; KPMAS = Kaiser Permanente Mid-Atlantic States. [1] Percentage potentially eligible. [2] Percentage of those confirmed eligible. [3] Percentage of those reached. [4] Percentage of those who completed Traceback testing. [5]—not applicable; result not indicated for cascade testing. [6]—defined as currently receiving care (Geisinger) or current member (KPWA, KPMAS). [7]—defined as study team spoke with person. KPMAS sent certified letters to all individuals attempted to reach.

Table 3. Genetic results, Traceback program (sites combined).

Genes with positive result [1] (Number of participants receiving result) (*n* = 10 participants)	Strong/definitive ovarian cancer association [3]: *ATM (2), BRCA2 (2), RAD51D (2), BRCA1* No evidence of ovarian cancer association [3]: *CHEK2 (3), TP53*
Variant of unknown significance (VUS) result genes [2] (*n* = 36 participants)	Strong/definitive ovarian cancer association [3]: *ATM* [4] *(3), BRCA2 (3), BRIP1(2), PALB2, RAD51D, SMARCA4,* No evidence of ovarian cancer association [3]: *NBN (3), POLE (3), BARD1(2), CDH1 (2), SDHA (2), APC, CHEK2, HOXB1, HOXB13, KIF1B, KIT, MSH3, NF1, NTHL1, PDGFRA, POLE/HOXB1, RAD50, RET, SDHC, TSC2*

[1] One participant received positive variants in two genes. [2] Three participants received multiple VUS. [3] Per Clinical Genome Resource or National Comprehensive Cancer Network as of January 2025. [4] One participant had an *ATM* VUS and *APC* VUS.

3.1. Traceback Participant Experience

We completed interviews with 47 probands, including 18 from Geisinger, 16 from KPWA, and 13 from KPMAS (Table 4). All participants reported female gender, and their median age was 69. Most participants (76.6%) were of white race, and none reported Hispanic ethnicity. More than half of the participants (57.4%) reported having a 4-year college degree or more education, and a quarter (25.5%) reported an annual household income of more than USD 75,000. Demographic characteristics were similar across study sites, though KPMAS had a larger proportion of Black participants (38.5%) compared with all sites combined (12.8%), and Geisinger had a smaller proportion of participants reporting incomes greater than USD 75,000 reflecting its mostly rural population . Though we looked for differences and similarities across the three study sites, we did not detect differences by site in how participants described their experiences of the Traceback program; therefore, the following results include data across all sites.

Participants generally had positive reactions to the Traceback program's invitation to have genetic testing, noting they trusted their health system and appreciated the opportunity to learn more about their genetic risk (Table 5). Only three participants reported surprise, confusion, or a negative reaction to the invitation, noting they did not understand why they were receiving the invitation many years after their initial cancer diagnosis.

Thirty of the forty-seven interview participants (63.8%) had accepted the invitation to receive genetic testing. Of the 17 participants who initially declined genetic testing, 5 changed their mind during the interview and agreed to receive testing. Reasons for accepting genetic testing included learning information that could affect participants' health and the health of family members, as well as an interest in contributing to research and helping future patients. Reasons for declining testing included the cost of testing, logistical barriers such as scheduling challenges, and a belief that the information would not be useful to them or their family (e.g., because the participant is elderly and/or already aware of an increased cancer risk).

Participants who received genetic counseling and testing generally reported positive experiences with the process, noting the sample collection process was easy, straightforward, and often could be completed as part of an unrelated clinic visit. The few instances of negative experiences were related to miscommunications or mix-ups in mailing genetic test kits (e.g., saliva kits mailed instead of buccal kits) that were later resolved. Participants also reported positive experiences talking with the genetic counselor, noting that the counselor was helpful, informative, and answered their questions. At KPMAS, participants reported finding value in the nurse coordinator calls, including the reminders, opportunities to have questions answered, and obtaining general information.

Table 4. Demographics of interview participants (n = 47).

	N	% or Range
Total interviewees	47	100%
Geisinger	18	38.3%
KPMAS	13	27.7%
KPWA	16	34.0%
Median age	69	(48–90)
Female	47	100%
Race/ethnicity		
Asian	4	8.5%
Black	6	12.8%
White	36	76.6%
Hispanic ethnicity	0	0.0%
4-year college graduate or more	27	57.4%
Annual income $50K or less	11	23.4%
Currently working for pay	8	17.0%
Currently married	26	55.3%
Completed Traceback genetic testing	30	63.8%

Abbreviations: KPMAS = Kaiser Permanente Mid-Atlantic States; KPWA = Kaiser Permanente Washington.

Table 5. Patient experiences with and reflections on Traceback program.

Category	Them	Example Quotes
Reaction to initial outreach	Positive reaction	My reaction is really positive. Because after like you looking out for me, you know, when you invite me to do genetic testing you're not only looking out for me, you're looking out for my family. So I feel good, you know, taking genetic testing is not only for me. KPWA00032 (Acceptor)
		I didn't have any qualms about it. I was happy that it was my. . . you know my Kaiser doctor. I trust them. I don't know how I would've felt had I gotten it from some other program. I don't know, but it was all through Kaiser so I didn't have any. . . any, any hesitation about it. KPMAS0008 (Acceptor)
	Negative reaction	First off I had ovarian cancer in 1973. I've been with multiple entities of what is now Kaiser forever and you're just now asking me this?. . . If you're going to tell me what you're looking for, if you're going to tell me how this is going to benefit and don't tell me it's free. Because Kaiser doesn't do nothing for free. . . To say it's free you're getting my DNA which is mine and ask me to give it to you for free, so that's about the only free there is. KPMAS0001 (Decliner)
		Well, what was going through my head was if you want me to come down there and get another needle you are nuts, that was going through my head before she even said anything. GE90692 (Decliner)
	Surprised by outreach	I was somewhat surprised, and I guess I felt as though oh what is it about me that makes them want me to do this? GE00345 (Decliner)
	Did not remember	I'm sorry, but I don't know if my husband threw that away. I don't remember seeing a letter if he answered the phone. GE00185 (Decliner)

Table 5. *Cont.*

Category	Them	Example Quotes
Genetic testing and counseling experience	Easy, straightforward testing process	It was pretty straightforward. I talked with the nurse and that was simple. The testing kit came as I was told it would and did the test. In an era of COVID testing this just seemed like one more set of swabs and containers. So it didn't really seem very exotic. KPMAS0002 (Acceptor)
	Informative conversations with genetic testing staff	I think it was great. Everybody was really nice and really informative and answered all the questions. And explained what was going in really clear terms so that I completely understood it. KPWA00059 (Acceptor)
Benefits of program	Prompt to get genetic testing that otherwise would not have pursued	I received the letter and I dismissed it at first because I thought I don't feel I need to do it. . . .Then one day I thought you know I have a child, I have a daughter. And I thought perhaps I should do this and gather more information for her. KPMAS0005 (Acceptor)
	Provided relevant risk information to family members	I encouraged them both to get tested. You know, and like I say, one daughter did, she did not have that genetic mutation. And hopefully the other one doesn't have it either, but she has not completed the genetic testing. KPWA00131 (Acceptor)
	Motivation to pursue preventive health actions	I'm going to get yearly mammograms and yearly breast exams. And just, you know, do what we can to catch it early if it's going to happen. Hopefully it won't. KPWA00059 (Acceptor)
Suggestions for improvement	Personalize outreach to proband's health history	So I think the only confusion that I had was that I already had the genetic testing, so I wasn't quite sure why they were reaching out to me again. KPWA00208 (Acceptor)
		I don't recall that there was any. . . any indication of awareness of how long in the past my cancer diagnosis was which you know it's kind of historical information. KPMAS0002 (Acceptor).
		The only question I have is what took them so long to decide they needed to do a genetic testing. . . Because a lot of people have died from ovarian cancer in 20 some years. KPWA00089 (Decliner)
	Explicitly explain benefits and potential impacts of testing	I would have appreciated maybe some more up-front information about how this can affect peoples' lives if they discover things in their genes that are--could be passed on. So, you know, I think just a little more education with why you're--why we need to do this. KPWA00037 (Acceptor)
		I think they could have identified, you know, these are the things that we might find out that would be helpful. You know, whether you could pass this along, whether somebody else in your family or immediate family might already have this, or they should get a particular test, or things like that. I have a couple of nieces out there, grandnieces, but because they're not direct genetically, I just assumed that that wouldn't affect them. So, if there is something that could be useful to them, then that would be good for me to pass along. KPWA00035 (Decliner)
		Well, people always want to contribute, you know. If somehow taking my blood would be greatly helpful in say prevention or detection, then that's fine. So I'd have to be convinced that it's worth my time, worth my effort. KPWA00066 (Decliner)
	Provide more information about cost and coverage of testing	I had no idea what it would cost, so I couldn't decide whether I wanted to go ahead and do it anyway because nobody could help me and tell me what out of pocket costs I would be looking at. So, therefore, I did not do the genetic testing. GE00345 (Decliner)
	Improve readability and clarity	Well, the letters were sort of wordy and vague. I mean, I understood what was going on, but I could see how they could confuse people who are less scientifically literate. KPWA00070 (Acceptor)
	Offer ways to address logistical barriers	I did read through it, and I found it a little difficult on how I could possibly participate. I have no means of getting, you know, to where I can do bloodwork or anything. KPWA00089 (Decliner)

Abbreviations: GE = Geisinger; KPMAS = Kaiser Permanente Mid-Atlantic States; KPWA = Kaiser Permanente Washington.

Participants reported several benefits of the Traceback program, noting that it prompted them to obtain genetic testing they may not have otherwise pursued, motivated them to pursue preventive care such as cancer screenings, and provided an opportunity to share the information with family members so they could learn more about their potential cancer risk. No participants reported any harms or negative consequences of their participation in the Traceback program.

Most participants, even those who did not undergo genetic testing themselves, expressed support for the idea of Traceback programs. Suggestions for improving the process included making the initial invitation more personalized to a patient's history with cancer and prior genetic testing, providing more explicit guidance about the benefits and potential impacts of testing, better explaining costs and insurance coverage for genetic testing, and improving the readability and clarity of outreach and informational materials.

3.2. Implementer Perspective

We spoke with 17 implementers within the three clinical sites who were involved in administering the Traceback program. The roles of the implementers included nurses ($n = 5$), genetic counselors ($n = 7$), genetic counseling assistants ($n = 4$), and a medical geneticist ($n = 1$).

All implementers endorsed (via "Agree" response) the domains of both the acceptability (AIM) and appropriateness (IAM) measures. Overall, implementers were supportive of the Traceback program. They felt the program improved care quality by offering genetic testing to previously untested ovarian cancer survivors. Implementers felt that genetic testing offered valuable information to survivors and families that could improve cancer detection and management. Implementers attributed the lack of relatives completing cascade testing to the small numbers of tested patients with positive or pathogenic results. They felt that even receiving negative results was of value to participants and did not detract from the value of Traceback programs (Table 6).

Table 6. Implementer perspectives on Traceback program.

Theme	Exemplar Quotes
Perceptions of the traceback program	I think the benefits of a program like this, something that comes to mind is we often, in my opinion, miss this specific patient population group for many reasons. They either are diagnosed a bit later in life, and they pass away fairly after their diagnosis or even if they do seek treatment, sometimes they are way too overwhelmed and busy to fit genetics in or maybe their physicians are not recommending genetics for whatever reason, so I think there's many reasons why they don't come to see us. So, I think a program like this pulls in this specific patient group where genetic testing is really beneficial for them. Genetic Counselor, Geisinger I think if we find better paths to go back and revisit some of these patients that have had really difficult diseases, that's part of our mission. Geneticist, KPWA
Traceback program value	I think a lot of people who are diagnosed with cancer, they don't realize that this [genetic testing] is an option for them, especially because they meet guidelines and that this could be a really great thing for their family and for kind of getting an answer. Maybe for future screening and future preventative care as well. So I think knowledge is power. So, you know, even if it's late knowledge in there, in some people's minds, it might be kind of late finding out. But I think any knowledge at any point is great…So I just think it's a great thing overall if we can explain it in a way that suits them and let them kind of decide what they want to do. Genetic Counseling Assistant, Geisinger I think the cascade testing which we talked about was the biggest benefit…I think that's why most people wanted to do testing and there was a huge benefit. Genetic Counselor, KPWA
Acceptability of outreach timing	…When someone is diagnosed with cancer and facing a cancer diagnosis, there's so many people contacting them, there are so many things they have to do, and then this was great because things had died down, it was much later, and they could focus on it, and possibly hear it more, so I really like this approach…it could really help. There may be people who decline, that's fine, but I think it could really help. Genetic Counselor, Geisinger

Table 6. *Cont.*

Theme	Exemplar Quotes
Perceptions of patient motivation to test	I ended up seeing a lot of people who had had an outdated genetic test, like the BRCA only or maybe a smaller panel than what we now have. I think though, overall, the most common reason that people decided to go forward with testing is to if they had children or other at-risk relatives, that was their strongest motivation. Genetic Counselor, KPWA Both of the patients I saw were elderly. . .so their ovarian cancer diagnoses were between 25 and 30 years prior, when genetic testing wasn't necessarily on the market, as widespread, their oncology team at the time might not have known that's something that can be done, what insurance and cost look like. . .But, I think for both of them, their main benefit of testing for them was their children and their family. Both had children, males and females, they were wanting [a] great opportunity to impact their care, both recognizing it probably didn't mean too much for them at their current age and health status. Genetic Counselor, Geisinger
Traceback workflow feedback	I didn't really notice any differences in the patients at all or the work I was conducting. It all seemed pretty similar to other patients I would see through the cancer clinic. Genetic Counseling Assistant, Geisinger I would say that the only thing that would make this more perfect would be to have them come in, get a cheek swab in the office and leave so they don't have to deal with sending the specimen back. Nurse Coordinator, KPMAS
Traceback program sustainability	I think working side by side with cancer genetics—that already has a really good process of getting these patients seen—it is perfect. I mean, this is just another way to capture a group of people that might be missed or might not know that they have this option. Genetic Counseling Assistant, Geisinger . . .one of the things that I learned from Traceback is that it is essential that we have training for all providers who interact with patients with this type of cancer or any type of serious cancer that we, give them the correct information. And if a provider can't do that directly, is to know that you collaborate with other providers to do that work. Geneticist, KPWA
Cost of testing	I got a lot of positive feedback from patients about how happy and excited they were that this was offered to them free of charge. Nurse Coordinator, KPMAS There seemed to be a good amount of times when they expected that this should be covered because I was contacted and it should be free, so that was the big difference I saw. Genetic Counselor, Geisinger
Future directions	I mean, even if there were something like population screening, I think it would still be valid to go back and recontact individuals and check on how they're doing. Genetic Counselor, KPWA I would be interested in maybe researching more into how we can make the genetic counseling model a little bit more inclusive to other groups. Genetic Counseling Assistant, KPWA

Implementers reported that patients who completed Traceback testing were cancer survivors with biological relatives who could benefit from testing, survivors seeking to update their outdated genetic testing, and survivors who had previously refused testing but ultimately decided to undergo testing. They also reported that some patients expressed concern about the cost of both testing and follow-up visits, particularly for at-risk relatives.

Implementers felt that caring for Traceback program participants fit well into their regular clinical practice for genetic testing, follow-up, and recommendations for cascade testing in at-risk relatives. However, they noted several inefficiencies in the identification of Traceback-eligible individuals, such as the unexpected time and clinical expertise needed for manual chart review to confirm eligibility for Traceback testing.

Traceback program implementers noted several areas of focus for Traceback program improvement and sustainability. Firstly, more collaboration between clinical departments and physician education could help streamline the program and support a multidisciplinary approach. At KPMAS, where mailed genetic testing kits were used, implementers suggested that laboratory-based testing could reduce the amount of nudging needed to encourage kit return. Future directions for Traceback suggested by implementers included taking a public health approach, where program staff could handle eligibility assessment, outreach, and facilitation of testing. Such an approach could limit clinician time to testing and follow-up. Additionally, implementers suggested expanding the Traceback program to cover additional clinical conditions, increasing the potential clinical benefits. Several implementers mentioned the need for equitable access to Traceback programs for populations underrepresented in genomics research and clinical practice.

4. Discussion

We conducted a multisite pilot implementation study of Traceback genetic testing for individuals with ovarian, peritoneal, or fallopian cancer at three integrated health systems in the United States. We found that Traceback testing programs reached individuals with prior cancer diagnosis but no previous genetic testing (phase 1 of the Traceback framework) and were effective in identifying and communicating pathogenic variants to tested individuals (phase 2 of the Traceback framework). Programs were acceptable to both participants and implementers and thought to be of high clinical value. Implementation was feasible in this pilot study, but adequate amount of time and resources required to identify Traceback testing eligible individuals using administrative data and tumor registries are likely determinants of future sustainability. The pilot program did not result in cascade testing of at-risk relatives, despite including strategies of education about free cascade testing, reminder calls to probands (with seven pathogenic variants) at one site and offers to directly contact at-risk relatives at another site with two pathogenic variants. However, the small number of pathogenic variants identified limited the number of at-risk relatives eligible for cascade testing.

These findings contribute important early evidence about implementation of Traceback testing programs in U.S. settings. In the GRACE study, another Traceback program conducted in two U.S. health systems, participation was similar to our experience; in total, 26% of eligible living individuals completed testing, compared to 22% in our study. Using a custom gene panel, their rate of pathogenic variant detection was 21% of people completing testing, while ours was 8% [19]. Both studies found incidental findings from panel testing in genes with no evidence of association with ovarian cancer risk. Contributors to this lower yield may include unmeasured differences between acceptors and decliners of Traceback testing (e.g., family history of ovarian cancer, stage at diagnosis); differences in the testing panels; or ineligibility for Traceback testing because of high prior probability of positive result. Rates of VUS in our study (27%) were similar to those reported in other screening studies [20,21].

Traceback programs in other countries suggest additional strategies that could improve testing uptake. In a Traceback program conducted in Australia, in which archival tissue or stored blood from deceased probands with ovarian cancer was tested for pathogenic variants in 10 genes, 84/787 (11%) of samples were positive. When families were contacted directly with the result, 19/29 (66%) of relatives, who were not previously aware of the variant in the proband, accepted cascade testing [22]. In a Traceback study in Sweden, 302 women with breast cancer were identified using national health statistics and invited to Traceback testing, with 56% of those invited completing testing and a 9.7% detection rate of pathogenic variants based on testing for variants in eight genes [23]. Non-clinical workload (identifying and contacting probands, including reminders) was estimated at 4.6 min per invited woman and 8.2 min per woman who underwent genetic testing. Survey measures of acceptability were generally high.

We were surprised that no at-risk relatives received cascade testing using the free testing provided by the testing laboratory. It is possible that some at-risk relatives received testing through their own physician or other testing companies, or were not informed of the free testing by the Traceback participant. However, since no participants provided consent for us to contact at-risk relatives directly, we cannot evaluate the impact of that strategy. Cascade testing rates are generally low; a 2022 meta-analysis of at-risk relatives from over 87 studies found that only 41% of relatives completed cascade testing [24]. Direct contact of relatives is not a current standard of care at our clinical sites, which limits our ability to influence cascade testing behavior, particularly in at-risk relatives who receive care in other health systems. Although we used evidence-based strategies to improve reach to relatives, including offering direct contact of relatives at one site and reminder calls to probands to support their own sharing at another, we did not reach any relatives. It is possible that because of the length of time between diagnosis and Traceback testing, familial communication or cascade testing may have already occurred at the time of our outreach. Different, more intensive strategies may be warranted to directly reach at-risk relatives in the Traceback setting [25].

Our study provides important new data relevant to the implementation of Traceback programs. Phases 1 and 2 of our Traceback program (proband identification and testing of those not previously tested) reached more than half of eligible individuals, and was acceptable to people who were reached, whether or not they chose to be tested. Both participants and implementers were primarily driven by the potential clinical benefits to patients and families. It is worth noting that Traceback testing and follow-up did not appear to stress existing clinical workflows. However, identification of eligible individuals was unexpectedly time-consuming and would likely be a determinant of sustainability and scale-up to other cancers or health systems without adaptation. Having reminder calls was important to encouraging testing uptake and test completion for people who received mailed testing kits. Implementers universally saw the value in Traceback approaches for improving clinical outcomes and care quality and suggested applying Traceback methods to other tumor types for which genetic testing is indicated.

Strengths of our study include its multisite, real-world health system setting, the focus on recruiting diverse populations, the design of materials and outreach procedures based on patient preferences and local context at each site, and its focus on evaluating implementation outcomes, learning from both clinical stakeholders involved in implementation and individuals invited to Traceback testing. Our study team's ancillary publications from this study also examined patient and family preferences for Traceback communications [12], legal aspects of using the U.S. public health exception for genetic testing programs like Traceback [26], and the cost-effectiveness of Traceback testing (in progress). The three health systems in this study represented a wide range of geographic, insurance coverage, and population demographics, increasing the potential generalizability of the results to other U.S. settings. However, results may not be generalizable to all health systems, such as safety-net health systems. Also, our study was limited to living cancer survivors, and so does not offer evidence on Traceback programs including relatives of deceased probands, such as consent and use of stored samples or testing of individuals who have a relative with ovarian cancer in the absence of test results in a proband [22,27]. Finally, the low prevalence of detected pathogenic variants in probands and different, context-specific strategies between sites limits our exploration of strategies for cascade testing in at-risk relatives, such as direct contact of relatives.

Future research could explore ways to directly engage at-risk relatives in the Traceback setting and more efficient ways to accurately identify Traceback-eligible individuals. Given the perceived value of Traceback programs, expanding the scope to more common cancers (e.g., breast or colorectal cancer) would allow observation of test uptake and identification of probands and cascade testing outcomes in larger sample sizes. This would also allow the study of the clinical and psychosocial outcomes of receiving Traceback results, including VUS, and of clinical intervention such as risk-reducing surgery in people with pathogenic variants identified though Traceback testing. Furthermore, Traceback approaches could be compared to population-based strategies such as population screening or population management approaches where untested probands and at-risk relatives continue to be offered testing at specific intervals.

5. Conclusions

This study demonstrates the feasibility, reach, acceptability, and potential effectiveness of Traceback testing for ovarian, fallopian, and peritoneal cancers. Further research is warranted on the sustainability of such programs, practices for facilitating cascade testing of at-risk relatives, and effectiveness of Traceback screening for other conditions.

Supplementary Materials: The following supporting information can be downloaded at: https://www.mdpi.com/article/10.3390/cancers17071154/s1, Semi-structured interview guide.

Author Contributions: Conceptualization, A.K.R., N.B.H. and M.C.J.; methodology, A.K.R., N.B.H., M.C.J. and P.R.B.; software, A.R., S.S.B. and J.B.; validation, A.R., S.S.B., K.L., P.R.B. and A.S.; formal analysis, D.H., J.H., P.R.B., Y.H., N.B.H. and M.C.J.; investigation, P.R.B., E.B., E.K., K.L., T.L., K.M.R., S.P., M.K., R.S., J.G., A.K.R., N.B.H., M.C.J., J.B., S.G., K.H., M.O., P.S., A.N.D., M.L. and D.C.; data curation, A.R., S.S.B., A.S., T.L., A.N.D. and J.B.; writing—original draft preparation, N.B.H., P.R.B. and M.C.J.; writing—review and editing, N.B.H., P.R.B., A.S., A.R., K.L., E.B., E.K., A.H.B., T.L., K.M.R. and P.S.; visualization, N.B.H., A.K.R. and M.C.J.; supervision, N.B.H., A.K.R., M.C.J., P.R.B., A.H.B., P.S. and K.L.; project administration, A.S., T.L., A.N.D., D.C., I.L., H.M.R. and M.O.; funding acquisition, A.K.R., N.B.H. and M.C.J. All authors have read and agreed to the published version of the manuscript.

Funding: This research was funded by the National Institutes of Health, National Cancer Institute U01 CA240747 (Rahm, Henrikson, Jonas).

Institutional Review Board Statement: The study was conducted according to the guidelines of the Declaration of Helsinki. KPWA study activities were approved by the Kaiser Permanente Interregional Institutional Review Board. FWA Number: 00002344 (Expires 11 October 2029); IORG Number: 0000241. IRB Number: 00013581. Study ID: IRBNet# 1821438. Approved: 08 December 2021 (current approval expiration 27 October 2025). Geisinger and KPMAS study activities were approved by the Geisinger Institutional Review Board, Geisinger Health System (IRB #2020-0281). IRB Number: 2020-0281. Approved: 25 March 2020 (no expiration).

Informed Consent Statement: Informed consent was obtained from all subjects involved in the study.

Data Availability Statement: The raw data supporting the conclusions of this article will be made available by the authors on request.

Acknowledgments: The study team acknowledges the study participants; Kathryn Carpenter and Emily Newman (Newman Carpenter, L.L.C.) for educational video design and production; Edward D. Esplin, FACMG, CGAF, FACP, Labcorp Genetics (formerly Invitae) and team for partnering on cascade testing outcome assessment; Selvi Sriranganathan (KP Mid-Atlantic States) for expertise and genetic counseling support; and Melissa Rabelhofer (KP Washington) for assistance preparing the manuscript.

Conflicts of Interest: Jessica Goehringer has a family member with equity in Genomic Life. All other authors declare no conflicts of interest. The funders had no role in the design of the study; in the collection, analyses, or interpretation of data; in the writing of the manuscript; or in the decision to publish the results.

References

1. Siegel, R.L.; Giaquinto, A.N.; Jemal, A. Cancer statistics, 2024. *CA Cancer J. Clin.* **2024**, *74*, 12–49. [CrossRef] [PubMed]
2. Eccles, D.M.; Balmaña, J.; Clune, J.; Ehlken, B.; Gohlke, A.; Hirst, C.; Potter, D.; Schroeder, C.; Tyczynski, J.E.; Gomez Garcia, E.B. Selecting Patients with Ovarian Cancer for Germline BRCA Mutation Testing: Findings from Guidelines and a Systematic Literature Review. *Adv. Ther.* **2016**, *33*, 129–150. [CrossRef]

3. Kuchenbaecker, K.B.; Hopper, J.L.; Barnes, D.R.; Phillips, K.A.; Mooij, T.M.; Roos-Blom, M.J.; Jervis, S.; van Leeuwen, F.E.; Milne, R.L.; Andrieu, N.; et al. Risks of Breast, Ovarian, and Contralateral Breast Cancer for BRCA1 and BRCA2 Mutation Carriers. *JAMA* **2017**, *317*, 2402–2416.
4. Daly, M.B.; Pal, T.; Maxwell, K.N.; Churpek, J.; Kohlmann, W.; AlHilli, Z.; Arun, B.; Buys, S.S.; Cheng, H.; Domchek, S.M.; et al. NCCN Guidelines(R) Insights: Genetic/Familial High-Risk Assessment: Breast, Ovarian, and Pancreatic, Version 2.2024. *J. Natl. Compr. Canc Netw.* **2023**, *21*, 1000–1010. [PubMed]
5. Rebbeck, T.R.; Kauff, N.D.; Domchek, S.M. Meta-analysis of risk reduction estimates associated with risk-reducing salpingo-oophorectomy in BRCA1 or BRCA2 mutation carriers. *J. Natl. Cancer Inst.* **2009**, *101*, 80–87.
6. Samimi, G.; Bernardini, M.Q.; Brody, L.C.; Caga-Anan, C.F.; Campbell, I.G.; Chenevix-Trench, G.; Couch, F.J.; Dean, M.; de Hullu, J.A.; Domchek, S.M.; et al. Traceback: A proposed framework to increase identification and genetic counseling of BRCA1 and BRCA2 mutation carriers through family-based outreach. *J. Clin. Oncol.* **2017**, *35*, 2329–2337. [CrossRef]
7. Samadder, N.J.; Riegert-Johnson, D.; Boardman, L.; Rhodes, D.; Wick, M.; Okuno, S.; Kunze, K.L.; Golafshar, M.; Uson, P.L.S., Jr.; Mountjoy, L.; et al. Comparison of Universal Genetic Testing vs Guideline-Directed Targeted Testing for Patients with Hereditary Cancer Syndrome. *JAMA Oncol.* **2021**, *7*, 230–237. [PubMed]
8. Roberts, M.C.; Dotson, W.D.; DeVore, C.S.; Bednar, E.M.; Bowen, D.J.; Ganiats, T.G.; Green, R.F.; Hurst, G.M.; Philp, A.R.; Ricker, C.N.; et al. Delivery of cascade screening for hereditary conditions: A scoping review of the literature. *Health Aff.* **2018**, *37*, 801–808.
9. DiNucci, A.; Henrikson, N.B.; Jonas, M.C.; Basra, S.; Blasi, P.; Brown, J.; Esplin, E.D.; Hassen, D.; Hao, J.; Hu, Y.; et al. Feasibility and Assessment of a Cascade Traceback Screening Program (FACTS): Protocol for a Multisite Study to Implement and Assess an Ovarian Cancer Traceback Cascade Testing Program. *J. Pers. Med.* **2021**, *11*, 543. [CrossRef]
10. Damschroder, L.J.; Reardon, C.M.; Widerquist, M.A.O.; Lowery, J. The updated Consolidated Framework for Implementation Research based on user feedback. *Implement. Sci.* **2022**, *17*, 75.
11. Brothers, K.B.; Bennett, R.L.; Cho, M.K. Taking an antiracist posture in scientific publications in human genetics and genomics. *Genet. Med.* **2021**, *23*, 1004–1007. [PubMed]
12. Romagnoli, K.M.; Kulchak Rahm, A.; Jonas, M.C.; Schwiter, R.; Klinger, T.; Ladd, I.; Salvati, Z.; DiNucci, A.; Blasi, P.R.; Sheridan, L.; et al. Human-Centered Design Study to Inform Traceback Cascade Genetic Testing Programs at Three Integrated Health Systems. *Public Health Genom.* **2023**, *26*, 45–57.
13. Bonini, K.E.; Smith, H.S.; Bonkowski, E.S.; Berkman, B.E.; Jamal, L. Modern Family: An Ethical Justification for System-Led Contact of Relatives Eligible for Cascade Screening in the United States. *Public Health Genom.* **2025**, *28*, 19–33.
14. Henrikson, N.B.; Wagner, J.K.; Hampel, H.; DeVore, C.; Shridhar, N.; Williams, J.L.; Donohue, K.E.; Kullo, I.; Prince, A.E.R. What guidance does HIPAA offer to providers considering familial risk notification and cascade genetic testing? *J. Law Biosci.* **2020**, *7*, lsaa071.
15. Crabtree, B.F.; Miller, W.L. Using codes and code manuals: A template organizing style of interpretation. In *Doing Qualitative Research*, 2nd ed.; Crabtree, B.F., Miller, W.L., Eds.; Sage Publications: Thousand Oaks, CA, USA, 1999; pp. 163–177.
16. Brooks, J.; McCluskey, S.; Turley, E.; King, N. The Utility of Template Analysis in Qualitative Psychology Research. *Qual. Res. Psychol.* **2015**, *12*, 202–222. [PubMed]
17. Braun, V.; Clarke, V. Using thematic analysis in psychology. *Qual. Res. Psychol.* **2006**, *3*, 77–101.
18. Weiner, B.J.; Lewis, C.C.; Stanick, C.; Powell, B.J.; Dorsey, C.N.; Clary, A.S.; Boynton, M.H.; Halko, H. Psychometric assessment of three newly developed implementation outcome measures. *Implement. Sci.* **2017**, *12*, 108.
19. White, L.L.; Sawyer, J.K.; Zepp, J.M.; Prado, Y.K.; Reyes, A.A.; Maiyani, M.; Shuster, E.; Zucker, R.; Henrikson, N.B.; Rope, A.F.; et al. Genetic Testing Uptake among Ovarian Cancer Survivors in the Genetic Risk Analysis in Ovarian Cancer (GRACE) Study. *Cancers* **2024**, *16*, 2563. [CrossRef]
20. Agrawal, N.Y.; Thawani, R.; Edmondson, C.P.; Chen, E.Y. Estimating the Time Toxicity of Contemporary Systemic Treatment Regimens for Advanced Esophageal and Gastric Cancers. *Cancers* **2023**, *15*, 5677. [CrossRef]
21. Makhnoon, S.; Levin, B.; Ensinger, M.; Mattie, K.; Volk, R.J.; Zhao, Z.; Mendoza, T.; Shete, S.; Samiian, L.; Grana, G.; et al. A multicenter study of clinical impact of variant of uncertain significance reclassification in breast, ovarian and colorectal cancer susceptibility genes. *Cancer Med.* **2023**, *12*, 2875–2884.
22. Delahunty, R.; Nguyen, L.; Craig, S.; Creighton, B.; Ariyaratne, D.; Garsed, D.W.; Christie, E.; Fereday, S.; Andrews, L.; Lewis, A.; et al. TRACEBACK: Testing of Historical Tubo-Ovarian Cancer Patients for Hereditary Risk Genes as a Cancer Prevention Strategy in Family Members. *J. Clin. Oncol.* **2022**, *40*, 2036–2047. [CrossRef] [PubMed]
23. Augustinsson, A.; Loman, N.; Ehrencrona, H. Retrospective genetic testing (Traceback) in women with early-onset breast cancer after revised national guidelines: A clinical implementation study. *Breast Cancer Res. Treat.* **2024**, *205*, 599–607. [PubMed]
24. Frey, M.K.; Ahsan, M.D.; Bergeron, H.; Lin, J.; Li, X.; Fowlkes, R.K.; Narayan, P.; Nitecki, R.; Rauh-Hain, J.A.; Moss, H.A.; et al. Cascade Testing for Hereditary Cancer Syndromes: Should We Move Toward Direct Relative Contact? A Systematic Review and Meta-Analysis. *J. Clin. Oncol.* **2022**, *40*, 4129–4143. [PubMed]

25. Jones, L.K.; Campbell-Salome, G.; Walters, N.L.; Brangan, A.; Morgan, K.M.; Tricou, E.P.; Lindsey Mills, Z.T.; McGowan, M.P.; Gidding, S.S.; Johns, A.M.; et al. IMPACT-FH Study for Implementing Innovative Family Communication and Cascade Testing Strategies for Familial Hypercholesterolemia. *JACC Adv.* **2024**, *3*, 101198.
26. Wagner, J.K.; Tanniru, J.K.; Chane, C.A.; Meyer, M.N. Exploring access to genomic risk information and the contours of the HIPAA public health exception. *J. Law Biosci.* **2022**, *9*, lsac034.
27. Kauffman, T.L.; Prado, Y.K.; Reyes, A.A.; Zepp, J.M.; Sawyer, J.; White, L.L.; Martucci, J.; Salas, S.B.; Vertrees, S.; Rope, A.F.; et al. Feasibility of a Traceback Approach for Using Pathology Specimens to Facilitate Genetic Testing in the Genetic Risk Analysis in Ovarian Cancer (GRACE) Study Protocol. *J. Pers. Med.* **2021**, *11*, 1194. [CrossRef]

cancers

MDPI

Article

Genetic Testing Uptake among Ovarian Cancer Survivors in the Genetic Risk Analysis in Ovarian Cancer (GRACE) Study

Larissa L. White [1,*], Jennifer K. Sawyer [1], Jamilyn M. Zepp [2], Yolanda K. Prado [2], Ana A. Reyes [2], Mahesh Maiyani [1], Elizabeth Shuster [2], Rachel Zucker [1], Nora B. Henrikson [3], Alan F. Rope [2,4], Sheila Weinmann [2], Heather S. Feigelson [1] and Jessica Ezzell Hunter [2,5]

1 Institute for Health Research, Kaiser Permanente Colorado, 16601 East Centretech Parkway, Aurora, CO 80011, USA; jennifer.k.sawyer@kp.org (J.K.S.); mahesh.maiyani@kp.org (M.M.); rachel.x.zucker@kp.org (R.Z.); heather.s.feigelson@kp.org (H.S.F.)

2 Center for Health Research, Department of Translational and Applied Genomics, Kaiser Permanente Northwest, 3800 North Interstate Avenue, Portland, OR 97227, USA; jamilyn.m.zepp@kpchr.org (J.M.Z.); yolanda.k.prado@kpchr.org (Y.K.P.); ana.a.reyes@kpchr.org (A.A.R.); elizabeth.shuster@kpchr.org (E.S.); alan@genomemedical.com (A.F.R.); sheila.weinmann@kpchr.org (S.W.); jehunter@rti.org (J.E.H.)

3 Kaiser Permanente Washington Health Research Institute, 1730 Minor Avenue, Seattle, WA 98101, USA; nora.b.henrikson@kp.org

4 Genome Medical, 701 Gateway Boulevard, South San Francisco, CA 94080, USA

5 Genomics, Ethics, and Translational Research Program, RTI International, 3040 East Cornwallis Road, Research Triangle Park, NC 27709, USA

* Correspondence: larissa.l.white@kp.org

Simple Summary: The improvement in genetic testing uptake for ovarian cancer faces barriers, such as deficiency in access to genetic testing and counseling, lack of referral, and poor follow-up to completion. Genetic risk information can be utilized to guide risk reduction strategies and management during ovarian cancer treatment. The Genetic Risk Analysis in ovarian CancEr (GRACE) study aimed to evaluate the feasibility of retrospective identification ("Traceback") of both individuals diagnosed with ovarian cancer and their families who could benefit from genetic risk information. The GRACE study findings can assist health care systems to implement and increase genetic testing for survivors of ovarian cancer and other hereditary cancers.

Abstract: Background: Recommendations state all people with ovarian cancers (OCs) receive genetic counseling, but testing uptake is only between 15 and 31%. Those with a prior diagnosis of OC who have not received genetic testing represent a missed opportunity for life-saving genetic risk information. The Genetic Risk Analysis in ovarian CancEr (GRACE) study aimed to evaluate the feasibility of the retrospective identification ("Traceback") of individuals diagnosed with OC. Methods: This nonrandomized intervention study within two integrated health care systems identified participants with a history of OC between 1998 and 2020 who did not have genetic testing or testing limited to BRCA1/2. Participants received clinical genomic sequencing via a custom 60 gene panel. This study measured the feasibility of the Traceback methodology in OC survivors. Results: The initial cohort included 929 individuals, of which 57% had no prior genetic testing. Of the 302 eligible for recruitment, 88 consented to participate. We were able to outreach 97% of the eligible population using contact information from medical records. The stage at diagnosis was the only factor associated with consent. Of the 78 who returned their saliva sample, 21% had pathogenic/likely pathogenic variants, and 79% had negative results. Conclusion: The GRACE study resulted in a 29% uptake of genetic testing in OC survivors. The time since diagnosis did not have an impact on consent or ability to contact. GRACE can inform the implementation of future Traceback programs, providing guidance on how to prevent and mitigate the burden of OC and other hereditary cancers.

Keywords: ovarian cancer; cancer prevention; gynecologic cancer risk reduction

1. Introduction

Approximately 23% of all ovarian cancers have a hereditary component, with pathogenic variants in *BRCA1/2* being the most common among epithelial ovarian cancers [1–4]. In 2024, it is estimated that there will be 19,680 new cases of ovarian cancer and an estimated 12,740 people will die of this disease [5]. Only 30.8% of diagnosed patients survive five years or more due to a late-stage diagnosis and a high recurrence rate due to chemoresistance, making ovarian cancer the leading cause of cancer death of the female reproductive system [5–7]. Late-stage ovarian cancer diagnosis is largely due to the absence of safe and effective routine screening options and nonspecific symptomology [8].

Since 2016, National Comprehensive Cancer Network (NCCN) recommendations have stated that all ovarian cases should have a referral to genetic counseling, but the range of testing uptake is only between 15 and 31% [9–11]. Barriers to uptake include lack of referral and access to genetic testing and counseling and poor follow-up to completion; this suggests factors impacting uptake occur at both the patient and provider level, extending to gaps in health care system infrastructure [9,12,13]. To address the growing body of evidence that genetic risk extends beyond *BRCA1/2*, NCCN recommended, in 2020, multi-gene panel testing for all individuals diagnosed with epithelial non-mucinous ovarian cancer, including fallopian tube and primary peritoneal cancers [14,15].

The benefits of genetic testing for ovarian cancer extend beyond the diagnosis itself. Genetic testing can identify individual and family risk for related cancers, such as breast cancer and Lynch syndrome [13,16]. This risk information can then be utilized in a provider's clinical strategy for risk reduction and management, such as surgery or chemo-preventive therapies [12,16]. Those with a prior diagnosis of ovarian cancer who have not received genetic testing represent a missed opportunity to receive life-saving genetic risk information for themselves and their families. The Genetic Risk Analysis in ovarian CancEr (GRACE) study aimed to evaluate the feasibility of the retrospective identification ("Traceback") of individuals diagnosed with ovarian cancer and their families who could benefit from genetic risk information [17].

2. Materials and Methods

2.1. Study Setting

This nonrandomized intervention study took place within two integrated health care systems, Kaiser Permanente Northwest (KPNW) and Kaiser Permanente Colorado (KPCO). KPCO provides comprehensive health care for approximately 516,000 members in Colorado, and KPNW serves over 600,000 members in Oregon and southwest Washington. The membership of KPNW and KPCO reflect the population in each catchment area. Attrition is low among members, with current annual retention rates of 88% at KPNW and KPCO. Both health systems have tumor registries that track a patient's entire course of care after a cancer diagnosis and have access to archived pathology specimens from biopsies and surgeries dating back at least 10 years. The electronic health record (EHR) system at both sites includes comprehensive information on diagnoses; medical procedures including surgical pathology; laboratory and imaging tests; genetic testing results; pharmaceutical orders; and notes from clinical encounters. KPNW and KPCO's EHR was established in 1998. Further details are published elsewhere [17]. The project and all materials were reviewed and approved by the Kaiser Permanente Interregional Institutional Review Board (IRB #1702425-29) and informed consent was required for each participant included in the study. The study covered all genetic testing costs; participants did not receive any additional compensation for their participation.

2.2. Study Population

We searched tumor registry data at KPNW and KPCO and identified health plan members with a history of ovarian cancer diagnosed between 1998 and 2020 who either did not have genetic testing or had genetic testing limited to BRCA1/2 and could benefit from testing with a comprehensive cancer gene risk panel. KPCO and KPNW used International

Classification of Diseases for Oncology (ICD-O) codes to match inclusion criteria for the cohort. The following codes documented in the EHR between 1998 and 2020 were considered eligible: C481 (Malignant neoplasm of specified parts of peritoneum), C482 (Malignant neoplasm of peritoneum, unspecified), C488 (Malignant neoplasm of overlapping sites of retroperitoneum and peritoneum), C569 (Malignant neoplasm of unspecified ovary), and C570 (Malignant neoplasm of fallopian tube). The following subtypes were considered eligible: endometrioid, clear cell, serous, mucinous, carcinoid, sex cord (stromal), malignant Brenner, peritoneal, and fallopian tube tumors, as well as bilateral teratoma (with family history of teratoma). Other inclusion criteria included female sex in the EHR; no evidence of prior genetic testing, or prior genetic testing for *BRCA1/2* only with a negative result or variant of uncertain significance (VUS); age 18 or older; and a KPNW or KPCO member at the time of ovarian cancer diagnosis. Exclusion criteria included prior diagnosis of a hereditary cancer syndrome, unable to consent in English (KPCO only), unable to provide informed consent, or opted out of research activities. Additionally, the following subtypes were excluded: germ cell, neuroendocrine, benign Brenner, and Krukenberg tumors, as well as sarcomas, uterine cancer in ovary, and lymphoma/leukemia. Borderline ovarian cancer and ovarian cancer of Low Malignant Potential (LMP), including LMP with microinvasion, were also excluded.

2.3. Chart Review

Trained research staff conducted a manual chart review to verify the inclusion criteria. Confirmation of ovary, fallopian, and peritoneal cancer was accomplished by identifying the corresponding surgical pathology report based on the tumor registry data. The surgical pathology report was reviewed by the study's genetic counselor, along with consultation with the study's physician, to ensure the tumor subtype was eligible for inclusion. The research staff looked for any prior diagnosis of a hereditary cancer syndrome and any prior genetic testing results, which are typically entered into the EHR as scanned documents and queried records where prior genetic testing results may have been captured. Chart review details, contact information, and eligibility status were tracked in REDCap, version 14.4.1 [18,19].

2.4. Recruitment Procedure

Starting in March 2021, potential eligible participants were sent a recruitment letter and fact sheet about the study offering genetic testing and counseling for themselves and their families, by both mail and email (if available), which included a personalized link to an electronic consent in REDCap, version 14.4.1 [20]; see Figure S1. Potential participants were also given the option to receive a paper copy of the consent form by mail depending on their preference. Trained research staff followed up by phone with potential participants after two weeks for those who did not respond to the recruitment outreach via mail or e-mail. We defined "active decliners" as people who declined participation during any of the follow-up calls and defined people as "passive decliners" after three failed phone call attempts to reach them.

2.5. Genetic Testing

After obtaining informed consent, saliva self-collection kits were mailed from the commercial genetic testing laboratory directly to participants. Participants collected their sample at home and mailed the kit back to the commercial laboratory. As a reminder, study staff called participants who did not return saliva kits after two weeks. The participant was marked as a passive decliner if the saliva kit was not returned after three attempts to reach the participant by phone. A custom 60 gene panel, comparable to the panel patients receive for genetic cancer risk assessment in the clinical setting, was used for clinical genomic sequencing (Table 1). The commercial laboratory performed genetic testing, variant confirmation, and variant interpretation. The results report included both pathogenic and likely pathogenic (P/LP) variants; variants of unknown significance (VUS) were not reported at the request of the study team.

Table 1. Custom gene panel.

Gene List					
APC	*CDH1*	*KIT*	*NBN*	*PTCH1*	*SDHD*
ATM	*CDK4*	*MAX*	*NF1*	*PTEN*	*SMAD4*
AXIN2	*CDKN2A*	*MEN1*	*NF2*	*RAD51C*	*SMARCA4*
BAP1	*CHEK2*	*MET*	*NTHL1*	*RAD51D*	*SMARCB1*
BARD1	*DICER1*	*MITF*	*PALB2*	*RB1*	*STK11*
BMPR1A	*EPCAM*	*MLH1*	*PDGFRA*	*RET*	*TEME127*
BRCA1	*FH*	*MSH2*	*PMS2*	*SDHA*	*TP53*
BRCA2	*FLCN*	*MSH3*	*POLD1*	*SDHAF2*	*TSC1*
BRIP1	*GREM1*	*MSH6*	*POLE*	*SDHB*	*TSC2*
CDC73	*HOXB13*	*MUTYH*	*PRKAR1A*	*SDHC*	*VHL*

2.6. Results Disclosure

Positive test results were returned by phone by the study genetic counselor. The genetic counselor collected family history and relatives' information during the results disclosure to discuss any changes in care and who would be eligible for cascade testing of the familial variant. The genetic counselor offered to directly contact relatives regarding cascade testing and/or provide a letter the participant could share with their relative to contact the study team. After the call, a letter summarizing the follow-up recommendations and a copy of the test results were mailed or emailed to the participant. Negative results were returned by letter along with a copy of the genetic testing results. All results, both positive and negative, were placed in the participant's medical record if they were a current or former member of the health plan. The study's genetic counselor coordinated the next steps in the participant's care as appropriate, and followed up with family members who expressed interest in testing for the familial variant. Future publications from this study will report on cascade testing among relatives.

2.7. Data Analysis

We descriptively measured feasibility using the following metrics: (1) accuracy of tumor registries and pathology reports to identify potential participants with a correct diagnosis, (2) number of eligible participants with available contact information in the EHR, (3) whether time since diagnosis had an impact on available contact information and consent rate, and (4) factors associated with genetic testing uptake (consent). We measured factors associated with uptake using Fisher's exact tests to account for small cell sizes, and a p-value of 0.05 or less was considered statistically significant. All analyses were conducted using SAS, version 9.4 (Cary, NC, USA).

3. Results

Initial eligibility assessment included 929 individuals, of which 43% (n = 404) were ineligible due to prior genetic testing, 18% (n = 170) had an ineligible tumor type, and 6% had missing contact information (n = 2), were cognitively impaired (n = 17), or had a prior diagnosis of hereditary cancer syndrome (n = 34; Figure 1). Of the 929 initially assessed, 45% had prior panel testing, 11% had *BRCA1/2* testing, 1% had other genetic testing, and 43% had no prior genetic testing. Of the 929 initially assessed, 33% (n = 302) were eligible for participation in the study (Figure 1).

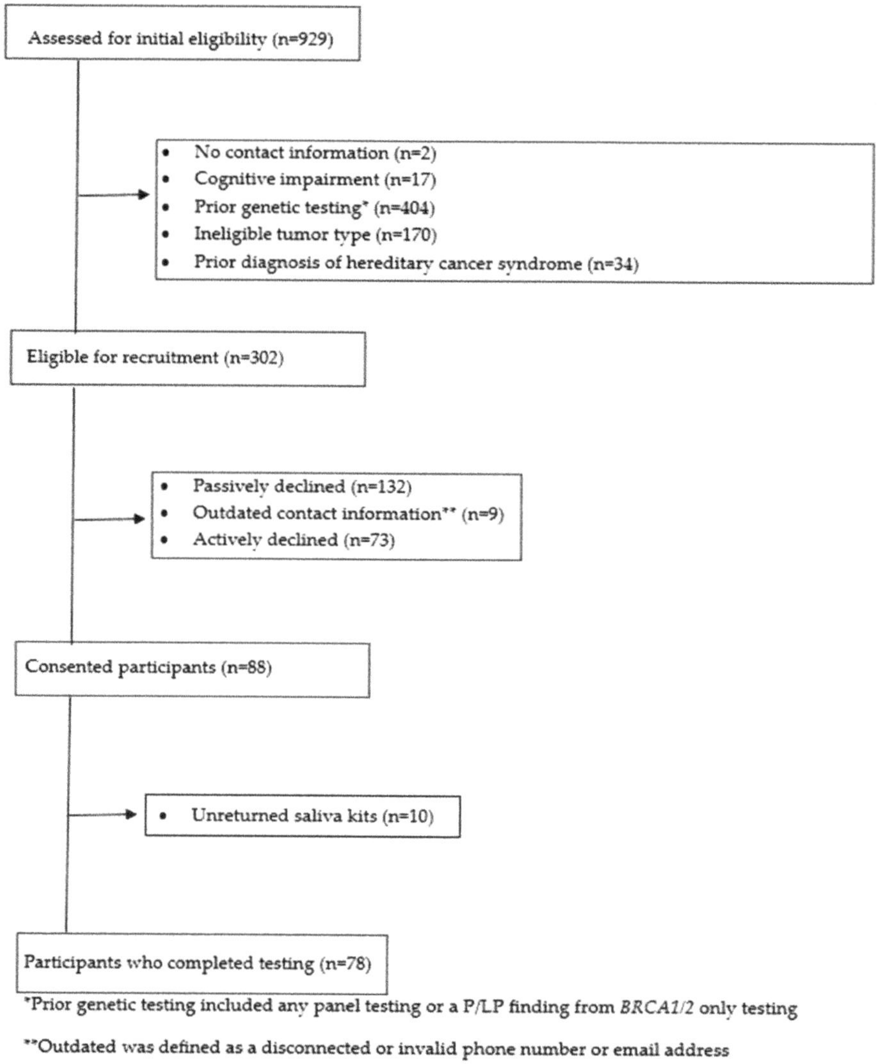

Figure 1. STrengthening the Reporting of OBservational studies in Epidemiology (STROBE) diagram.

Of the 302 eligible for recruitment, 29% (n = 88) consented to participate in the study, 44% (n = 132) passively declined, 24% (n = 73) actively declined, 3% (n = 9) had outdated contact information, and 3% (n = 10) of consenting participants did not return their saliva kits (Figure 1). Overall, we were able to outreach 97% (n = 293) of the eligible population using contact information from the EHR. Reasons for active decline were not systematically captured, but eligible recruits reported to research staff that they had no interest/non-specific reason, no family/biological children, they were managing a serious medical condition or recurrence of cancer either personally or as a caregiver at the time of recruitment, and had privacy concerns as it relates to genetic information.

Of the 302 eligible for recruitment, 61% were diagnosed between the ages of 45 and 64, 52% were diagnosed at stage 1, 56% were 65 years or older at the time of recruitment, 79% were White, and 87% were non-Hispanic/Latino. The eligible and consenting populations were proportionally similar to the overall population of survivors identified in the tumor registry (Table 2). Of those eligible, stage at diagnosis was the only statistically significant factor associated with consent ($p \leq 0.0001$; Table 2).

Table 2. Study population characteristics.

Characteristic	Overall (n = 929) N (%)	Eligible (n = 302) N (%)	Consented (n = 88) N (%)	p *
Age at diagnosis				0.43
<18	10 (1%)	1 (<1%)	0 (0%)	
18–44	216 (23%)	60 (20%)	20 (23%)	
45–64	481 (52%)	184 (61%)	48 (54%)	
≥65	222 (24%)	57 (19%)	20 (23%)	
Age at recruitment				0.12
<18	0 (0%)	0 (0%)	0 (0%)	
18–44	87 (9%)	19 (6%)	9 (10%)	
45–64	357 (39%)	113 (38%)	28 (32%)	
≥65	485 (52%)	170 (56%)	51 (58%)	
Race				0.69
American Indian/Alaskan Native	2 (<1%)	1 (<1%)	0 (0%)	
Native Hawaiian/Other Pacific Islander	1 (<1%)	0 (0%)	0 (0%)	
Asian	34 (4%)	11 (4%)	3 (3%)	
Black/African American	25 (3%)	14 (5%)	4 (5%)	
White	764 (82%)	239 (79%)	75 (85%)	
Multiple Races	32 (3%)	5 (2%)	0 (0%)	
Other Race	20 (2%)	7 (2%)	1 (1%)	
Unknown/Not Reported	51 (6%)	25 (8%)	5 (6%)	
Hispanic/Latino				0.58
Yes	89 (10%)	29 (10%)	6 (7%)	
No	821 (88%)	262 (87%)	79 (90%)	
Unknown/Not Reported	19 (2%)	11 (3%)	3 (3%)	
Stage at diagnosis				<0.0001
0	2 (<1%)	0 (0%)	0 (0%)	
1	374 (40%)	157 (52%)	59 (67%)	
2	83 (9%)	37 (12%)	5 (6%)	
3	190 (20%)	58 (19%)	13 (15%)	
4	71 (8%)	11 (4%)	0 (0%)	
Unknown/Unstaged	209 (23%)	39 (13%)	11 (12%)	

* p-value comparing the characteristics between the eligible and consenting populations.

The most successful contact method for recruitment was a combination of letter and email, with a 33% consent rate (Table 3). About 25% of the eligible population preferred to consent electronically. While the prevalence of consent, passive decline, and active decline varied somewhat between eligible participants diagnosed from 1998 to 2020, time since diagnosis was not associated with consent (Figure 2). Time since diagnosis was also not associated with ability to contact, with contact information being available for most eligible participants between 1998 and 2020 (Figure 2).

Table 3. Consent status by contact method of eligible population.

Contact Method *	Consented (Paper)	Consented (Electronic)	Did Not Consent	Total
Letter (row %)	3 (10%)	1 (3%)	27 (87%)	31
Email (row %)	0 (0%)	12 (23%)	40 (77%)	52
Phone (row %)	0 (0%)	0 (0%)	1 (100%)	1
Letter + Email (row %)	8 (4%)	64 (29%)	146 (67%)	218
Total (row %)	11 (4%)	77 (25%)	214 (71%)	302

* Contact method was based on availability of contact information in the electronic health record (EHR).

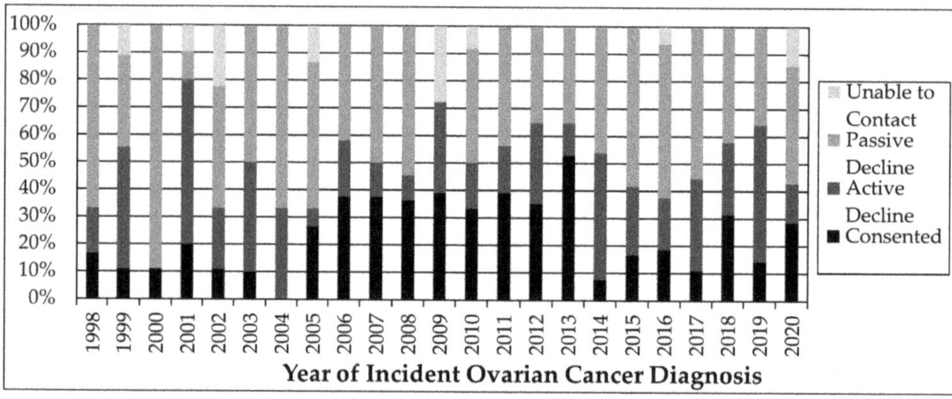

Figure 2. Time since diagnosis, ability to contact, and consent status.

Of the 78 who returned their saliva sample to perform genetic testing, 21% (n = 16) had a P/LP variant result and 79% (n = 62) had a negative result. *BRCA1/2* were the most common variants identified from the comprehensive panel, resulting in a positive rate of 13% for genes historically associated with ovarian cancer only (Table 4).

Table 4. Total newly identified pathogenic/likely pathogenic (P/LP) variants.

Gene	Variant	Reference Sequence [c]	Diagnosis Year	Diagnosis Age
ATM	p.M1? (c.3G>A)	NM_000051.4	2006	65
APC [a,b]	p.I1307K (c.3920T>A)	NM_000038.6	2018	34
BARD1 [b]	c.2300_2301delTG	NM_000465.2	2013	61
BRCA1	c.3756_3759delGTCT	NM_007294.4	2013	54
BRCA1	p.V1736A (c.5207T>C)	NM_007294.4	2006	48
BRCA2	c.5073dupA	NM_000059.4	2010	57
BRCA2	c.3847_3848delGT	NM_000059.4	2011	60
CHEK2	p.H143Y (c.427C>T)	NM_007194.4	2007	55
DICER1	p.R688* (c.2062C>T)	NM_177438.3	2013	23
HOXB13 [a,b]	p.G84E (c.251G>A)	NM_006361.6	2007	47
MITF [a,b]	p.E318K (c.952G>A)	NM_000248.3	2007	51
MUTYH [a,b]	p.G396D (c.1187G>A)	NM_001128425.1	2006	49
MUTYH [a,b]	p.G396D (c.1187G>A)	NM_001128425.1	2013	68
PMS2	c.1040delA	NM_000535.7	2013	52
RAD51D	c.896_*505del761insT	NM_002878.4	2006	83
SDHB [a,b]	p.L111V (c.331C>G)	NM_003000.3	2009	62

[a] Genetic test result not related to personal or family cancer history; [b] gene not historically associated with ovarian cancer risk; [c] matched annotation from the National Center for Biotechnology Information (NCBI) and European Molecular Biology Laboratory-European Bioinformatics Institute (EMBL-EBI) (MANE) reference sequence, version 1.3.

4. Discussion

The GRACE study aimed to recruit ovarian cancer survivors who had not yet received genetic testing, representing 57% of our initial study cohort. GRACE study recruitment resulted in a 29% uptake of genetic testing in individuals who initially did not know their genetic risk status, thereby providing potentially life-saving information to their relatives. Recruitment was via email and letter, and saliva kits were mailed directly to participants; thus, no in-person visits were required for participation. Among those who consented, 89% returned their saliva samples, representing a successful genetic testing methodology for ovarian cancer survivors. Contact information was available for 97% of the eligible study population, which included participants who were no longer insured by the health system, and time since diagnosis did not have an impact on consent or ability to contact, further strengthening support for Traceback methods in ovarian cancer survivors. Our findings from this feasibility study translate to several recommendations for future Traceback program implementation.

Our largest proportion of dropout was due to passive decline, showcasing a potential barrier of the Traceback approach. Consent rate was highest for participants who received both electronic and mailed study information, but depending on availability of information in the EHR, some participants received letter or email only. Notes collected on reasons for active decline suggest potential participants had no living family or biological children or had no interest in knowing their genetic risk status, in which case, changes to recruitment methodology are unlikely to have an impact on consent rate. However, it is also possible that participants who passively declined were not expecting to receive an offer for genetic testing, nor were they expecting to consider their past cancer diagnosis, especially those who were no longer insured via KPNW/KPCO. A qualitative study [21] interviewed 70 participants on Traceback genetic testing messaging and uptake. Participants preferred directed, personalized communication and the use of multiple, repetitive modes of outreach, primarily from their physician over other clinicians [21]. Furthermore, the study team [21] hypothesized that direct forms of recruitment may lessen the degree of surprise that participants experience with the Traceback approach. Given these findings and the frequency of passive decline in GRACE, retention for future Traceback studies may be improved by embedding recruitment outreach and reminders via the health system's respective clinical care unit to improve study corroboration and visibility.

Time since diagnosis did not impact consent rate or ability to contact, which highlights the utility of Traceback approaches to address the health care gap in genetic testing uptake for ovarian cancer survivors, especially those who were diagnosed prior to NCCN guideline implementation. However, the time and staff necessary for manual chart review to identify survivors who are eligible for genetic testing may not be feasible for many health systems. To facilitate genetic testing uptake and to close the gap for genetic testing referral and completion, health systems should consider the implementation of an auto-referral and reminder system to improve genetic testing uptake at or near the time of diagnosis [22,23]. An additional suggestion is for clinicians to reach out to individuals soon after active treatment is complete to inquire about testing at that time.

Relatedly, stage at diagnosis was associated with genetic testing uptake in those who were eligible. While this study cannot address why participants diagnosed at an early stage decided to consent more than those diagnosed at a late stage, about 64% of eligible participants were diagnosed at stages one and two compared to 23% at stages three and four. Given that this is a survivor's cohort, it is likely that more late stage participants were initially excluded due to vital status. Furthermore, participants diagnosed at an early stage met eligibility criteria more often than those diagnosed at a late stage, suggesting that late-stage participants were excluded due to prior genetic testing or ineligible tumor type. There were no significant differences in other participant-specific factors between those who were eligible and those who consented, indicating that the Traceback approach is unlikely to exacerbate existing health disparities, such as inequities by race/ethnicity.

Our custom gene panel resulted in newly identified P/LP findings for 21% of the tested population, with 13% of P/LP findings specific to ovarian cancer (including *BRCA1/2* and *ATM*). Given that approximately 15% of ovarian cancers are heritable through *BRCA1/2* mutation, our results are slightly lower compared to the current literature on this topic [24,25]. An additional Traceback study [26] saw similar results to the current literature, with 29 consented participants, or 14% of their sample, receiving P/LP findings. However, this study [26] utilized a panel consisting of *BRCA1/2*, *PALB2*, *CHEK2*, and *ATM* only, whereas the GRACE panel was more extensive and included genes not traditionally associated with ovarian cancer, which explains our higher prevalence of P/LP variants overall. Furthermore, the original NCCN guidelines focused more on *BRCA1/2* than the more recently detected variants like *CHEK2*, and it is possible that genetic testing had already occurred for older cases [14,15]. It is unlikely that the exclusion of other ovarian tumors, such as neuroendocrine and germ cell, significantly impacted the identification of P/LP variants, given their rarity [27]. Depending on the aims of future Traceback programs and the state of the literature, researchers and clinicians may want to consider the inclusion of genes associated with cancers other than ovarian to offer a fuller profile of genetic risk assessment and management, especially individuals diagnosed with hereditary breast and ovarian cancer syndrome and Lynch syndrome [4].

Our custom gene panel included both high- and moderate-risk genes for ovarian cancer, as well as other cancers. For many of these genes, there are limited data on cancer risk with no guidelines on clinical actionability [9,14,15,28]. Therefore, it may be difficult to use a singular P/LP variant to assign risk. In many cases, the information from testing for moderate-risk genes does not change clinical management compared to that based on family history alone [9,14,15,28]. Our study's approach relied on the skills and knowledge of our genetic counselor to interpret a participant's risk and next steps, given known barriers to genetic risk assessment in clinical care. We highly recommend that other studies and future Traceback programs include a genetic counselor on their team to assist participants in navigating the clinical process after the return of the results.

This study had several strengths. The sample size was sufficient for a rare outcome such as ovarian cancer, making our results statistically generalizable to similar health system populations. We verified ovarian cancer and genetic testing history using both tumor registry data and manual chart review, improving our case definition and application of inclusion/exclusion criteria. Saliva testing was performed by a Clinical Laboratory Improvement Amendments (CLIA)-certified laboratory via third party testing, which is comparable to the current clinical testing procedure for individuals with ovarian cancer in the health care systems in which this study was performed.

We also note some limitations. We could not confirm whether all passive declines were true declines or outdated contact information. However, we did record if contact information was confirmed as outdated. We did not measure differences between the original sampling frame and those who eventually consented, but given the homogeneity of our eligible population vis-à-vis demographics and cancer characteristics, it is unlikely that there are meaningful differences between those who were ineligible and those who consented. However, we recommend future studies examine any potential disparities that the Traceback approach may cause for uninsured/underinsured individuals, as well as those with lower socioeconomic status. Future publications from this study will report on cascade testing among relatives, as well as efforts to provide genetic risk information to families of patients who are deceased.

5. Conclusions

The GRACE study findings illustrate the promise of Traceback approaches to provide life-saving information to individuals and their families at increased genetic risk for ovarian cancer. The GRACE study can inform the broad implementation of future Traceback programs across health care systems, providing guidance on how to prevent and mitigate the burden of ovarian and other hereditary cancers.

Supplementary Materials: The following supporting information can be downloaded at: https://www.mdpi.com/article/10.3390/cancers16142563/s1, Figure S1: Fact sheet.

Author Contributions: Conceptualization, J.E.H.; methodology, L.L.W., N.B.H., H.S.F. and J.E.H.; software, M.M. and E.S.; formal analysis, L.L.W. and E.S.; investigation, L.L.W., J.K.S., J.M.Z., A.A.R., A.F.R., H.S.F. and J.E.H.; data curation, M.M., E.S. and J.E.H.; writing—original draft preparation, L.L.W. and J.K.S.; writing—review and editing, L.L.W., J.K.S., J.M.Z., Y.K.P., A.A.R., M.M., E.S., R.Z., N.B.H., A.F.R., S.W., H.S.F. and J.E.H.; visualization, L.L.W. and E.S.; supervision, N.B.H., S.W., H.S.F. and J.E.H.; project administration, Y.K.P. and R.Z.; funding acquisition, J.E.H. All authors have read and agreed to the published version of the manuscript.

Funding: This research was funded by the National Cancer Institute, grant number XU01CA244323 (PI: Hunter).

Institutional Review Board Statement: This study was conducted in accordance with the Declaration of Helsinki and was approved by the Institutional Review Board (or Ethics Committee) of Kaiser Permanent Northwest Institutional Review Board (protocol code 1702425-29 approved 18 February 2021).

Informed Consent Statement: Informed consent was obtained from all subjects involved in the study.

Data Availability Statement: The data underlying this article cannot be shared publicly due to the privacy of individuals and health system members that participated in the study. The derived data will be shared on reasonable request to the corresponding author.

Acknowledgments: The authors would like to thank Tia L. Kauffman, Katrina Goddard, Michelle Wrenn, Jessica Martucci, Sandra Soo-Jin Lee, Mika Caruncho, Aaron Scrol, Suzanne Bianca Salas, and Sarah Vertrees for their past contributions to this study.

Conflicts of Interest: The authors declare no conflicts of interest. The funders had no role in the design of the study; in the collection, analyses, or interpretation of data; in the writing of the manuscript; or in the decision to publish the results.

References

1. National Cancer Institute. Ovarian, Fallopian Tube, and Primary Peritoneal Cancer—Health Professional Version. Available online: https://www.cancer.gov/types/ovarian/hp (accessed on 15 May 2024).
2. Zhang, S.; Royer, R.; Li, S.; McLaughlin, J.R.; Rosen, B.; Risch, H.A.; Fan, I.; Bradley, L.; Shaw, P.A.; Narod, S.A. Frequencies of BRCA1 and BRCA2 Mutations among 1342 Unselected Patients with Invasive Ovarian Cancer. *Gynecol. Oncol.* **2011**, *121*, 353–357. [CrossRef]
3. American Cancer Society. Key Statistics for Ovarian Cancer. Available online: https://www.cancer.org/cancer/types/ovarian-cancer/about/key-statistics.html (accessed on 15 May 2024).
4. Toss, A.; Tomasello, C.; Razzaboni, E.; Contu, G.; Grandi, G.; Cagnacci, A.; Schilder, R.J.; Cortesi, L. Hereditary ovarian cancer: Not only BRCA 1 and 2 genes. *BioMed Res. Int.* **2015**, *2015*, 341723. [CrossRef]
5. National Cancer Institute. SEER Cancer Stat Facts: Ovarian Cancer. Available online: https://seer.cancer.gov/statfacts/html/ovary.html (accessed on 15 May 2024).
6. Prat, J. Ovarian carcinomas: Five distinct diseases with different origins, genetic alterations, and clinicopathological features. *Virchows Arch.* **2012**, *460*, 237–249. [CrossRef]
7. Doubeni, C.A.; Doubeni, A.R.; Myers, A.E. Diagnosis and Management of Ovarian Cancer. *Am. Fam. Physician* **2016**, *93*, 937–944.
8. Fantone, S.; Piani, F.; Olivieri, F.; Rippo, M.R.; Sirico, A.; Di Simone, N.; Marzioni, D.; Tossetta, G. Role of SLC7A11/xCT in Ovarian Cancer. *Int. J. Mol. Sci.* **2024**, *25*, 587. [CrossRef]
9. Samimi, G.; Bernardini, M.Q.; Brody, L.C.; Caga-Anan, C.F.; Campbell, I.G.; Chenevix-Trench, G.; Couch, F.J.; Dean, M.; de Hullu, J.A.; Domchek, S.M.; et al. Traceback: A Proposed Framework to Increase Identification and Genetic Counseling of BRCA1 and BRCA2 Mutation Carriers through Family-Based Outreach. *J. Clin. Oncol.* **2017**, *35*, 2329–2337. [CrossRef]
10. Moss, H.A.; Samimi, G.; Havrilesky, L.J.; Sherman, M.E.; Myers, E.R. Estimating the Number of Potential Family Members Eligible for BRCA1 and BRCA2 Mutation Testing in a "Traceback" Approach. *Genet. Epidemiol.* **2018**, *42*, 117–122. [CrossRef]
11. Kurian, A.W.; Ward, K.C.; Howlader, N.; Deapen, D.; Hamilton, A.S.; Mariotto, A.; Miller, D.; Penberthy, L.S.; Katz, S.J. Genetic Testing and Results in a Population-Based Cohort of Breast Cancer Patients and Ovarian Cancer Patients. *J. Clin. Oncol.* **2019**, *37*, 1305–1315. [CrossRef]
12. Meyer, L.A.; Anderson, M.E.; Lacour, R.A.; Suri, A.; Daniels, M.S.; Urbauer, D.L.; Nogueras-Gonzalez, G.M.; Schmeler, K.M.; Gershenson, D.M.; Lu, K.H. Evaluating women with ovarian cancer for BRCA1 and BRCA2 mutations: Missed opportunities. *Obstet. Gynecol.* **2010**, *115*, 945–952. [CrossRef]
13. Febbraro, T.; Robison, K.; Wilbur, J.S.; Laprise, J.; Bregar, A.; Lopes, V.; Legare, R.; Stuckey, A. Adherence patterns to National Comprehensive Cancer Network (NCCN) guidelines for referral to cancer genetic professionals. *Gynecol. Oncol.* **2015**, *138*, 109–114. [CrossRef]
14. Daly, M.B.; Pilarski, R.; Yurgelun, M.B.; Berry, M.P.; Buys, S.S.; Dickson, P.; Domchek, S.M.; Elkhanany, A.; Friedman, S.; Garber, J.E.; et al. NCCN Guidelines Insights: Genetic/Familial High-Risk Assessment: Breast, Ovarian, and Pancreatic, Version 1.2020. *J. Natl. Compr. Cancer Netw. JNCCN* **2020**, *18*, 380–391. [CrossRef]
15. Daly, M.B.; Pal, T.; Berry, M.P.; Buys, S.S.; Dickson, P.; Domchek, S.M.; Elkhanany, A.; Friedman, S.; Goggins, M.; Hutton, M.L.; et al. Genetic/Familial High-Risk Assessment: Breast, Ovarian, and Pancreatic, Version 2.2021, NCCN Clinical Practice Guidelines in Oncology. *J. Natl. Compr. Cancer Netw. JNCCN* **2021**, *19*, 77–102. [CrossRef]
16. Temkin, S.M.; Bergstrom, J.; Samimi, G.; Minasian, L. Ovarian Cancer Prevention in High-risk Women. *Clin. Obstet. Gynecol.* **2017**, *60*, 738–757. [CrossRef]

17. Kauffman, T.L.; Prado, Y.K.; Reyes, A.A.; Zepp, J.M.; Sawyer, J.; White, L.L.; Martucci, J.; Salas, S.B.; Vertrees, S.; Rope, A.F.; et al. Feasibility of a Traceback Approach for Using Pathology Specimens to Facilitate Genetic Testing in the Genetic Risk Analysis in Ovarian Cancer (GRACE) Study Protocol. *J. Pers. Med.* **2021**, *11*, 1194. [CrossRef]
18. Harris, P.A.; Taylor, R.; Thielke, R.; Payne, J.; Gonzalez, N.; Conde, J.G. Research electronic data capture (REDCap)—A metadata-driven methodology and workflow process for providing translational research informatics support. *J. Biomed. Inform.* **2009**, *42*, 377–381. [CrossRef]
19. Harris, P.A.; Taylor, R.; Minor, B.L.; Elliott, V.; Fernandez, M.; O'Neal, L.; McLeod, L.; Delacqua, G.; Delacqua, F.; Kirby, J.; et al. The REDCap consortium: Building an international community of software platform partners. *J. Biomed. Inform.* **2019**, *95*, 103208. [CrossRef]
20. Lawrence, C.E.; Dunkel, L.; McEver, M.; Israel, T.; Taylor, R.; Chiriboga, G.; Goins, K.V.; Rahn, E.J.; Mudano, A.S.; Roberson, E.D.; et al. A REDCap-based model for electronic consent (eConsent): Moving toward a more personalized consent. *J. Clin. Transl. Sci.* **2020**, *4*, 345–353. [CrossRef]
21. Romagnoli, K.M.; Kulchak Rahm, A.; Jonas, M.C.; Schwiter, R.; Klinger, T.; Ladd, I.; Salvati, Z.; DiNucci, A.; Blasi, P.R.; Sheridan, L.; et al. Human-Centered Design Study to Inform Traceback Cascade Genetic Testing Programs at Three Integrated Health Systems. *Public Health Genom.* **2023**, *26*, 45–57. [CrossRef]
22. McGee, J.; Peart, T.M.; Foley, N.; Bertrand, M.; Prefontaine, M.; Sugimoto, A.; Ettler, H.; Welch, S.; Panabaker, K. Direct Genetics Referral Pathway for High-Grade Serous Ovarian Cancer Patients: The "Opt-Out" Process. *J. Oncol.* **2019**, *2019*, 6029097. [CrossRef]
23. Petzel, S.V.; Vogel, R.I.; McNiel, J.; Leininger, A.; Argenta, P.A.; Geller, M.A. Improving referral for genetic risk assessment in ovarian cancer using an electronic medical record system. *Int. J. Gynecol. Cancer* **2014**, *24*, 1003–1009. [CrossRef]
24. Norquist, B.M.; Harrell, M.I.; Brady, M.F.; Walsh, T.; Lee, M.K.; Gulsuner, S.; Bernards, S.S.; Casadei, S.; Yi, Q.; Burger, R.A.; et al. Inherited Mutations in Women With Ovarian Carcinoma. *JAMA Oncol.* **2016**, *2*, 482–490. [CrossRef]
25. Alsop, K.; Fereday, S.; Meldrum, C.; deFazio, A.; Emmanuel, C.; George, J.; Dobrovic, A.; Birrer, M.J.; Webb, P.M.; Stewart, C.; et al. BRCA mutation frequency and patterns of treatment response in BRCA mutation-positive women with ovarian cancer: A report from the Australian Ovarian Cancer Study Group. *J. Clin. Oncol. Off. J. Am. Soc. Clin. Oncol.* **2012**, *30*, 2654–2663. [CrossRef]
26. Augustinsson, A.; Nilsson, M.P.; Ellberg, C.; Kristoffersson, U.; Olsson, H.; Ehrencrona, H. Genetic testing in women with early-onset breast cancer: A Traceback pilot study. *Breast Cancer Res. Treat.* **2021**, *190*, 307–315. [CrossRef]
27. Randall, T.C.; Armstrong, K. Health care disparities in hereditary ovarian cancer: Are we reaching the underserved population? *Curr. Treat. Options Oncol.* **2016**, *17*, 39. [CrossRef]
28. Sekine, M.; Isobe, M.; Nishino, K.; Adachi, S.; Suda, K.; Yoshihara, K. Challenges for clinical application of "TRACEBACK" study: Testing of historical Tubo-Ovarian cancer patients for hereditary risk genes. *Ann. Transl. Med.* **2023**, *11*, 295. [CrossRef]

cancers MDPI

Systematic Review

Imiquimod Is Effective in Reducing Cervical Intraepithelial Neoplasia: A Systematic Review and Meta-Analysis

Balázs Hamar [1,2,*], Brigitta Teutsch [1,3], Eszter Hoffmann [1,2], Péter Hegyi [1,3,4], Andrea Harnos [5], Péter Nyirády [2,6], Zsombor Hunka [2], Nándor Ács [1,2], Ferenc Bánhidy [1,2] and Zsolt Melczer [1,2]

1 Centre for Translational Medicine, Semmelweis University, 1088 Budapest, Hungary; teutschbrigitta@gmail.com (B.T.); h.eszter@icloud.com (E.H.); hegyi2009@gmail.com (P.H.); acsnandor@gmail.com (N.Á.); banhidyferenc@hotmail.com (F.B.); melczerzsolt@gmail.com (Z.M.)
2 Department of Obstetrics and Gynecology, Semmelweis University, 1088 Budapest, Hungary; nyiradyp@gmail.com (P.N.); hunka.zsombi@gmail.com (Z.H.)
3 Institute for Translational Medicine, Medical School, University of Pécs, 7621 Pécs, Hungary
4 Institute of Pancreatic Diseases, Semmelweis University, 1088 Budapest, Hungary
5 Department of Biostatistics, University of Veterinary Medicine, 1078 Budapest, Hungary; harnos.andrea@gmail.com
6 Department of Urology, Semmelweis University, 1082 Budapest, Hungary
* Correspondence: hamar.balazs@semmelweis.hu; Tel.: +36-30-613-0965

Simple Summary: There are publications on the use of Imiquimod in cervical intraepithelial neoplasia (CIN) and HPV clearance; however, the literature is not consistent about its efficacy. Moreover, in cervical precancers, surgical solutions are widely accepted therapies, despite their association with increased obstetrical complications, such as miscarriage and preterm birth. Therefore, a conservative solution is needed. Topical Imiquimod reduced CIN and enhanced HPV clearance, though surgical intervention conization was found to be more effective than Imiquimod treatment. Side effects were common, though mostly mild. Topical Imiquimod could be a valuable therapeutic option for CIN patients, especially for women who have future pregnancy desires. Imiquimod should be incorporated into guidelines as evidence shows it is effective and safe.

Abstract: Introduction: Topical Imiquimod is an immune response modifier approved for the off-label use of vulvar intraepithelial neoplasia. We conducted this systematic review and meta-analysis to investigate the efficacy and safety of Imiquimod in treating cervical intraepithelial neoplasia (CIN) and human papillomavirus (HPV)-positive patients. Methods: The study was prospectively registered (CRD420222870) and involved a comprehensive systematic search of five medical databases on 10 October 2022. We included articles that assessed the use of Imiquimod in cervical dysplasia and HPV-positive patients. Pooled proportions, risk ratios (RRs), and corresponding 95% confidence intervals (CIs) were calculated using a random effects model to generate summary estimates. Statistical heterogeneity was assessed using I^2 tested by the Cochran Q tests. Results: Eight articles reported on 398 patients who received Imiquimod out of 672 patients. Among CIN-2–3 patients, we observed a pooled regression rate of 61% (CI: 0.46–0.75; I^2: 77%). When compared, Imiquimod was inferior to conization (RR: 0.62; CI: 0.42–0.92; I^2: 64%). The HPV clearance rate in women who completed Imiquimod treatment was 60% (CI: 0.31–0.81; I^2: 57%). The majority of side effects reported were mild to moderate in severity. Conclusions: Our findings indicate that topical Imiquimod is safe and effective in reducing cervical intraepithelial neoplasia and promoting HPV clearance. However, it was found to be inferior compared to conization. Imiquimod could be considered a potential medication for high-grade CIN patients and should be incorporated into guidelines for treating cervical dysplasia.

Keywords: Imiquimod; gynecologic cancer risk reduction; HPV; cervical cancer

1. Introduction

Cervical intraepithelial neoplasia (CIN) 2–3 is the precursor lesion of cervical cancer, one of the leading causes of cancer-related death in women. In 2020, there were 604,000 new cases and 342,000 deaths worldwide, according to GLOBOCAN [1]. High-risk human papillomavirus (HPV) is the predominant factor (99.7%) responsible for cervical cancer [2].

While only a minority of cases progress to invasive cancer after years of persistence, most patients experience regression of CIN to a normal condition [3]. For histologic high-grade intraepithelial lesions (HSILs), excisional treatment is preferred according to the 2019 American Society for Colposcopy and Cervical Pathology (ASCCP) Risk-Based Management Consensus Guideline [4]. However, these procedures may impact pregnancy outcomes, such as preterm delivery, premature rupture of membranes, and low birth weight [5,6]. Moreover, the persistence of HPV has been linked to an increased recurrence rate following surgical intervention [7]. Consequently, alternative conservative therapies are necessary to reduce the frequency of surgical interventions and associated complications.

Topical Imiquimod has been approved by the US Food and Drug Administration (FDA) for treating external genital and perianal warts, basal cell carcinoma, and actinic keratoses [8]. This compound is believed to activate immune cells as a Toll-like receptor-7 agonist. It exerts its antiviral effects by activating dendritic cells and inducing cytokines such as tumor necrosis factor-alpha (TNF-α), interferon-alpha (IFN-α), and interleukins (ILs) [9]. Multiple studies have shown that Imiquimod could be a potential conservative treatment for precursor cervical lesions by accelerating viral clearance [10–12]. However, some other publications have found that it is ineffective in reducing CIN [13]. There were no meta-analyses on the subject to answer this important question.

In this present study, based on the available literature, we aimed to determine the efficacy and safety of topical Imiquimod therapy in reducing the incidence of cervical intraepithelial neoplasia (CIN) and its impact on HPV clearance.

2. Materials and Methods

We followed the PRISMA 2020 (Table S1) for conducting a systematic review and meta-analysis, as recommended by the Cochrane Handbook for Systematic Reviews of Interventions [14,15] (see Figure S1). The study design and protocol were registered in PROSPERO (CRD420222870), and we adhered to them completely.

2.1. Search Strategy

The complete search key is provided in the Supplementary Materials. During the systematic search, the following search strategy was used: 'Imiquimod', 'cervical intraepithelial neoplasia', 'cervical dysplasia', 'human papillomavirus'.

2.2. Literature Search and Eligibility Criteria

A systematic search was conducted using five major databases: MEDLINE (through PubMed), Embase, Cochrane Central Register of Controlled Trials (CENTRAL), Scopus, and Web of Science until 10 October 2022. No restrictions or filters were applied during the search. We used two frameworks to describe the eligibility criteria in the articles. First, the CoCoPop framework was used in studies with no comparators to assess. We investigated women with cervical intraepithelial neoplasia (Population) who applied topical Imiquimod (Context). We determined cervical dysplasia regression, estimation of treatment success, assessment of HPV clearance, and adverse events (Condition). Afterwards, the PICO framework was used. We assessed women (P) with cervical dysplasia or who were HPV positive. In the intervention group (I), women had to receive topical Imiquimod products for their cervical disease. Patients in the comparator group (C) received the standard treatment, predominantly surgical solutions: conization, cryotherapy, laser therapy or expectant management. Outcome (O) parameters included the assessment of cervical dysplasia regression, assessment of HPV clearance, and adverse events [16]. Cervical dysplasia regression was defined as the absence of dysplasia or regression from CIN 2–3 to CIN 1. HPV clearance was effective when the original HPV types could not be detected after treatment. Cohorts, case-control studies, and randomized controlled trials (RCT) were accepted. Only studies with patient follow-up were included. We imposed no language restrictions; non-English articles were translated into English and evaluated afterwards.

2.3. Selection Process and Data Collection

Article selection was performed using the reference management program EndNote X9. Duplicate removal was conducted by two independent reviewers (B.H., H.E.) at each stage: after title and abstract selection and during full-text selection. Cohen's kappa coefficient (κ) was used to measure the level of agreement [17]. Disputed articles were resolved by a third independent reviewer (H.Z.S.).

Pre-defined variables were described in a Microsoft Excel spreadsheet (Windows 11 Pro 10) by two independent reviewers (B.H., E.H.). The following variables were extracted: first author, year of publication, digital object identifier, study type, study design, country, study period, centers, and duration of follow-up. The following were extracted for both the intervention group and the control group: patient numbers, patient age, pregnancy status, smoking status, number of sexual partners, histological findings (cervical intraepithelial neoplasia 2–3), and HPV status. The dose, duration, and application form were recorded in the intervention group. Outcomes were collected in two-by-two tables. Whenever possible, risk ratios (RRs) were extracted directly. Intention-to-treat (ITT) and per-protocol (PP) data were collected from RCTs. Response rate data were recorded separately when available. Adverse events were collected using the National Cancer Institute (NIH) website's Common Terminology Criteria for Adverse Event protocols [18]. Adverse events were graded from 1 to 4 for the following: fatigue, headache, myalgia, flu-like symptoms, fever, abdominal pain, vaginal pruritus, vaginal discharge, vaginal bleeding, and inflammation. A third reviewer (Z.S.H.) resolved the conflict in case of disagreements.

2.4. Risk of Bias and Quality Assessment of Included Articles

Assessing bias and quality risk of bias and quality assessment depended on study type. RCTs were evaluated using the Risk of Bias II (ROB II) tool, while non-randomized interventions used the Risk of Bias in Non-Randomized Studies (ROBINS I) [19,20]. Response rates without a control group were assessed with the Joanna Briggs Institute (JBI) Critical Appraisal Checklists [21]. GRADE was applied to grade evidence, and a Summary of Findings Table was formulated using GradePro [22]. Two reviewers (B.H., E.H.) conducted assessments, with disputes resolved by a third reviewer (Z.S.H.).

2.5. Synthesis Methods

In data synthesis, both qualitative and quantitative analyses utilized R statistical programming language (R version 4.3). Quantitative synthesis required a minimum of three studies, presented in forest plots. Subgroup analyses were conducted based on article type and cervical dysplasia grade. RCTs' ITT data were analyzed, while other study types were grouped as cohorts. For cervical dysplasia, subgroups included studies without CIN, CIN 1–2–3, and CIN 2–3. Per-protocol data were analyzed for RCTs with complete treatment. Risk ratios (RR) with 95% confidence intervals (CI) assessed effect sizes. The Clopper-Pearson method calculated CIs. Statistically significant results excluded the null value within pooled CI [23]. Forest plots summarized meta-analysis findings. Higgins & Thompson's I^2 assessed heterogeneity, with τ^2 indicating variance [24].

Heterogeneity levels were categorized: 0–40% possibly not important, 30–60% moderate, 50–90% substantial, 75–100% considerable. Subgroups used a fixed effects "plural" model. Cochrane Q test evaluated subgroup differences [25]. To assess the difference between the subgroups, a "Cochrane Q" test was used between subgroups (Harrer et al. 2021). The null hypothesis was rejected on a 5% significance level [26].

Publication bias was not assessed due to limited studies (<10).

3. Results

3.1. Search and Selection

Our systematic search identified 3141 articles from five databases. After removing duplicates, 2218 articles were analyzed for title and abstract selection. In the full-text selection, 13 eligible articles were screened. Finally, we found eight eligible articles for quantitative and qualitative synthesis (see Figure 1).

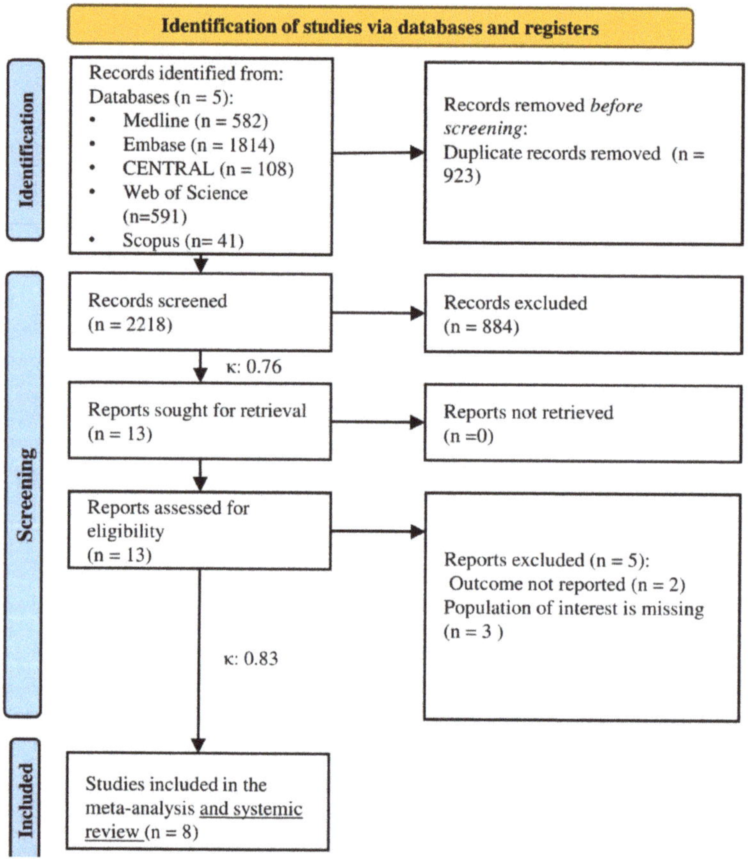

Figure 1. PRISMA flowchart of selection n: number of studies, κ: Cohen's kappa coefficient.

3.2. Basic Characteristics of Included Studies

The eligible articles were published between 2012 and 2022. Regarding demographics, the mean age of women included in the studies was 30.41 years (±2.15). The mean follow-up time was 18.62 (±12.00) months. In six studies, women had histologically proven CIN 2–3; in the seventh study, both cytology and histology were used [27]. Quantitative synthesis was possible only for a subpopulation with HPV status. HPV tests were performed in seven articles. Details about the doses and application of Imiquimod can be found in Table S2.

Altogether, 672 patients were included from the eight studies [10–13,27–30], with 398 women receiving Imiquimod treatment. Detailed baseline characteristics can be found in Table **??**.

Table 1. Basic characteristics of included studies.

Author, Years	Study Type	Region	Follow Up Time (Months)	Number of Patients in Intervention	Age (Mean) Intervention, SD	Number of Patients in Control	Age (Mean) Control, SD	CIN [B]	CIN2/CIN3 Ratio	HPV [D] Type	Dose of Imiquimod/Patient	Intervention of Control Group	Adverse Event Reporting	Dropout of Patients
Grimm et al., 2012 [11]	RCT [A]	Austria	5	30	29.2 ± 6.1	29	31.8 ± 7.8	CIN 2–3	1.73	HPV 16/18, other HR [E] HPV	243.75 mg	observation	CTCAE [F] 3.0	6.70%
Hendriks et al., 2022 [12]	Non-randomized interventional	The Netherlands	6	61	33.3 ± 9.1	62	35.2 ± 7	CIN 2–3	0.69	HPV 16/18, other HR HPV	300 mg	conization	VAS [G]	22.90%
Cokan et al., 2021 [28]	RCT	Slovenia	6	52	28.3 ± 4.2	52	26 ± 4.6	CIN 2–3	0.79	NA	600 mg	conization	CTCAE 5.0	17.30%
Lin et al., 2012 [29]	Retrospective cohort analysis	Taiwan	33.4	72	51.75 [B]	20	50 [B]	NA [C]	NA	persistent HR-HPV	150 mg	observation	NA	NA
Fonseca et al., 2021 [10]	RCT	Brazil	24	45	32 [B]	45	36 [B]	CIN 2–3	0.4	NA	150 mg	observation	CTCAE 4.0	15.60%
Pachman et al., 2012 [13]	RCT	USA	37.2	28	30 ± 8.9	28	29 ± 9.7	CIN 1–2–3	1.42	HR-HPV	12.5 mg	conization, laser, cryotherapy	CTCH [H] 2.0	7.14%
Polterauer et al., 2022 [30]	RCT	Austria	24	51	31.4 [B]	42	30.1 [B]	CIN 2–3	0.28	HPV 16/18, other HR HPV	243.75 mg	conization	CTCAE 3.0	9.80%
Kim et al., 2019 [27]	retrospective cohort analysis	Republic of Korea	13.4	55	30 [B]	NA	NA	CIN 2–3	0.74	HPV 16/18, other HR HPV	100 mg	NA	NA	1.80%

[A] randomised control trial; [B] median cervical intraepithelial neoplasia; [C] Not applicable; [D] Human papillomavirus; [E] High-risk; [F] Common Terminology Criteria for Adverse Events; [G] Visual Analog Scale; [H] Common Toxicity Criteria.

3.3. CIN 2–3 Regression

A total of 294 women received topical Imiquimod treatment for CIN 2–3 [10,12,27,28,30]. These patients showed a regression rate of 61% (CI: 0.46–0.75; I^2: 77%) to CIN 1 or no disease after topical Imiquimod therapy (see Figure 2). The subgroup analysis based on study type revealed a histologic regression rate of 59% (CI: 0.47–0.70) in the ITT-RCTs, while the response rate in the cohort studies was 64% (CI: 0–1.00 I^2: 94) (see Figure 2). In the PP population, which consisted of 155 patients, the regression rate was 67% (CI: 0.54–0.78; I^2: 0%) (see Figure S1).

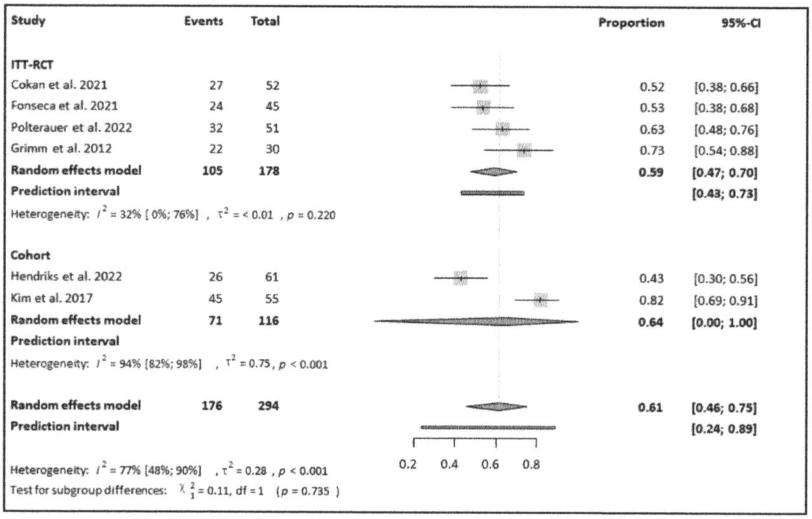

Figure 2. Forest plot of studies representing Imiquimod and CIN 2–3 regression based on study type [10–12,27,28,30]. ITT: intention to treat, RCT: randomized control trial. Effect Estimate: The effect estimate for each study is represented by a grey square, located along the x-axis. Confidence Interval (CI): A line extending from the effect estimate represents the confidence interval. This indicates the range within which the true effect size is likely to lie, with the point estimate positioned at the center of the bar. Overall Estimate: A summary effect estimate, represented by a blue diamond at the bottom of the plot, combines the results of all studies included in the meta-analysis. The center of the diamond represents the point estimate, and the width of the diamond represents the confidence interval around the summary estimate. I^2 (I-squared): A measure of heterogeneity in meta-analysis, indicating the proportion of total variation across studies that is due to heterogeneity rather than chance. p-value (Probability value): The probability of obtaining test results at least as extreme as the observed results, indicating the significance of the heterogeneity test. χ^2 (Chi-square): A test for subgroup differences, evaluating whether the observed differences between subgroups are statistically significant. df (degrees of freedom): The degrees of freedom, which represent the number of independent pieces of information used to estimate a parameter. τ^2 (Tau-squared): The variance of true effects in a random-effects model, reflecting the variability of effect sizes across studies beyond sampling error.

Two articles investigated the efficacy of topical Imiquimod [10,11]. In both studies, the RR for CIN regression was higher when Imiquimod was compared to no treatment (RR: 1.87; CI: 1.12–3.10 and RR: 2.37; CI: 1.25–4.48, respectively) (see Figure S2).

In the experimental group, 196 women received Imiquimod treatment, while 196 women were in the control group who underwent conization [10,12,28,30]. Women who underwent conization had a 38% decrease in the risk of persistence or progression in CIN compared to the women who applied Imiquimod (RR: 0.62; CI: 0.42–0.92; I^2: 64%) (see Figure 3). The subgroup analysis showed a randomized clinical trial where conization was superior to Imiquimod and had a 44% decrease in the risk of unsuccessful treatment (RR: 0.56; CI: 0.43–0.74) (see

Figure 3) [28]. Likewise, in the PP analysis, Imiquimod was not a superior intervention to conization (RR: 0.78; CI: 0.56–1.07; I^2: 0%) (see Figure S3).

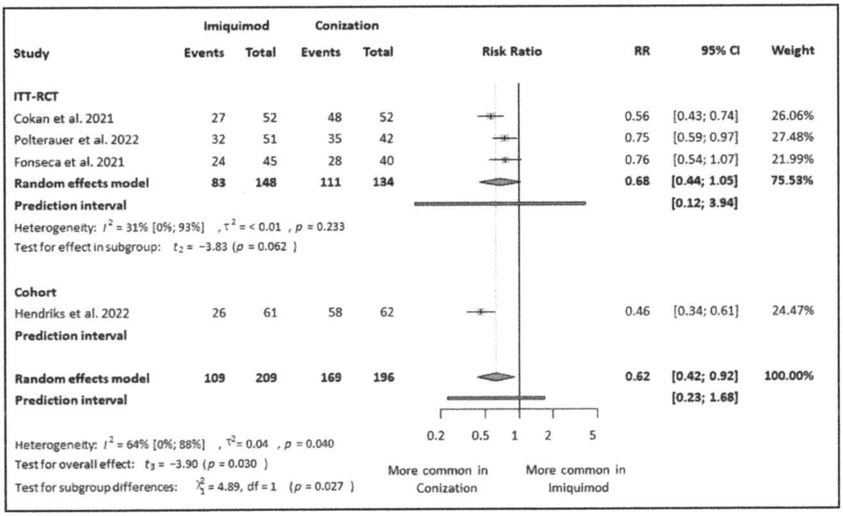

Figure 3. Forest plot of studies representing the Imiquimod group compared to conization on CIN 2–3 regression based on study type [10,12,28,30]. ITT: intention to treat, RCT: randomized control trial. Effect Estimate: The effect estimate for each study is represented by a grey square, located along the x-axis. Confidence Interval (CI): A line extending from the effect estimate represents the confidence interval. This indicates the range within which the true effect size is likely to lie, with the point estimate positioned at the center of the bar. Overall Estimate: A summary effect estimate, represented by a blue diamond at the bottom of the plot, combines the results of all studies included in the meta-analysis. The center of the diamond represents the point estimate, and the width of the diamond represents the confidence interval around the summary estimate. I^2 (I-squared): A measure of heterogeneity in meta-analysis, indicating the proportion of total variation across studies that is due to heterogeneity rather than chance. p-value (Probability value): The probability of obtaining test results at least as extreme as the observed results, indicating the significance of the heterogeneity test. χ^2 (Chi-square): A test for subgroup differences, evaluating whether the observed differences between subgroups are statistically significant. df (degrees of freedom): The degrees of freedom, which represent the number of independent pieces of information used to estimate a parameter. τ^2 (Tau squared): The variance of true effects in a random-effects model, reflecting the variability of effect sizes across studies beyond sampling error. Test for overall effect (t_3): A statistical test used to assess whether the observed effect size is significantly different from zero, indicating the presence of an overall effect in the meta-analysis.

3.4. Imiquimod on HPV Clearance

Among the 254 patients who received Imiquimod treatment, 50% (CI: 0.35–0.64; I^2: 64) of women had HPV clearance (see Figure 4) [10,13,27,29,30]. We performed a subgroup analysis according to the grade of cervical dysplasia (see Figure 4). When CIN 2–3 was the diagnosis, the HPV clearance rate was 42% (CI: 0.29–0.56; I^2: 49%); when CIN 1–3 was the diagnosis, the HPV clearance rate was 68% (CI: 0.48–0.84). Finally, when there was no CIN, only HPV positivity, the HPV clearance rate was 65% (CI: 0.44–0.83). However, we had only one study for each outcome. The subgroup analysis regarding the study types showed a 56% (CI: 0.28–0.80; I^2: 59%) HPV clearance in the ITT-RCTs and 44% (CI: 0.17–0.75; I^2: 73%) in the cohort studies (see Figure S4). Moreover, in the PP population of 100 patients, the HPV clearance rate was higher at 60% (CI: 0.35–0.84; I^2: 57%) (see Figure S5). The subgroup analysis in the PP for CIN 2–3 showed a 54% HPV clearance (CI: 0.06–0.96; I^2: 47%), and for CIN 1–3, the HPV clearance was 73%

(CI: 0.52–0.88) (see Figure S6). The HPV tests were conducted on average 2.33 (SD: ±1.91) months after finishing Imiquimod treatment.

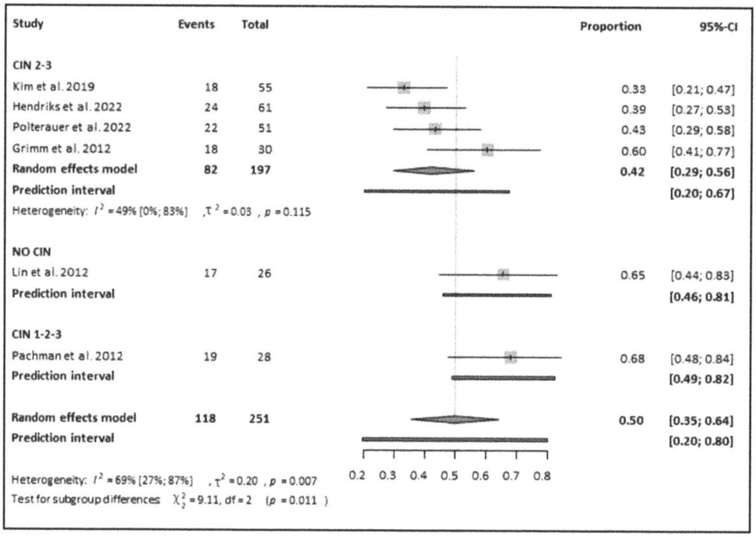

Figure 4. Forest plot of studies representing Imiquimod on HPV clearance based on the CIN status [11–13,27,29,30]. CIN: cervical intraepithelial neoplasia. Effect Estimate: The effect estimate for each study is represented by a grey square, located along the x-axis. Confidence Interval (CI): A line extending from the effect estimate represents the confidence interval. This indicates the range within which the true effect size is likely to lie, with the point estimate positioned at the center of the bar. Overall Estimate: A summary effect estimate, represented by a blue diamond at the bottom of the plot, combines the results of all studies included in the meta-analysis. The center of the diamond represents the point estimate, and the width of the diamond represents the confidence interval around the summary estimate. I^2 (I-squared): A measure of heterogeneity in meta-analysis, indicating the proportion of total variation across studies that is due to heterogeneity rather than chance. p-value (Probability value): The probability of obtaining test results at least as extreme as the observed results, indicating the significance of the heterogeneity test. χ^2 (Chi-square): A test for subgroup differences, evaluating whether the observed differences between subgroups are statistically significant. df (degrees of freedom): The degrees of freedom, which represent the number of independent pieces of information used to estimate a parameter. τ^2 (Tau-squared): The variance of true effects in a random-effects model, reflecting the variability of effect sizes across studies beyond sampling error.

The Imiquimod group had 196 patients, while the control arm had 180 patients. Imiquimod treatment did not result in better HPV clearance compared to that of the control group (RR: 1.29; CI: 0.52–3.21; I^2: 80%) (see Figure 5). In the control group, the treatment differed between the studies. When the control group received conization, it was more effective than Imiquimod (RR: 0.67; CI: 0.46–0.99) [30]. However, when no intervention was implemented in the control group and HPV infection was persistent, Imiquimod was more effective (RR: 4.20; CI: 1.62–10.89) [11]. In another study, when there was only persistent HPV positivity and no cervical dysplasia, and the control arm received no intervention, Imiquimod was more effective (RR: 2.18; CI: 1.06–4.05) [29]. In one study, the control arm had surgical interventions (conization, cryotherapy, laser) [13]. We found that in this article, Imiquimod treatment did not result in better HPV clearance than the control group (RR: 1.19; CI: 0.79–1.79) [13]. When we examined the PP group, we concluded that Imiquimod treatment did not result in a higher HPV clearance rate compared to the control group (see Figure S7).

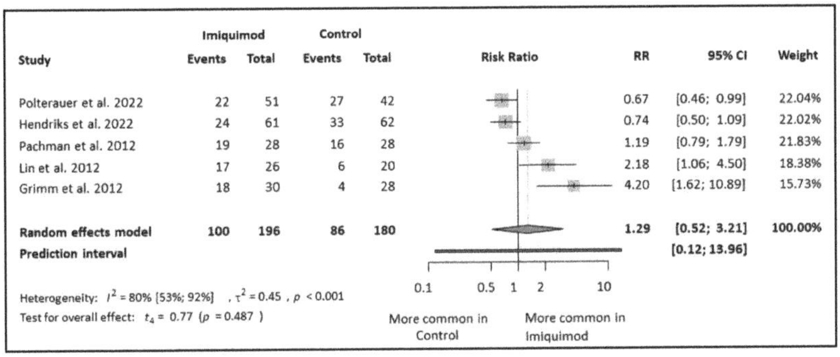

Figure 5. Forest plot of studies representing the Imiquimod group compared to control on HPV clearance [11–13,29,30]. Effect Estimate: The effect estimate for each study is represented by a grey square, located along the x-axis. Confidence Interval (CI): A line extending from the effect estimate represents the confidence interval. This indicates the range within which the true effect size is likely to lie, with the point estimate positioned at the center of the bar. Overall Estimate: A summary effect estimate, represented by a blue diamond at the bottom of the plot, combines the results of all studies included in the meta-analysis. The center of the diamond represents the point estimate, and the width of the diamond represents the confidence interval around the summary estimate. I^2 (I-squared): A measure of heterogeneity in meta-analysis, indicating the proportion of total variation across studies that is due to heterogeneity rather than chance. p-value (Probability value): The probability of obtaining test results at least as extreme as the observed results, indicating the significance of the heterogeneity test. τ^2 (Tau-squared): The variance of true effects in a random-effects model, reflecting the variability of effect sizes across studies beyond sampling error. Test for overall effect (t_4): A statistical test used to assess whether the observed effect size is significantly different from zero, indicating the presence of an overall effect in the meta-analysis.

Among the 186 patients who received Imiquimod, we investigated HPV 16/18 clearance compared to the clearance of other high-risk HPV (HR-HPV) types [11,12,27,30]. Our findings indicate no significant difference between HPV 16/18 clearance and clearance of other HR-HPV types (RR: 0.89; CI: 0.58–1.37; I^2: 0) (see Figure S8).

3.5. Adverse Events

In five studies, we were able to quantitatively synthesize the adverse events in patients who received Imiquimod, as they all used a very similar grading system (see Table ??) [10,11,13,28,30]. They graded side effects on a scale of one to five, where the grades ranged from mild to moderate, serious, life-threatening, and death (see Figure 6 and Figures S9–S18).

The most common systemic side effects were flu-like symptoms and myalgia, while from the local side effects, vaginal pruritus was the most frequent.

Grade 3 side effects occurred 8 times, with two articles [10,28] reporting abdominal pain and two [28,30] reporting headache. The remaining four occurrences of grade 3 side effects were one flu-like symptom [11], one fever [30], one myalgia [30], and one vaginal inflammation [28].

3.6. Risk of Bias Assessment and GRADE

For randomized controlled studies, the ROB2 showed some concern for the risk of bias in five outcomes and a low risk of bias in two outcomes. In non-randomized clinical trials, the ROBINSON tool showed a moderate and serious risk of bias in two outcomes. The latter study was a retrospective cohort analysis that used a historical control group for comparison and had methodological concerns. When we analyzed the response rates according to the JBI critical appraisal checklist, we found that the most frequent issue was regarding sample size (see Figures S19–S21).

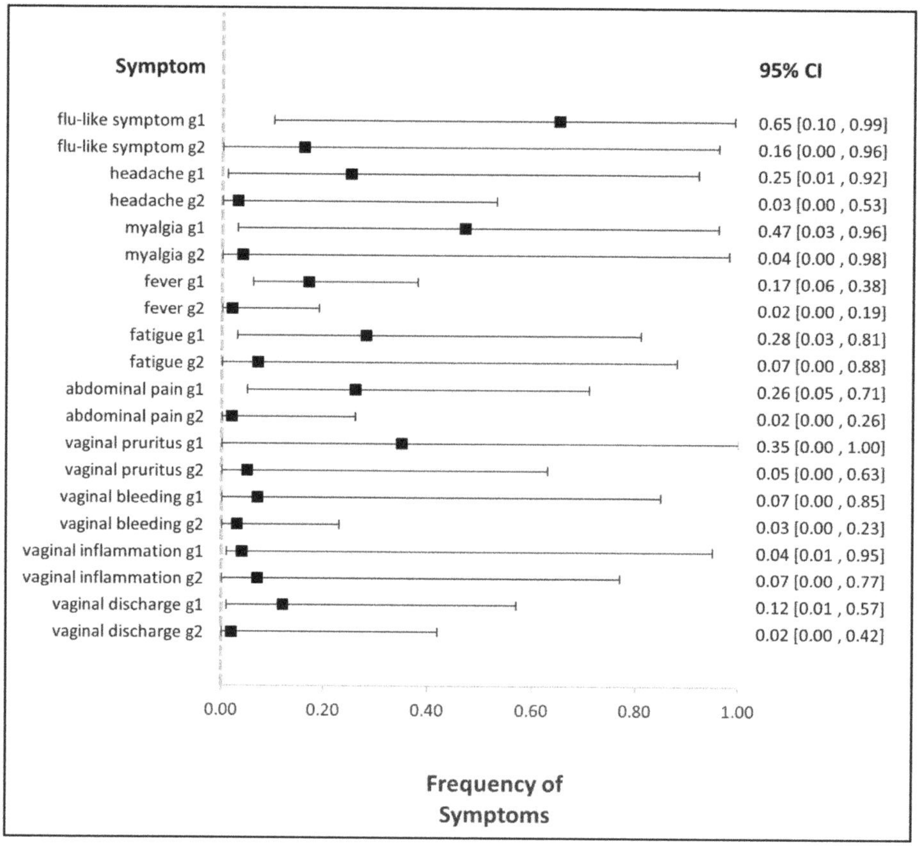

Figure 6. Forest plot of studies representing the frequency of all symptoms. g1: grade1, g2: grade2, CI: confidence interval.

Our summary of findings consisted of five outcomes where we included a control group. The quality of evidence was assessed as high for two outcomes, while it was deemed low for the remaining three outcomes (see Tables S3–S5).

4. Discussion

We investigated the safety and efficacy of topical Imiquimod on cervical intraepithelial neoplasia and HPV clearance. After Imiquimod treatment in CIN 2–3 patients, the regression rate was 61%. Regarding efficacy, we analyzed the biopsies of CIN 2–3 patients and concluded that women who received Imiquimod treatment had a higher rate of CIN regression compared to those who did not receive Imiquimod. We found that conization is more effective than Imiquimod in treating CIN 2–3 patients, as there was a 38% increase in successful treatment in the conization arm.

Imiquimod treatment in HPV-positive women showed a 50% HPV clearance rate and was 60% for women who completed the treatment. Overall, Imiquimod treatment did not result in better HPV clearance compared to the control group's treatment. When Imiquimod was compared to conization in one study, HPV clearance was higher in the conization group [30]. However, in another study comparing Imiquimod to placebo, HPV clearance was higher in the Imiquimod arm [11].

Regarding side effects, most side effects were mild, and hospitalization was not required in the majority of cases.

Increasing the dosage of Imiquimod did not result in a higher rate of CIN 2–3 regression. Regardless of the dose of Imiquimod, women who completed the treatment had a similar rate of dysplasia regression. In two studies, the remission rates were higher than in other studies [11,27]. In one study, the higher CIN 2/CIN 3 ratio could explain this. CIN 2 has a higher rate of spontaneous regression than CIN 3 and is considered a milder lesion [31,32]. The other study used cytological confirmation of CIN regression [27]. Although cytology is not a reproducible method of detecting cervical dysplasia [33], Imiquimod has been shown to be effective in reducing vulvar intraepithelial neoplasia and vaginal intraepithelial neoplasia [34,35]. The American College of Obstetricians and Gynecologists recognizes the off-label use of Imiquimod for vulvar intraepithelial neoplasia [36]. Our findings on Imiquimod in cervical dysplasia regression align with previous findings on other lower genital intraepithelial neoplasia [34,35]. Imiquimod is effective in reducing cervical CIN.

When comparing Imiquimod with the already existing surgical therapy for CIN 2–3, we found Imiquimod inferior. However, when the topical immunomodulator was used before conization, the positive margins of the resected tissue were lower than the average in the literature [10,37]. A logical explanation could be that Imiquimod reduced the lesion's depth and width and made it more suitable for surgical excision [10]. For selected patients, Imiquimod can be a choice; for example, women with future pregnancy desires have a higher demand for conservative treatment [38], given that conization increases the risk of miscarriage and preterm birth by causing cervical incompetence [5]. The ASCCP guideline recommends diagnostic evaluation of the cervix after six months in case of positive surgical excision margins [4]. Implementing Imiquimod in positive margin cases could lower the need for additional surgical excision. This could be highly desirable in women considering future pregnancy, as repeated surgical intervention of the cervix increases the risk of preterm birth compared to one surgical excision of the cervix [39]. A recent study showed that patients who respond to Imiquimod treatment could be selected priorly with an immunohistochemical method, as the immune microenvironment predicts whether the patient will respond to Imiquimod treatment [40]. This could personalize Imiquimod treatment, as not all patients respond to it.

When examining the HPV clearance for HPV 16 and 18 and other HR-HPV types, we conclude that the clearance rate is not worse for HPV 16 and 18. This is interesting since HPV 16 and 18 are known to be more aggressive and accountable for 70% of all cervical cancer [41]. We investigated the best three HPV clearance rates in the studies. We found that in one article there were only persistent HPV infections without CIN lesions [29]. In another study, CIN 2/CIN 3 rate was the highest among all publications [11]. In the third study, the CIN 2/CIN 3 rate was high, and CIN 1 also occurred [13]. Moreover, in this study, the HPV tests were taken after 6 months of Imiquimod discontinuation, which is a long time considering the natural clearance of HPV [42]. Patients with CIN 2+ lesions are known to be, molecularly, a vastly heterogeneous population; with progression, the cellular changes are more extensive, and the spontaneous regression of CIN and HPV declines [43]. Our findings support this observation, as when more CIN 3 lesions occur, HPV clearance rates are lower with Imiquimod. Higher Imiquimod doses did not result in a higher HPV clearance rate. When Imiquimod was compared to the control group, we did not experience a better HPV clearance rate. However, it should be mentioned here that the control differed between studies: surgical excision of the HPV-infected area [30] and only expectant management with no intervention [11]. In the previous case, the surgical solution was comparable or more effective than Imiquimod; however, in the latter case, Imiquimod was more effective than expectant management.

Systemic and local side effects were frequent but mostly mild, and the symptoms could have been reduced with non-steroid anti-inflammatory drugs [11]. This variability can be attributed to the systemic side effects associated with Imiquimod, which other common infections or health-related conditions can influence. Additionally, local side effects caused by Imiquimod are commonly observed in general gynecological practice. This variability in side effects can help explain the differences observed among the studies.

Dropouts can happen for several reasons (long travel, financial reasons, dissatisfaction); however, it should be mentioned that in the two studies where the highest dose of Imiquimod was implemented, the two highest rates of dropout were also observed [12,28]. Severe side effects occurred just eight times, with high doses of Imiquimod used in seven of these cases. In two studies, besides the lower doses of Imiquimod used, the application was implemented by doctors. Accordingly, direct application could contribute to the low and mild side effects [10,27].

4.1. Strengths and Limitations

This meta-analysis is the first to synthesize the findings on Imiquimod use in cervical dysplasia and HPV-positive patients.

However, several limitations should be noted. Firstly, many studies had poor patient recruitment, and efforts were made to enroll more women. Secondly, the patients were mostly selected based on specific criteria, limiting the generalizability of the study's implications to all cervical dysplasia patients. Thirdly, in some cases, the comparison was made with control groups that used different methods. Fourthly, many studies lacked longer follow-up intervals, raising questions about the durability of dysplasia remission. Fifthly, the endpoint and timing of different outcome measures were often inconsistent, which is problematic given the spontaneous tendency of these lesions to regress. Sixthly, the clinical and statistical heterogeneity was substantial in several cases.

4.2. Implications for Practice and Research

Our findings show Imiquimod is safe and effective in reducing CIN and facilitating HPV clearance. While Imiquimod is inferior to conization, it could still be considered for use in selected patients, particularly after positive margins of conization, to avoid subsequent surgical excision of the cervix.

Personalized treatment strategies hold promise for enhancing the therapeutic efficacy of Imiquimod, as the immune environment serves as a predictive factor for treatment success [40]. Further refinement of immunohistochemical methods to identify specific biomarkers could lead to even greater therapeutic responses, potentially revolutionizing the approach to CIN treatment. Investigating the microenvironment and molecular profiles in greater detail through focused studies has the potential to transform the utilization of Imiquimod in clinical practice. Additionally, exploring combined therapies involving Imiquimod and other agents, such as 5-Fluorouracil, may synergistically augment the reduction of CIN lesions [44]. This approach not only offers the possibility of reducing Imiquimod dosage but also of lowering potential side effects, thereby improving medication adherence. Determining the optimal dosage of Imiquimod is paramount, with further studies warranted to elucidate this aspect. Such investigations are motivated by the desire to minimize costs associated with Imiquimod, an expensive medication, while simultaneously mitigating side effects. Prior to the incorporation of Imiquimod into treatment guidelines, conducting cost–effectiveness analyses is imperative. Assessing the cost–benefit ratio compared to traditional treatments is essential for informing healthcare resource allocation decisions. A comprehensive understanding of these factors will be instrumental in optimizing the utilization of Imiquimod in the management of cervical dysplasia. Further interventional studies are needed in this field to better understand how Imiquimod can reduce the burden of cervical dysplasia. Particularly, investigations with extended follow-up periods are warranted, as current studies often lack prolonged monitoring. Consequently, elucidating the duration of Imiquimod's effects remains an unanswered question that requires attention.

5. Conclusions

In conclusion, Imiquimod is not a substitute for cone biopsy; however, it can be a valuable treatment option for high-grade cervical dysplasia.

Supplementary Materials: The following supporting information can be downloaded at https://www.mdpi.com/article/10.3390/cancers16081610/s1, Figure S1: Forest plot of studies representing Imiquimod and CIN 2–3 regression in the PP analysis; Figure S2: Forest plot of studies representing Imiquimod compared to no intervention on CIN 2–3 regression; Figure S3: Forest plot of studies representing Imiquimod group compared to conization on CIN 2–3 regression in the PP analysis; Figure S4: Forest plot of studies representing Imiquimod on HPV clearance according to study type; Figure S5: Forest plot of studies representing Imiquimod on HPV clearance in the PP analysis; Figure S6: Forest plot of studies representing Imiquimod on HPV clearance according to subgroup analysis of CIN status in the PP analysis; Figure S7: Forest plot of studies representing the Imiquimod group compared to control on HPV clearance in the PP analysis; Figure S8: Forest plot of studies representing the HPV 16/18 clearance compared to other HR-HPV clearance in the Imiquimod group; Figure S9: Forest plot of studies representing the occurrence of headaches in patients treated with Imiquimod; Figure S10: Forest plot of studies representing the occurrence of myalgia in patients treated with Imiquimod; Figure S11: Forest plot of studies representing the occurrence of fatigue in patients treated with Imiquimod; Figure S12: Forest plot of studies representing the occurrence of flu-like symptoms in patients treated with Imiquimod; Figure S13: Forest plot of studies representing the occurrence of fever in patients treated with Imiquimod; Figure S14: Forest plot of studies representing the occurrence of abdominal pain in patients treated with Imiquimod; Figure S15: Forest plot of studies representing the occurrence of vaginal pruritus in patients treated with Imiquimod; Figure S16: Forest plot of studies representing the occurrence of vaginal bleeding in patients treated with Imiquimod; Figure S17: Forest plot of studies representing the occurrence of vaginal discharge in patients treated with Imiquimod; Figure S18: Forest plot of studies representing the occurrence of inflammation of the vagina in patients treated with Imiquimod; Figure S19: Risk of bias assessment of randomized control studies; Figure S20. Risk of bias assessment of non-randomized controlled studies; Figure S21: JBI Critical Appraisal Checklist for response rate outcomes; Table S1: PRISMA 2020 checklist; Table S2: Imiquimod administration; Table S3: Grade for outcomes that assessed Imiquimod compared to conization; Table S4: Grade for outcomes that assessed Imiquimod compared to control on HPV clearance; Table S5: Grade on HPV 16/18 clearance compared to other HR-HPV clearance. References [10–13,27–30] are cited in the Supplementary Materials.

Author Contributions: B.H.: conceptualization, writing, project administration; B.T.: project administration, methodology, writing; E.H.: data curation; P.H.: project administration, methodology; A.H.: data analysis; P.N.: conceptualization; Z.H.: data curation; N.Á.: conceptualization, supervision; F.B.: conceptualization; Z.M.: conceptualization, writing, review and editing. All authors have read and agreed to the published version of the manuscript.

Funding: Funding was provided by the ÚNKP-22-3 New National Excellence Program of the Ministry for Innovation and Technology from the source of the National Research, Development and Innovation Fund (to B.T.—ÚNKP-22-3-I-PTE-1693).

Data Availability Statement: The dataset utilized in this meta-analysis is available in the full-text articles included in this systematic review.

Conflicts of Interest: The authors declare no conflicts of interest.

References

1. Sung, H.; Ferlay, J.; Siegel, R.L.; Laversanne, M.; Soerjomataram, I.; Jemal, A.; Bray, F. Global Cancer Statistics 2020: GLOBOCAN Estimates of Incidence and Mortality Worldwide for 36 Cancers in 185 Countries. *CA Cancer J. Clin.* **2021**, *71*, 209–249. [CrossRef] [PubMed]
2. Okunade, K.S. Human papillomavirus and cervical cancer. *J. Obstet. Gynaecol.* **2020**, *40*, 602–608. [CrossRef] [PubMed]
3. Loopik, D.L.; Bentley, H.A.; Eijgenraam, M.N.; IntHout, J.; Bekkers, R.L.M.; Bentley, J.R. The Natural History of Cervical Intraepithelial Neoplasia Grades 1, 2, and 3: A Systematic Review and Meta-analysis. *J. Low. Genit. Tract. Dis.* **2021**, *25*, 221–231. [CrossRef] [PubMed]

4. Perkins, R.B.; Guido, R.S.; Castle, P.E.; Chelmow, D.; Einstein, M.H.; Garcia, F.; Huh, W.K.; Kim, J.J.; Moscicki, A.B.; Nayar, R.; et al. 2019 ASCCP Risk-Based Management Consensus Guidelines for Abnormal Cervical Cancer Screening Tests and Cancer Precursors. *J. Low. Genit. Tract. Dis.* **2020**, *24*, 102–131. [CrossRef] [PubMed]
5. Weinmann, S.; Naleway, A.; Swamy, G.; Krishnarajah, G.; Arondekar, B.; Fernandez, J.; Myers, E. Pregnancy Outcomes after Treatment for Cervical Cancer Precursor Lesions: An Observational Study. *PLoS ONE* **2017**, *12*, e0165276. [CrossRef] [PubMed]
6. Lieb, J.A.; Mondal, A.; Lieb, L.; Fehm, T.N.; Hampl, M. Pregnancy outcome and risk of recurrence after tissue-preserving loop electrosurgical excision procedure (LEEP). *Arch. Gynecol. Obstet.* **2023**, *307*, 1137–1143. [CrossRef] [PubMed]
7. Bogani, G.; Di Donato, V.; Sopracordevole, F.; Ciavattini, A.; Ghelardi, A.; Lopez, S.; Simoncini, T.; Plotti, F.; Casarin, J.; Serati, M.; et al. Recurrence rate after loop electrosurgical excision procedure (LEEP) and laser Conization: A 5-year follow-up study. *Gynecol. Oncol.* **2020**, *159*, 636–641. [CrossRef]
8. Food and Drug Administration (FDA). Available online: https://www.fda.gov/drugs/drug-and-biologic-approval-and-ind-activity-reports/2021-first-generic-drug-approvals (accessed on 6 July 2023).
9. Stary, G.; Bangert, C.; Tauber, M.; Strohal, R.; Kopp, T.; Stingl, G. Tumoricidal activity of TLR7/8-activated inflammatory dendritic cells. *J. Exp. Med.* **2007**, *204*, 1441–1451. [CrossRef] [PubMed]
10. Fonseca, B.O.; Possati-Resende, J.C.; Salcedo, M.P.; Schmeler, K.M.; Accorsi, G.S.; Fregnani, J.H.T.G.; Antoniazzi, M.; Pantano, N.P.; Santana, I.V.V.; Matsushita, G.M.; et al. Topical Imiquimod for the Treatment of High-Grade Squamous Intraepithelial Lesions of the Cervix: A Randomized Controlled Trial. *Obstet. Gynecol.* **2021**, *137*, 1043–1053. [CrossRef]
11. Grimm, C.; Polterauer, S.; Natter, C.; Rahhal, J.; Hefler, L.; Tempfer, C.B.; Heinze, G.; Stary, G.; Reinthaller, A.; Speiser, P. Treatment of cervical intraepithelial neoplasia with topical imiquimod: A randomized controlled trial. *Obstet. Gynecol.* **2012**, *120*, 152–159. [CrossRef]
12. Hendriks, N.; Koeneman, M.M.; Van De Sande, A.J.M.; Penders, C.G.J.; Piek, J.M.J.; Kooreman, L.F.S.; Van Kuijk, S.M.J.; Hoosemans, L.; Sep, S.J.S.; De Vos Van Steenwijk, P.J.; et al. Topical Imiquimod Treatment of High-grade Cervical Intraepithelial Neoplasia (TOPIC-3): A Nonrandomized Multicenter Study. *J. Immunother.* **2022**, *45*, 180–186. [CrossRef]
13. Pachman, D.R.; Barton, D.L.; Clayton, A.C.; McGovern, R.M.; Jefferies, J.A.; Novotny, P.J.; Sloan, J.A.; Loprinzi, C.L.; Gostout, B.S. Randomized clinical trial of imiquimod: An adjunct to treating cervical dysplasia. *Am. J. Obstet. Gynecol.* **2012**, *206*, 42.e1–42.e7. [CrossRef] [PubMed]
14. Page, M.J.; McKenzie, J.E.; Bossuyt, P.M.; Boutron, I.; Hoffmann, T.C.; Mulrow, C.D.; Shamseer, L.; Tetzlaff, J.M.; Akl, E.A.; Brennan, S.E.; et al. The PRISMA 2020 statement: An updated guideline for reporting systematic reviews. *BMJ* **2021**, *372*, n71. [CrossRef] [PubMed]
15. Cumpston, M.; Li, T.; Page, M.J.; Chandler, J.; Welch, V.A.; Higgins, J.P.; Thomas, J. Updated guidance for trusted systematic reviews: A new edition of the Cochrane Handbook for Systematic Reviews of Interventions. *Cochrane Database Syst. Rev.* **2019**, *10*, Ed000142. [CrossRef]
16. Munn, Z.; Stern, C.; Aromataris, E.; Lockwood, C.; Jordan, Z. What kind of systematic review should I conduct? A proposed typology and guidance for systematic reviewers in the medical and health sciences. *BMC Med. Res. Methodol.* **2018**, *18*, 5. [CrossRef] [PubMed]
17. McHugh, M.L. Interrater reliability: The kappa statistic. *Biochem. Med.* **2012**, *22*, 276–282. [CrossRef]
18. National Cancer Institute (NIH). Available online: https://ctep.cancer.gov/protocoldevelopment/electronic_applications/ctc.htm (accessed on 8 July 2023).
19. Sterne, J.A.C.; Savović, J.; Page, M.J.; Elbers, R.G.; Blencowe, N.S.; Boutron, I.; Cates, C.J.; Cheng, H.Y.; Corbett, M.S.; Eldridge, S.M.; et al. RoB 2: A revised tool for assessing risk of bias in randomised trials. *BMJ* **2019**, *366*, l4898. [CrossRef]
20. Sterne, J.A.; Hernán, M.A.; Reeves, B.C.; Savović, J.; Berkman, N.D.; Viswanathan, M.; Henry, D.; Altman, D.G.; Ansari, M.T.; Boutron, I.; et al. ROBINS-I: A tool for assessing risk of bias in non-randomised studies of interventions. *BMJ* **2016**, *355*, i4919. [CrossRef]
21. Joanna Briggs Institute (JBI). Critical Appraisal Checklist. 2017. Available online: https://jbi.global/sites/default/files/2019-05/JBI_Critical_Appraisal-Checklist_for_Prevalence_Studies2017_0.pdf (accessed on 7 August 2023).
22. GradePro. Available online: https://www.gradepro.org (accessed on 11 May 2023).
23. Clopper, C.J.; Pearson, E.S. The Use of Confidence or Fiducial Limits Illustrated in the Case of the Binomial. *Biometrika* **1934**, *26*, 404–413. [CrossRef]
24. Higgins, J.P.; Thompson, S.G. Quantifying heterogeneity in a meta-analysis. *Stat. Med.* **2002**, *21*, 1539–1558. [CrossRef]
25. Borenstein, M.; Hedges, L.; Higgins, J.; Rothstein, H. An Introduction to Meta-Analysis. In *Introduction to Meta-Analysis*; Wiley: Hoboken, NJ, USA, 2009; Volume 19.
26. Harrer, M.; Cuijpers, P.; Furukawa, T.; Ebert, D. *Doing Meta-Analysis with R: A Hands-On Guide*; CRC Press: Boca Raton, FL, USA, 2021.
27. Kim, J.H.; Kim, D.Y. Imiquimod as an alternative option for young women with high-grade cervical intraepithelial neoplasia. *Eur. J. Gynaecol. Oncol.* **2019**, *40*, 943–947.
28. Cokan, A.; Pakiž, M.; Serdinšek, T.; Dovnik, A.; Kodrič, T.; Fokter, A.R.; Kavalar, R.; But, I. Comparison of conservative treatment of cervical intraepithelial lesions with imiquimod with standard excisional technique using lletz: A randomized controlled trial. *J. Clin. Med.* **2021**, *10*, 5777. [CrossRef] [PubMed]

29. Lin, C.T.; Qiu, J.T.; Wang, C.J.; Chang, S.D.; Tang, Y.H.; Wu, P.J.; Jung, S.M.; Huang, C.C.; Chou, H.H.; Jao, M.S.; et al. Topical imiquimod treatment for human papillomavirus infection in patients with and without cervical/vaginal intraepithelial neoplasia. *Taiwan. J. Obstet. Gynecol.* **2012**, *51*, 533–538. [CrossRef] [PubMed]
30. Polterauer, S.; Reich, O.; Widschwendter, A.; Hadjari, L.; Bogner, G.; Reinthaller, A.; Joura, E.; Trutnovsky, G.; Ciresa-Koenig, A.; Ganhoer-Schimboeck, J.; et al. Topical imiquimod compared with conization to treat cervical high-grade squamous intraepithelial lesions: Multicenter, randomized controlled trial. *Gynecol. Oncol.* **2022**, *165*, 23–29. [CrossRef] [PubMed]
31. Chan, J.K.; Monk, B.J.; Brewer, C.; Keefe, K.A.; Osann, K.; McMeekin, S.; Rose, G.S.; Youssef, M.; Wilczynski, S.P.; Meyskens, F.L.; et al. HPV infection and number of lifetime sexual partners are strong predictors for 'natural' regression of CIN 2 and 3. *Br. J. Cancer* **2003**, *89*, 1062–1066. [CrossRef] [PubMed]
32. Tainio, K.; Athanasiou, A.; Tikkinen, K.A.O.; Aaltonen, R.; Cárdenas, J.; Glazer-Livson, S.; Jakobsson, M.; Joronen, K.; Kiviharju, M.; Louvanto, K.; et al. Clinical course of untreated cervical intraepithelial neoplasia grade 2 under active surveillance: Systematic review and meta-analysis. *BMJ* **2018**, *360*, k499. [CrossRef] [PubMed]
33. Kitchener, H.C.; Castle, P.E.; Cox, J.T. Chapter 7: Achievements and limitations of cervical cytology screening. *Vaccine* **2006**, *24* (Suppl. S3), 63–70. [CrossRef] [PubMed]
34. Terlou, A.; van Seters, M.; Ewing, P.C.; Aaronson, N.K.; Gundy, C.M.; Heijmans-Antonissen, C.; Quint, W.G.; Blok, L.J.; van Beurden, M.; Helmerhorst, T.J. Treatment of vulvar intraepithelial neoplasia with topical imiquimod: Seven years median follow-up of a randomized clinical trial. *Gynecol. Oncol.* **2011**, *121*, 157–162. [CrossRef] [PubMed]
35. Tranoulis, A.; Laios, A.; Mitsopoulos, V.; Lutchman-Singh, K.; Thomakos, N. Efficacy of 5% imiquimod for the treatment of Vaginal intraepithelial neoplasia-A systematic review of the literature and a meta-analysis. *Eur. J. Obstet. Gynecol. Reprod. Biol.* **2017**, *218*, 129–136. [CrossRef]
36. Committee Opinion No. 675 Summary: Management of Vulvar Intraepithelial Neoplasia. *Obstet. Gynecol.* **2016**, *128*, 937–938.
37. O'Shea, A.S.; Stockdale, C.K. The impact of LEEP margin status on subsequent abnormal cervical cytology. *Proc. Obstet. Gynecol.* **2014**, *4*, 1–8. [CrossRef]
38. Koeneman, M.M.; Essers, B.A.; Gerestein, C.G.; van de Sande, A.J.M.; Litjens, R.; Boskamp, D.; Goossens, M.F.J.; Beekhuizen, H.J.; Kruitwagen, R.; Kruse, A.J.; et al. Treatment of Cervical Intraepithelial Neoplasia: Patients Preferences for Surgery or Immunotherapy with Imiquimod. *J. Immunother.* **2017**, *40*, 148–153. [CrossRef] [PubMed]
39. Kyrgiou, M.; Athanasiou, A.; Kalliala, I.E.J.; Paraskevaidi, M.; Mitra, A.; Martin-Hirsch, P.P.; Arbyn, M.; Bennett, P.; Paraskevaidis, E. Obstetric outcomes after conservative treatment for cervical intraepithelial lesions and early invasive disease. *Cochrane Database Syst. Rev.* **2017**, *11*, Cd012847. [CrossRef] [PubMed]
40. Abdulrahman, Z.; Hendriks, N.; Kruse, A.J.; Somarakis, A.; van de Sande, A.J.; van Beekhuizen, H.J.; Piek, J.M.; de Miranda, N.F.; Kooreman, L.F.; Slangen, B.F.; et al. Immune-based biomarker accurately predicts response to imiquimod immunotherapy in cervical high-grade squamous intraepithelial lesions. *J. Immunother. Cancer* **2022**, *10*, e005288. [CrossRef] [PubMed]
41. Da Silva, R.L.; da Silva Batista, Z.; Bastos, G.R.; Cunha, A.P.A.; Figueiredo, F.V.; de Castro, L.O.; Dos Anjos Pereira, L.; da Silva, M.; Vidal, F.C.B.; Barros, M.C.; et al. Role of HPV 16 variants among cervical carcinoma samples from Northeastern Brazil. *BMC Women's Health* **2020**, *20*, 162. [CrossRef] [PubMed]
42. Gravitt, P.E.; Winer, R.L. Natural History of HPV Infection across the Lifespan: Role of Viral Latency. *Viruses* **2017**, *9*, 267. [CrossRef] [PubMed]
43. Steenbergen, R.D.; Snijders, P.J.; Heideman, D.A.; Meijer, C.J. Clinical implications of (epi)genetic changes in HPV-induced cervical precancerous lesions. *Nat. Rev. Cancer* **2014**, *14*, 395–405. [CrossRef]
44. Rahangdale, L.; Lippmann, Q.K.; Garcia, K.; Budwit, D.; Smith, J.S.; van Le, L. Topical 5-fluorouracil for treatment of cervical intraepithelial neoplasia 2: A randomized controlled trial. *Am. J. Obstet. Gynecol.* **2014**, *210*, 314.e1–314.e8. [CrossRef]

cancers

MDPI

Article

Overcoming Cultural Barriers and Building Healthcare Trust: A Mixed Methods Study on Cervical Cancer Prevention Among Arab American Women Across Major U.S. Cities

Lilah Khoja [1], Manar Zoulfikar [1], Layla Hak [2], Sabrina Yousif [1], Manar Aljebori [1], Matthew Stiffler [3], Madiha Tariq [4], Sarah Burgard [5], Nancy L. Fleischer [1], Lauren P. Wallner [1,6] and Celeste Leigh Pearce [1,*]

1 Department of Epidemiology, University of Michigan School of Public Health, 1415 Washington Heights, Ann Arbor, MI 48109, USA; lkhoja@umich.edu (L.K.); zmanar@umich.edu (M.Z.); sabyou@umich.edu (S.Y.); mjebori@umich.edu (M.A.); nancyfl@umich.edu (N.L.F.); lwallner@umich.edu (L.P.W.)
2 College of Human Medicine, Michigan State University, 804 Service Rd Suite A112, East Lansing, MI 48824, USA; haklayla@msu.edu
3 Center for Arab Narratives at the Arab American National Museum, ACCESS, 6450 Maple St., Dearborn, MI 48126, USA; mstiffler@accesscommunity.org
4 Oakland County Michigan Government, 2100 Pontiac Lake Road, Waterford Township, MI 48328, USA; tariqm@oakgov.com
5 Department of Sociology, University of Michigan, 500 S State St., Ann Arbor, MI 48109, USA; burgards@umich.edu
6 Department of Internal Medicine, University of Michigan, 1500 E Medical Center Dr, Ann Arbor, MI 48109, USA
* Correspondence: lpearce@umich.edu

Simple Summary: Cervical cancer is preventable through a combination of vaccination, screening, and treatment of pre-cancerous lesions. However, not all groups are up to date on their cervical cancer screening. One such group is Arab American women. This study evaluated Arab American women's knowledge and attitudes toward HPV vaccination, cervical cancer screening in general, and HPV self-sampling as a screening approach. A mixed-method approach was employed, using focus groups and an electronic questionnaire. Results revealed varying levels of knowledge about cervical cancer and HPV. Many women who participated in the focus groups were only aware of Pap tests after marriage or immigration. Attitudes towards HPV vaccination were generally positive. Participants expressed trust in healthcare providers for sample collection, doubting self-sampling accuracy. This study highlights the need for culturally tailored education and interventions involving healthcare providers to improve screening rates within this population.

Abstract: Background: Understanding why Arab American women have lower adherence to cervical cancer screening compared to other racial/ethnic groups is important. The study aimed to understand attitudes and knowledge of cervical cancer prevention and HPV vaccination among Arab American women. Methods: A mixed-method approach was employed, including nine focus groups and an online questionnaire. Demographic characteristics, medical history, screening practices, and attitudes towards HPV vaccination and HPV self-sampling for cervical cancer screening were assessed. Results: Focus group participants (n = 22) demonstrated varying levels of knowledge about cervical cancer and HPV, including limited awareness of the Papanicolaou (Pap) test. Participants expressed mixed feelings about HPV self-sampling. Among questionnaire respondents (n = 25), who on average had a higher socioeconomic status than focus group participants, 73.9% had undergone a Pap test, with 94% up to date on screening. While 59% preferred self-sampling at home, almost two in five cited concerns about sample accuracy. Conclusions: Our study demonstrates the variability in attitudes and experiences towards cervical cancer screening among Arab American women, potentially driven by socioeconomic disparities. Our qualitative results suggest the need for targeted, culturally tailored health education in the Arab American community. Further research should explore effective strategies to engage this underserved group and enhance adherence to prevention programs.

Keywords: cervical cancer; self-collection; HPV; cancer screening; Arab American; women's health

1. Introduction

Cervical cancer prevention is one of the biggest public health success stories. The Papanicolaou test ("Pap test") was shown to be efficacious as a screening method in 1941 [1] and, since 1975, has been credited with the dramatic decline in cervical cancer incidence and mortality in the United States [2]. In 2014, the United States Food and Drug Administration (FDA) approved a human papillomavirus (HPV) DNA test for use as a primary cervical cancer screening test [3]. Both screening methods are now approved by the United States Preventive Services Task Force, though HPV testing is not recommended for people under 30 [4]. Recently, the FDA approved self-collection of cervical samples for HPV testing as a form of cervical cancer screening [5].

In the United States (U.S.), cervical cancer screening is higher in the non-Hispanic White population compared to non-White groups [6]. Based on results from the 2014, 2016, and 2018 Behavioral Risk Factor Surveillance System data [7], Black and Hispanic women are less likely to be screened compared to their non-Hispanic White counterparts. The disparities in screening mirror differences in cervical cancer incidence in the U.S.; between 2000 and 2018, cervical cancer incidence rates were 13.8 per 100,000 for Hispanic women and 11.8 per 100,000 for Black women, compared to 6.8 per 100,000 for non-Hispanic White women [7].

Rates of cervical cancer screening are not known for all groups in the U.S. Historically, Arab Americans have been categorized as non-Hispanic White, thus obscuring their health trends due to the larger proportion of non-Hispanic Whites who are predominantly of European descent. However, because the Arab American population in Michigan is large, the Behavioral Risk Factor Surveillance System has conducted two Arab Michigan Behavioral Risk Factor Surveys [8,9]. Based on data from the most recent survey (2016), cervical cancer screening rates in Arab American women are lower than for non-Hispanic, non-Arab White women in Michigan, with only 64.2% of Arab American women in Michigan ever having a Pap test compared to 90.5% of non-Hispanic, non-Arab White women in the state [9].

Given the concerningly low cervical cancer screening rates among Arab American women in Michigan, we previously conducted focus groups in the state to better understand attitudes and beliefs around cervical cancer screening and, more broadly, cervical cancer prevention [10]. Several themes were revealed, notably that there exists a dearth of health literacy, knowledge, and materials related to women's health, an ambivalence towards vaccination, and areas of potential intervention. We found a need for the creation of appropriate health materials and education about women's health within the community.

However, the Arab American population is heterogeneous [11,12], and it is unclear whether these themes reflect the experiences of Arab American women across the US or if there exist geographic differences in attitudes and beliefs about cervical cancer screening. To address this gap, we carried out an additional qualitative study among Arab American women residing in Southern California or the Chicago and New York metropolitan areas to understand their views and knowledge about cervical cancer screening. Questions about

HPV vaccination were also included to provide information more broadly about cervical cancer prevention.

Research conducted in recent years suggests that HPV self-sampling is a feasible approach to reach populations that are under-screened for cervical cancer [13–15], like the Arab American population in Michigan. During our Michigan focus groups, we discussed HPV self-sampling, and while participants voiced concerns about self-efficacy, they were also cautiously interested. We therefore conducted a study to ascertain more detailed quantitative data on cervical cancer prevention and risk factors in the Arab American populations in Michigan, Southern California, Chicago, and the New York metropolitan areas. Herein, the results of our qualitative and quantitative studies are reported.

2. Materials and Methods

A mixed-method study with both quantitative and qualitative components, the Study of Arab American Health Attitudes-Human Papillomavirus (SAHA-HPV), was designed to evaluate the acceptability and feasibility of HPV self-sampling as an alternative to traditional forms of cervical cancer screening. SAHA-HPV is an exploration of Arab American women's knowledge and attitudes toward cervical cancer, traditional cervical cancer screening, self-collection as a form of screening, and the HPV vaccine. The qualitative data collection and analysis were completed prior to the launch of the quantitative component. As such, the study is de facto an emergent exploratory sequential mixed-methods study. Recruitment into the qualitative and quantitative studies was independent; thus, there was little to no overlap in participants. We cannot explicitly determine overlap because identifying information on participants in the qualitative study is not available by design.

2.1. Qualitative Arm: Study Population and Procedures

Women who identified as Arab American and Chaldean American aged 30–65 years old and who lived in Southern California or the New York or Chicago metropolitan areas were eligible to participate. The qualitative component of the study consisted of focus groups conducted remotely by Zoom (n = 9). Participants were recruited between October 2022 and January 2023 by word of mouth, informational flyers, and through community-based partners. The informational flyers were shared on Arabic-language and Arab-American-interest-focused social media, inviting women who resided in the three metropolitan areas to participate. For two of the focus groups, participants were recruited through The Arab American Family Support Center, a community-based partner located in Brooklyn, New York. Nine focus groups consisting of three to six women each were conducted between October 2022 and February 2023. All study materials were available in Arabic and English. The number of focus groups was chosen to best achieve data saturation [16]. Ultimately, the decision was made to keep the focus groups small (with four participants or less) due to the sensitive nature of the research. This decision was further guided by guidelines and recommendations for conducting online focus groups [17].

All potential focus group participants were provided with a study informational sheet outlining the study goals, benefits, and risks in Arabic and English one week ahead of the scheduled focus group time. Upon the start of each focus group, the study team discussed the information sheet with each participant privately in a Zoom breakout room and ensured that each participant gave informed verbal consent prior to participating in the focus groups. Participants received a USD 75 gift card to a vendor of their choice (Amazon, Target, Visa) for participating.

After receiving verbal consent, participants were asked to complete a brief anchoring survey with demographic questions and a self-assessment of HPV knowledge. A focus group guide served as an overall framework for the focus group. The details of the guide have been previously published, but briefly, it covered beliefs about cervical cancer and HPV risk, attitudes towards prevention, including self-collection and vaccination, and barriers to screening.

Seven focus groups were completed in English or a mix of English and Arabic, and two were conducted in Arabic only. The two Arabic-exclusive focus groups consisted of participants recruited through our New York community-based partner. All but one participant completed the anchoring survey (n = 21). The focus groups were led by LK.

2.2. Qualitative Arm: Data Processing and Analysis

Each focus group was recorded, then transcribed and translated as needed. Transcripts were prepared without identifying information. Coding and analysis were conducted using Dedoose Version 9.0.017 [18]. Four members of the study team participated in the coding and analysis (LK, MZ, SY, and LH).

An iterative process drawing from grounded theory was used to analyze the focus groups. Grounded theory [19] is an iterative approach to the collection, analysis, and interpretation of data that is structured yet flexible and is appropriate when little or none is known about the phenomenon under study.

First, three of the nine focus groups were coded independently by four team members using an open coding process with descriptive codes. After a team discussion around the application of codes, a codebook was developed. This codebook was further refined upon more coding of the focus groups. All focus groups were ultimately coded with the final codebook by at least two coders. Code reports were then prepared. Thematic content and ethnographic analysis techniques were applied to identify patterns and themes across several team meetings and discussions. Inter-coder reliability was assessed qualitatively [20].

2.3. Quantitative Arm: Subject Recruitment

For the quantitative arm, women who identified as Arab American and Chaldean American, aged 30–65 years old, who lived in Michigan, Southern California, or the Chicago or New York metropolitan areas were eligible to participate. Women who had undergone hysterectomy or had previously been diagnosed with cervical cancer were ineligible to participate. All study materials were available in Arabic and English. Participants were recruited between August 2023 and June 2024, both in-person and remotely.

In-person recruitment took place primarily in Dearborn, Michigan, at Arab Community Center for Economic and Social Services (ACCESS), a community-based partner. Flyers were also dropped off and posted at local women's-only beauty salons and a local health clinic in Southwest Michigan. Participants were also recruited via word-of-mouth, through ACCESS referrals, and through various social media channels and influencers on Instagram, TikTok, and Facebook that are targeted to the Arabic-speaking populations within Michigan, Southern California, and the Chicago and New York metropolitan areas. Furthermore, influencers, defined as women with public-facing social media who had at least 1500 followers, living in the Chicago metropolitan area and catering to a predominantly Arab identifying population, also shared the IRB-approved fliers on their social media accounts. A USD 10 Visa gift card incentive for enrolling in the study and completing the questionnaire was offered.

2.4. Quantitative Arm: Study Procedures

When conducting in-person recruitment, women were verbally asked if they qualified and thus did not complete the written screening survey. The number of women approached in person was not documented. Most women approached through in-person recruitment were not eligible because they did not meet the minimum age requirement of 30. For those recruited in person, participants were taken to a private area and a member of the study team went over the informed consent form and made sure the participant fully understood the study procedures. After signing the consent, the participants were then given access to a computer by the study team to complete the questionnaire. Participants recruited remotely were prompted to review the informed consent hosted on Qualtrics after establishing eligibility; after signing the consent, participants were redirected to the questionnaire, which was also hosted on Qualtrics. All participants received an electronic copy of their consent form.

A total of 136 women completed at least part of the electronic screening questionnaire. Of these 136, 8 women did not fully complete it, and 69 were ineligible for at least one of the following reasons: not identifying as Arab or Chaldean (5.8%), falling outside of the eligible age range (11.0%), having had a previous cervical cancer diagnosis (13.2%), and having had a hysterectomy (15.4%). The remaining fifty-nine (43.4%) women were eligible to participate, but eight women, six from Illinois and two from California, declined to participate and did not ultimately sign the consent form. Of the 51 participants who signed the consent form and agreed to participate, 25 (49%) completed the questionnaire.

The self-administered questionnaire included questions about demographic characteristics, medical history, sexual practices, prior cervical cancer screening practices, cervical cancer risk factors, HPV vaccination practices, and interest in receiving a self-collection kit via mail. The questionnaire was modified from a previously used instrument assessing the feasibility and acceptability of HPV self-collection in minority communities, including Muslim communities, and in Guatemala and Thailand [21–23].

2.5. Quantitative Arm: Data Processing and Analysis

Results from questionnaire modules that covered topics parallel to the focus groups are presented with frequencies and proportions. Data were processed and analyzed using Microsoft Excel [24] and RStudio Version 04.24 [25].

2.6. Ethical Considerations

The qualitative component of the study was approved by the University of Michigan Health Sciences and Behavioral Sciences Institutional Review Board (IRB-HSBS HUM00150710) and granted a determination of exemption under Federal Exemption 2, which does not require signed informed consent. The quantitative arm of the study was initially approved by the University of Michigan Health Sciences and Behavioral Sciences Institutional Review Board (IRB-HSBS) and later transferred to and approved by the Institutional Review Boards of the University of Michigan Medical campus (IRBMED HUM00147147). The change was due to an institutional decision that all HPV self-sampling studies would fall under the purview of the IRBMED.

3. Results

3.1. Quantitative Study

Recruitment into the quantitative arm of this study proved very difficult despite the multiple recruitment strategies employed. Less than half of the women who consented to participate went on to complete the questionnaire. The descriptive characteristics of the 25 participants who completed the questionnaire are shown in Table 1. The average age of the participants in the quantitative study was 39.5 years (SD: 10.5). Over half were born in the U.S. (n = 14; 56%), and the majority of participants identified as Muslim (76%). In their households, most participants spoke either Arabic (n = 8) or a mixture of Arabic and English (n = 15); only two households were exclusively English-speaking. The participants were highly educated, and more than half indicated their household income was greater than USD 75,000 (see Table 1).

Table 1. General characteristics of questionnaire respondents (N = 25).

Measure	N (%)	
Age (mean/standard deviation)	39.5	10.5
Born in the United States	14	58.0%
Year of immigration *		
Between 1988 and 1997	1	4.0%
Between 1998 and 2007	4	16.0%
Between 2008 and 2012	2	8.0%
Between 2013 and 2019	2	8.0%
State of Residence		
California	8	32.0%
Illinois	6	24.0%
Michigan	9	36.0%
New York/New Jersey	2	8.0%
Religious Affiliation		
Christian	5	20.0%
Muslim	19	76.0%
Do not know/prefer not to answer	1	4.0%
Languages spoken at home		
Arabic	8	32.0%
Arabic and English	15	60.0%
English	2	8.0%
Arabic language fluency		
Can read and write	22	88.0%
Can read only	2	8.0%
Can neither read nor write	1	4.0%
Marital Status		
Married	14	56.0%
Single	7	28.0%
Divorced	3	12.0%
Education		
Advanced/post-graduate degree	12	48.0%
University/college 4 years or more (university/college graduate)	5	20.0%
University/college 1 year to 3 years (some university/college or technical school)	2	8.0%
Grade 12 or GED (high school graduate)	3	12.0%
Grades 9 through 11 (some high school)	1	4.0%
Income		
USD 75,000 or more	14	56.0%
USD 50,000–USD 74,999	3	12.0%
USD 35,000–USD 49,999	1	4.0%
USD 20,000–USD 34,999	1	4.0%
USD 15,000–USD 19,999	1	4.0%
USD 10,000–USD 14,999	2	8.0%
Less than USD 10,000	1	4.0%
Do not know/prefer not to answer	1	4.0%

* Complete data missing for two participants.

3.2. Quantitative: Understanding of HPV and Cervical Cancer and Their Risk Factors

Among questionnaire respondents, 40.1% indicated that all women aged 21–65 should be screened for cervical cancer regardless of their sexual history. Over a quarter of respondents endorsed that cervical cancer is genetic, with 26.1% of questionnaire respondents indicating that women with a family history of cervical cancer should be screened. (see Table 2). Most of the participants who completed the questionnaire had heard of HPV, which is in stark comparison to the focus group results, where knowledge of the virus appeared lacking.

Table 2. Questionnaire respondents' understanding of risk factors—questionnaire (N = 23 *).

Measure	N	(%)
Who should be screened for cervical cancer? **		
All women regardless of age	4	17.4%
All women beginning of puberty	6	26.1%
All women when they begin their sex lives	7	30.4%
All women ages 21–65 regardless of sexual history	9	39.1%
Only women who are more sexually active	3	13.0%
Women with a family history of cervical cancer	6	26.1%
Older women	2	8.7%
Menopausal women	2	8.7%
Have you heard of HPV?		
Yes	20	85.0%
No	3	15.0%
Which of the following increases your risk for HPV (human papillomavirus) infection?		
If you have many sexual partners	17	73.9%
If your partner has many sexual partners	13	56.5%
If you begin your sex life at an early age	3	13.0%
Birth control pills	4	17.4%
Smoking	3	13.0%
Excessive stress	6	26.1%
Poor nutrition	4	17.4%

* Two participants were missing data for this module. ** Will sum to greater than 22 because participants were able to choose more than one option.

Among the questionnaire respondents, 73.9% reported that having many sexual partners increases one's risk for an HPV infection. Fewer participants (56.5%), but still a majority, reported that one's partner having many sexual partners increases the risk for HPV.

3.3. *Quantitative: Preventive Health Behaviors and Attitudes*

3.3.1. Screening

Approximately three-quarters of the questionnaire respondents reported ever having a Pap test (73.9%), which was similar to the focus group participants, where most reported having been screened at least once. Among respondents, 94% were up to date on their screening. Overall, 73.9% of respondents indicated that a doctor's recommendation would be most encouraging to get screened for cervical cancer (see Table 3).

3.3.2. Vaccination Against HPV

The same findings from the focus groups are echoed in the questionnaire results: 78.3% of questionnaire respondents had heard of the HPV vaccine. While most questionnaire respondents were in favor of vaccinating and immunizing their children, 30% responded "no" or "don't know/prefer not to answer" (See Table 4).

Table 3. Questionnaire respondents' views on cervical cancer screening (N = 23 *).

Measure	N	(%)
Have you ever had a pap smear?		
Yes	17	73.90%
No	5	21.70%
Do not know/prefer not to answer	1	4.30%
How long has it been since you had your last pap smear?		
Within the past 5 years	1	5.90%
More than 5 years ago	16	94.10%
Which of the following would encourage you to get screened for cervical cancer? **		
Your doctor recommending it	17	73.90%
Being able to test yourself at home	8	34.80%
A friend or family member telling you why screening is important	3	13.00%
Being able to do it for free	8	34.80%

* Two participants were missing data for this module. ** Participants could select more than one response.

Table 4. Questionnaire respondents' views on HPV vaccination (N = 23 *).

Measure	N	(%)
Have you heard there is a vaccine against HPV (human papillomavirus)?		
Yes	18	78.30%
No	5	21.70%
Are you in favor of vaccinating and immunizing your children?		
Yes	16	69.60%
No	4	17.40%
Do not know/prefer not to answer	3	13.00%
Would you vaccinate your children against HPV?		
Yes	17	73.90%
No	3	13.00%
Do not know/prefer not to answer	3	13.00%

* Two participants were missing data for this module.

3.3.3. Self-Collection

Participants in the quantitative study largely shared the same attitudes regarding self-collection as participants in the qualitative component of the study. Collection at a doctor's office was preferable for 59% of the participants, and 68.2% stated that they would prefer the doctor to perform a swab for an HPV test as opposed to themselves. Like the focus group sentiments, participants who completed the questionnaire stated that they would be worried that they did not collect the sample properly (see Table 5).

The quantitative study provided a broad overview of participants' demographic characteristics, knowledge, attitudes, and behaviors regarding HPV and cervical cancer. However, additional insights are necessary to fully understand the nuances within this population. Thus, the qualitative focus groups were designed to explore in greater depth the themes emerging from the quantitative data, such as the barriers to screening and vaccination, the role of healthcare providers in influencing health behaviors, and the cultural and social stigmas associated with women's health. By conducting focus groups across different metropolitan areas, we aimed to capture a diverse range of perspectives and provide a richer, more detailed narrative that complements and contextualizes the findings of the quantitative study.

Table 5. Self-collection attitudes—questionnaire (N = 23 *).

Measure	N	(%)
Have you heard there is a vaccine against HPV (human papillomavirus)?		
Yes	18	78.3%
No	5	21.7%
Are you in favor of vaccinating and immunizing your children?		
Yes	16	69.6%
No	4	17.4%
Do not know/prefer not to answer	3	13.0%
Would you vaccinate your children against HPV?		
Yes	17	73.9%
No	3	13.0%
Do not know/prefer not to answer	3	13.0%

* Two participants were missing data for this module.

3.4. Qualitative Study

The Southern California focus groups (n = 3) consisted of six women, the New York focus groups (n = 3) consisted of nine women, and the Chicago focus groups (n = 3) consisted of seven women. Each focus group consisted of at least two participants. The mean age of participants was 37.6 years (standard deviation: 8.7), 61.9% were married, and 66.7% had children. The anchoring questionnaire was filled out in Arabic by 38% of participants. Approximately half of the participants described themselves as being "moderately informed" or "well informed" about cervical cancer (see Table 6).

Table 6. Focus group participant characteristics (N = 22 *).

Measure	N	(%)
Location		
New York metropolitan area	8	38.10%
Southern California	7	33.30%
Chicago metropolitan area	6	28.60%
Anchoring survey language		
English	13	61.90%
Arabic	8	38.10%
Age, mean and standard deviation	37.6	8.7
Marital status		
Married	13	61.90%
Not married	8	38.10%
How well would you say the financial needs are met in your household?		
Not met at all	4	19.10%
Barely making it	5	23.80%
Just sufficient	8	38.10%
More than sufficient	4	19.10%
How well-informed do you think you are about cervical cancer?		
Not informed at all	1	4.80%
Not well informed	9	42.90%
Moderately informed	8	38.10%
Well informed	3	14.30%

Table 6. *Cont.*

Measure	N (%)	
Have children		
Yes	14	66.70%
No	7	33.30%

* Data missing for one participant.

3.5. *Qualitative: Understanding of HPV and Cervical Cancer and Their Risk Factors*

The focus group discussions covered the understanding of HPV and cervical cancer, as well as their risk factors and who is at risk. Some participants in the focus groups described the number of sexual partners as a risk factor for cervical cancer, whereas others cited genetics and a lack of cleanliness (see Table 7). Additionally, fewer participants in the focus groups seemed aware of HPV; many had neither heard of it before nor knew that there was a test or vaccine for it.

Table 7. Focus group excerpts: understanding of HPV and cervical cancer risk factors.

"Anyone who is active with more than one partner."—CA #1
"Those who are not vaccinated against HPV."—IL #1
"Women who are of age to have sex."—CA #2
"You have a genetic predisposition"—CA #3
"It might be genetic"—IL #2
"It might be from a lack of cleanliness"—NY #2

3.6. *Qualitative: Preventive Health Behaviors and Attitudes*

3.6.1. Screening

Participants in the focus groups had a mixed understanding of cervical cancer screening, although the majority had undergone screening at least once. The majority of participants stated that they follow the guidelines and what is recommended to them by their physician. This was the driving factor behind their decision to undergo cancer screening (see Table 8).

Table 8. Focus group excerpts: screening.

"For me I just follow the guidelines like annually I do my mammogram and I do my Pap smear you know that the-the annual test I do for myself. Prevention."—CA #1
"My gynecologist she-she says that she recommends—whatever she doe-she says I have to do."—CA #3
"The doctor requests it from me, like as a check-up."—IL #2

The focus group leader described the Pap test; some participants were not aware of what a Pap test was and had never discussed it with their healthcare providers. Others stated that they first heard of Pap tests when they came to the U.S. Other participants stated that the first time they had a Pap and gynecologist visit was following marriage (see Table 9).

Table 9. Focus group excerpts: awareness of Pap tests.

"The first time I got it was when I was married."—IL #1
"This is the first time I hear of [it]. . . I went to many doctors, and nobody has requested this from me."—NY #1
"I don't remember anyone asking me to do it in the Arab countries—I lived in Damascus, Saudi Arabia, and Kuwait, and nobody ever brought this up to me [. . .] this was in approximately 2012."—IL #2
"I did bring it up to my mom once before I got married and she said only married girls get the test."—IL#1

3.6.2. Vaccination Against HPV

Most participants had heard of the HPV vaccine. Attitudes towards the HPV vaccine, and vaccination, broadly, were varied. Among focus group participants who had heard of the HPV vaccine, the majority expressed sentiments in favor of the vaccine and had either been vaccinated or had vaccinated their children. Focus group members who were in favor of vaccination, for themselves and for their children, discussed viewing it as a form of prevention. Several participants cited wanting to protect their children against future health problems (i.e., cancer in part due to an HPV infection).

However, there were several vaccine skeptic sentiments expressed by focus group members. Vaccine skeptic sentiments were expressed about both some and all vaccines and were mostly related to side effects. Concerns about side effects were mostly due to fears of infertility. Several participants described their families' vaccine skeptic attitudes. Among some participants, the HPV vaccine's link to promiscuity was mentioned as a reason for vaccine skepticism within their community and themselves.

Additionally, several women mentioned the HPV vaccine never being brought up to them or their children by their healthcare providers (see Tables 10 and 11).

Table 10. Focus group excerpts: attitudes towards vaccination.

In Favor of Vaccination	Vaccine Skeptic
"I can't prevent anything so I would rather be very safe and make sure that you know he gets any like all the vaccines especially this vaccine to save his life, you just never know."—IL #2	"Sometimes girls get [this thing] and become infertile."—NY #1
"In this case I believe in this vaccine definitely because nowadays uh teenagers they don't think, they're very into sex and you know so I think it's-it's a good idea to do that for them definitely."—CA #3	"I don't believe too much in the vaccine of uh, coronavirus."—CA #3
"A few of the women that I'm friends within community with have gotten HPV like after they got married because their husbands didn't know they had it either, so when I got engaged last year I was like you know what I was like I'm gonna vaccinate myself just in case to protect myself."—IL #1	"I had a younger cousin that took the vaccine and another person form the family said, "Oh there goes that, she's not going to have babies now."—IL #1
"I don't know how to say that but the temptations happen in life you—you—you always need to protected is what I'm trying to say and uh yeah so I'm all for the vaccine."—IL #3	"I believe it was offered to my daughter when maybe she was 16/17 and I struggled with it because you know we don't necessarily, I felt like this is probably important for uhh sexually active people and we tend—you know were—we're not, you know, unless you're married so—so I struggle with it."—CA #1

Table 11. Focus group excerpts: healthcare providers and vaccine encouragement.

"No to be honest, I take all my daughters to the doctor and he explains to me but I have never heard of a vaccine against cervical cancer. I ask him every time, "Do I have any vaccines?" and I ask about vaccines that are for protection and never in his life has told me about a vaccine for cervical cancer, no."—NY #2
"I haven't heard of it, no doctor told me about it. I don't know."—IL #2
"When I was growing up, that wasn't offered to me by my pediatrician"—CA #3

3.6.3. HPV Self-Collection

Focus group participants appeared to have ambivalent attitudes towards HPV self-collection. While the ease and convenience of HPV collection as an alternative to Pap tests was mentioned, participants seemed to place higher trust in results from a healthcare provider. Others noted that they would use it only in certain circumstances, such as when gender-concordant care is unavailable (see Table 12).

Table 12. Focus group excerpts: self-collection attitudes.

"If we can do this at home, it's nice for us to do it at home. If a woman is employed or if she has kids and she can't leave but at the same time if you have the time to go to your monthly appointments with the doctor, I would prefer to do it with the doctor."—NY #1
"I'd prefer to go do it at the doctor, if the doctor isn't available, or the female doctor, then I'd do it, I don't have a problem doing it, but I'd prefer to be in the office I feel—I don't know I feel it'd be more accurate."—IL #2
"Maybe I don't know. but it's better to do it at the female doctor."—NY #2
"If there's like only a mem—male provider at the time I would like yeah prefer using this."—NY #3
"I think it makes it a lot more accessible than having to make a doctor's appointment [...] you know, it's—it's just another tool that you can access yourself directly and administer yourself and be able to do that privately I think that's important."—CA #2

3.7. *Further Insights from the Focus Groups*

Within the focus group discussions, participants discussed the stigmas they grew up with related to women's health. Participants discussed how women's health in their communities is broadly ignored due to stigmas surrounding sexual behavior and virginity. Several participants, however, noted that there was a generational divide with respect to stigmas (see Table 13).

Table 13. Further insights from the focus groups: taboo and stigma.

Theme	Representative Excerpt
Taboo around women's health	"We don't have topics open among us. We have, that when a girl, she wants to get married, she doesn't know what's going on with her."—IL #2 "I wasn't personally even told about my period growing up. You know, I have a mom and many sisters, and I think there's a lot of cultural stigma around reproductive and sexual health."—CA #2

Table 13. *Cont.*

Theme	Representative Excerpt
Impact of taboo on health-seeking behavior	"I even had friends who had you know like were dealing with really horrifying health issues with extremely heavy periods or dealing with—with lots of pain um pain obviously somewhere in their ovaries or somewhere in their—in their kind of stomach area and their moms wouldn't allow them to go get even a Pap smear or get an examination and they were just living through this pain."—CA #2 "Still, you know this is their viewpoint, they feel like nice girls don't need this, they don't need—they don't need to take this vaccine."—CA #3 "These things are important in our communities. Sort of—really prevent care on a very significant level for women um and access to care just because of the sort of social stigmas around them and just lack of education like I think a lot of women don't even have conversations around safe sex and you know STIs and basic birth control conversations within our communities so all of that ties into this including this sort of preventative care."—CA #2
Inter-generational impact of stigma	"I've talked to some of my sisters about this and they're definitely, you know they—they've experienced the same thing, a lack of education and the lack of clarity around these things, um, and they've you know had to learn things in their own way."—CA #2

One participant even detailed her panic at seeing a cousin at the gynecologist's office, reflecting the stigma of gynecologic health:

> "You know the Arab community sometimes gossips and stuff, so I was super worried I don't know why I was so worried, but it was like my second cousin's daughter or something like that and she was pregnant and she was in the waiting room. I wasn't married, I was just getting my vaccines at the time and I was like... I was like trying to talk to the clerk on the way out, and I was trying to say it out loud like "Yeah, the vaccine"—just like, I don't know I was like trying to make sure she knows I'm not there for like an STI or other thing."

When asked what would have the biggest impact on improving understanding and awareness of women's health, to hopefully improve health outcomes, participants discussed the need for community education and open and honest conversations within their families, with their children, and within the community (see Table 14).

Table 14. Focus group excerpts: further insights related to improving awareness of women's health.

Theme	Representative Excerpts
Need for community education	"Trying to reach people. Like for example maybe like women's groups in mosques, if they have like female healthcare providers coming to talk to women in the spaces that they want to reach about like the risks and the importance of it."—NY #3 "I think reaching out to the communities you know through uh I—through mosque—mosques probably would be the first approach since—so because you know from my point of view, uh like, this ignorance is among people who do not want to educate themselves."—IL #3

Table 14. *Cont.*

Theme	Representative Excerpts
Open, honest conversations	"Having honest conversations with each other, not being ashamed to talk about like health problems probably. . . . I think just people who are willing to go up and be embarrassed and talk about these things in public." CA #3 "You want your family and your friends to be healthy so it's good to talk about it openly."—CA #3 "I've definitely had those conversations with people in my age group and people I'm close to um and I feel like people in my age group are more willing to speak about these things and also are the ones who are more informed and educated on it."—NY #3 "I think also talking about it more like for both its good and bad like Instagram, social media a lot of those things will talk about it more. I think people are willing to learn more about their struggles with their health publicly when they hear others talk about it."—NY #3 "You know some health concerns in the last few years with my cervical health so it's kind of become the backdrop for me to speak more openly with my family about it and especially now that I have lots of nieces—um—that, you know, my sisters and my sister-in-law have definitely discussed more preventative um care with them."—CA #2
Open conversations with children	"I have daughters and I've been talking to them any question they have I just answer I'm just very open with them. And I was just like I'm not gonna raise my daughters the way I was brought up its gonna be different this time."—IL #1 "Every mother [should] to educate her daughter."—NY #2

The quantitative and qualitative results of this study provide complementary perspectives. In our quantitative study, a significant majority of respondents (85%) had heard of HPV, in stark contrast to the qualitative findings where focus group participants often lacked awareness of HPV and its vaccine. Quantitatively, 73.9% of respondents reported ever having a Pap smear, aligning closely with the qualitative insights where most focus group participants also had undergone cervical cancer screening, motivated predominantly by physician recommendations, further emphasizing the critical role healthcare providers play in preventive health behaviors. Both datasets highlight concerns about self-collection, with 68.2% of questionnaire respondents preferring doctor-administered tests, echoed by focus group participants who preferred clinic-based screenings for accuracy and trust. Additionally, while the quantitative results indicate high levels of cervical cancer screening adherence, qualitative findings reveal underlying issues such as stigma and lack of education, which impact health-seeking behavior, especially among younger women and recent immigrants. Combining these insights provides a nuanced understanding of the barriers and motivators influencing cervical health behaviors in this population, pointing towards the need for community-based education and open conversations to overcome cultural taboos and enhance preventive care outcomes.

4. Discussion

The qualitative and quantitative findings combined provide us with a rich picture of Arab American women's attitudes towards awareness and understanding of cervical cancer, HPV, and their risk factors; attitudes towards preventative health and preventative health behaviors, including HPV self-collection; as well as important community insights and how to improve screening uptake within the Arab American community. The moderate awareness of HPV and cervical cancer risk factors indicates that there is work yet to be completed with regard to health education. Indeed, participants themselves cited a need for education within the community. For example, some focus group participants expressed a belief that cervical cancer is related to genetics. Generally, cervical cancer is not a hereditary disease, and any genetic component is likely related to a small fraction of the population who are more or less susceptible to HPV infection. Thus, clarifying this misconception is important as it may serve as a deterrent to screening if people perceive themselves as not at risk.

Contrary to what has been found in the literature, the majority of our questionnaire respondents were up to date on their cervical cancer screenings [8,9,26–29], though it is likely that this is not representative of Arab American women, as we know Arab American women in Michigan, for example, have much lower rates of being up to date on their cervical cancer screenings than their non-Hispanic, non-Arab White, Black, and Hispanic counterparts [9]. This discrepancy can likely be explained by the relatively higher socioeconomic status of the questionnaire respondents compared to the focus group participants. There exist socioeconomic disparities around income and education within the Arab American community that likely contribute to the discrepancy in observed rates [30].

Some facilitators and barriers to screening identified in this study are not dissimilar to what has been identified in other immigrant and minority groups, such as the impact of physician recommendations [31,32] and lack of knowledge [32,33]. Some differences exist, however. For example, barriers such as lack of insurance were not identified as a barrier to screening within our study populations, while it has been identified as one for many other underserved groups in the U.S. [34], and competing interests for time did not emerge as a primary concern or barrier either [33]. Perceived discrimination was also not identified as a barrier in our study [35]. However, the lack of nationally representative data on Arab Americans makes it hard to truly contextualize our findings.

An interesting finding that emerged in both the focus groups and the questionnaire findings is the hesitancy towards HPV self-collection as an alternative method of screening. This finding differs from other studies that have found broad acceptability of self-collection as an alternative method of screening [14,21–23,36–39] but is in line with what we found in our previous study conducted in Michigan [10]. However, our findings could differ due to perceptions of self-efficacy: several participants in the focus groups and a high percentage of participants in the survey indicated a reason for not wanting to use self-collection was due to the fear of collecting the sample incorrectly. This hesitancy towards self-collection sheds important light on the type of intervention that is likely to work within this community. Given the recent emphasis on self-collection for cervical cancer screening (e.g., the National Cancer Institute's "Last Mile" Initiative), exploring these barriers warrants further investigation, as self-collection may not improve rates in the populations of women who are at greatest risk of going unscreened.

Attitudes towards vaccination were positive in the quantitative and qualitative components of the study, though some vaccine skeptic sentiments were present. COVID-19 and other vaccine-hesitant sentiments have been previously documented in the Arab American community [40–42]. In this sample, vaccine skepticism appears to be linked to misinformation. Our findings are largely aligned with those from a cross-sectional study carried out in New York [43]. They found that the primary reasons for not vaccinating a child against HPV among Arab American mothers were a lack of awareness of the vaccine and a lack of provider recommendations [43]. Mixed attitudes towards HPV vaccination were also observed in our previous study conducted in Michigan in which vaccine hesitancy was noted [10].

A key theme that emerged in the focus groups and in our prior study [10] is the importance of physician recommendations to this population, for both vaccination and screening. This finding is in line with other research: a cross-sectional study conducted in New York found that Arab American mothers would be more likely to vaccinate their children after a doctor's recommendation [43]. Furthermore, in a study conducted in Cleveland, Ohio, among Arab refugee women, physician recommendation was found to be one of the top perceived facilitators of cervical cancer screening [44]. Thus, it seems that efforts and interventions that support physicians in communicating with patients about the benefits of screening and vaccination are warranted.

Physician's offices appear to be a safe space for Arab American women. Materials can be developed for physicians and other healthcare workers to help them encourage their Arab American patients to get screened for cervical cancer, regardless of their age and childbearing status. There are other avenues of intervention that are worth exploring within this population. For example, culturally sensitive health presentations at mosques were suggested by our participants. These would take place during women's-only events. This would engage the women in a safe space. Another safe space to engage the women in is at the physician's office. As inadequate knowledge was identified as a barrier for Arab American women, a friend-based intervention wherein women inform each other about cervical cancer and encourage each other to get screened could also be impactful.

Interestingly, per the focus groups, there did not appear to be any differences by geography with respect to attitudes toward prevention and cervical cancer, including our previously published work in Michigan [10]. However, two patterns were observed. First, the immigrant experience with respect to women's health markedly differed from that of subsequent U.S.-born generations. While their U.S.-born children may benefit from health curriculum in school and access to resources, our focus group findings illustrate that there is still an impact on their children's health knowledge and literacy. Second, our findings also suggest that socioeconomic status may contribute to differences in attitudes and behavior within this population. We were unable to assess differences in geography or socioeconomic status quantitatively due to insufficient numbers. The focus groups, relative to the quantitative data, allowed for the exploration of more nuanced topics. Participants discussed their communities' attitudes and stigmas towards women's health and ways to improve screening rates and health awareness within the community. In particular, participants discussed the taboo surrounding women's health within their communities. Participants cited that as a reason for poor community uptake but also as a major problem as the taboo impacts health-seeking behaviors. A study evaluating the impact of disclosure-oriented sexual health interviews with Arab American women found that 55% of participants had never discussed sexual health with their physician [45]. This is also in line with what we observed in our previously conducted Michigan study [10].

To the best of our knowledge, this is the first study of its kind: a mixed-method approach to understanding the awareness and knowledge of HPV and cervical cancer risk factors as well as attitudes towards self-collection and vaccination against HPV among Arab American women residing across four major metropolitan areas. The Arab American experience is diverse, owing to the diversity of the Arab world, heterogeneity of the immigrant experience, and socioeconomic status. The qualitative findings can help direct future research and develop interventions; the quantitative findings can help us understand important health patterns, including sexual behavior, which was not touched upon in the focus groups.

As this community is understudied, our study also helps shed light on successful recruitment efforts with the population. Unfortunately, recruitment was extremely difficult for the quantitative study despite attempts at various methods. Recruitment in person did not prove to be successful within this group, either. This is likely due to potential participants' competing interests when in public spaces. For example, those coming to the community center have appointments during normal business hours (i.e., 9-5 M-F) and are often not willing to engage before or after. Wider remote recruitment could be attempted, but also opens up the study to imposter participants, a growing concern for researchers [46]. Further investigation is needed into best practices in research in the Arab American population. Recruitment into the focus groups was easier. This may be a result of the anonymity provided by the focus groups or the higher incentive (USD 75 for the focus group compared to USD 10 for the questionnaire).

4.1. Limitations

The difficulties with recruitment and, as a result, the small sample size, are a limitation of the quantitative component of the study. The small sample size made it difficult to carry out more sophisticated analyses and made it difficult to generalize our findings. For example, where other estimates place the median income of a MENA American household was USD 67,000 [47], more than 50% of our questionnaire respondents reported a household income of over USD 75,000. This indicates that our sample has higher income and therefore they may have a different attitude and approach to healthcare, especially as socioeconomic status and health in the U.S. are linked [48].

The participants of the qualitative component may be a better reflection of the Arab American population as only 20% of focus group participants reported that their financial needs were more than sufficiently met. However, the primary objective of qualitative research is not to generalize findings to a larger population but to achieve in-depth insights into the subject. Data saturation can be achieved in 4–8 focus groups [16] for a narrowly defined objective; however, that is for a more homogenous population. While our focus group study had narrowly defined objectives, we had an extremely heterogeneous study population. Though we achieved data saturation within our focus groups, an expanded participant pool in other geographic areas could potentially provide additional perspectives. Regardless, we focused on recruiting participants from areas that are believed to be home to a majority of Arab Americans [49].

4.2. Future Directions and Applicability of the Research

Our findings are relevant for cancer control efforts in the U.S. First, as part of the effort to combat the increasing incidence of cervical cancer and lagging screening goals, in May 2024 the FDA approved HPV self-collection in healthcare settings [5]. Developing programs meant to tackle disparities in incidence and mortality will require an understanding of the attitudes and barriers to HPV self-collection specifically outside of a healthcare setting. Programs tailored to capture under-screened populations will need to address their specific needs. While some research has demonstrated its potential for implementation in the U.S. to reach under-screened groups [36,49,50], further research on how to best implement HPV self-collection in more underserved groups is needed. Our findings shed light on potential interventions to improve uptake in the Arab American community that can be applied to other immigrant and minority groups in the United States. For example, healthcare provider-centered interventions may be more successful at improving uptake than community-based interventions and warrant further study in this population and beyond.

Secondly, the Office of Management and Budget has recently revised the Standards for Maintaining, Collecting, and Presenting Federal Data on Race and Ethnicity, known as Directive 15, to include a "Middle Eastern or North African" minimum reporting category, which means that Arab Americans will finally be distinguished for federal reporting purposes [51]. This will require institutions and others that receive government funding to update their demographic forms. Federal agencies are expected to start implementing this new standard of its initial use in the 2030 Census. These findings can help direct future research efforts in a more targeted way.

There are several exciting avenues to further this research: continue recruitment for the quantitative arm of the study in order to capture a more representative population, pilot test a mailed HPV self-collection kit within this community, and explore alternative methods to evaluate screening rates through electronic health records and other large datasets. There is the potential to make a large impact on the health of an understudied and underserved population.

5. Conclusions

Our study provides important insights into the current understandings, attitudes, and behaviors of Arab American women regarding cervical cancer, HPV, and preventative health measures, particularly screening and HPV vaccination. Our findings illustrate that while there is moderate awareness of HPV and cervical cancer risk factors among the Arab American community, there is a dearth of targeted, culturally tailored health education and interventions for Arab American women. Our results suggest the necessity for provider-level education and interventions. Further research should explore effective strategies to engage this underserved group and enhance adherence to prevention programs. This insight points to the need for targeted education and communication efforts to improve self-efficacy and comfort with self-collection methods. Furthermore, physician recommendations emerged as a crucial influence on participants' health behaviors, underscoring the importance of strong provider–patient communication in increasing screening and vaccination rates.

Our findings also revealed challenges and opportunities in recruiting Arab American women for health research. Despite employing varied recruitment strategies, the quantitative component faced notable difficulties, which resulted in a smaller-than-expected sample size. This obstacle highlights the necessity for continued exploration of effective recruitment methods and the development of culturally sensitive and adaptable health interventions. Overall, our study emphasizes the importance of tailored approaches to health education and intervention within the Arab American community and sets a foundational basis for addressing health disparities and improving cervical cancer prevention efforts in this underserved population.

Author Contributions: Conceptualization, C.L.P., L.K., M.T., M.S.; Methodology, C.L.P., L.K.; Software, L.K., M.Z., M.A., S.Y., L.H.; Formal Analysis, L.K., M.Z., M.A., S.Y., L.H.; Investigation, C.L.P., L.K.; Resources, C.L.P.; Data Curation, L.K.; Writing—Original Draft Preparation, L.K., C.L.P.; Writing—Review and Editing, N.L.F., L.P.W., M.S., S.B.; Supervision, C.L.P.; Project Administration, L.K.; Funding Acquisition, C.L.P., L.K. All authors have read and agreed to the published version of the manuscript.

Funding: This project was supported by the University of Michigan Rogel Cancer Center grant number NIH grant number P30CA046592, and the Dr. Philip M. Kayal Fund for Arab American Research (2022).

Institutional Review Board Statement: The qualitative arm of the study was reviewed by the Institutional Review Board of Health Sciences and Behavioral Sciences at the University of Michigan and was determined to be Exempt category 2 determination. (HUM00150710; 10/16/2018). The quantitative arm of the study was reviewed by the Michigan Medicine Institutional Review Board (HUM00147147; 13 January 2020).

Informed Consent Statement: Informed written consent was obtained from all subjects involved in the quantitative arm of the study; in the qualitative arm of the study, obtaining informed consent was not required per the exempt determination per the Institutional Review Board Health Sciences and Behavioral Sciences.

Data Availability Statement: The data used in this study are not laboratory data. We collected human subject data using a survey and focus groups; the data we collected contains sensitive information related to sexual history and infection status. We do not have approval from our participants to share the data.

Acknowledgments: We extend our gratitude to all our participants for their time and candor. We also thank the Arab American Family Support Center, Xpress Pharmacy, Family Eye Physicians, Rula Thabata, Cedra Atassi, and ACCESS for their time and help in recruitment.

Conflicts of Interest: The authors declare no conflicts of interest.

References

1. Vilos, G.A. The History of the Papanicolaou Smear and the Odyssey of George and Andromache Papanicolaou. *Obstet. Gynecol.* **1998**, *91*, 479–483. [CrossRef] [PubMed]
2. Yang, D.X.; Soulos, P.R.; Davis, B.; Gross, C.P.; Yu, J.B. Impact of Widespread Cervical Cancer Screening: Number of Cancers Prevented and Changes in Race-Specific Incidence. *Am. J. Clin. Oncol.* **2018**, *41*, 289–294. [CrossRef] [PubMed]
3. FDA Approves First HPV Test for Primary Cervical Cancer Screening. The ASCO Post 2014. Available online: https://ascopost.com/issues/may-15-2014/fda-approves-first-hpv-test-for-primary-cervical-cancer-screening/ (accessed on 12 August 2024).
4. US Preventive Services Task Force. Screening for Cervical Cancer: US Preventive Services Task Force Recommendation Statement. *JAMA* **2018**, *320*, 674–686. [CrossRef] [PubMed]
5. FDA Approves HPV Self-Collection Solution. The ASCO Post Staff. Available online: https://ascopost.com/news/may-2024/fda-approves-hpv-self-collection-solution/ (accessed on 12 August 2024).
6. Sabatino, S.A.; Thompson, T.D.; White, M.C.; Villarroel, M.A.; Shapiro, J.A.; Croswell, J.M.; Richardson, L.C. Up-to-Date Breast, Cervical, and Colorectal Cancer Screening Test Use in the United States, 2021. *Prev. Chronic Dis.* **2023**, *20*, 230071. [CrossRef] [PubMed]
7. Cohen, C.M.; Wentzensen, N.; Castle, P.E.; Schiffman, M.; Zuna, R.; Arend, R.C.; Clarke, M.A. Racial and Ethnic Disparities in Cervical Cancer Incidence, Survival, and Mortality by Histologic Subtype. *J. Clin. Oncol.* **2023**, *41*, 1059–1068. [CrossRef] [PubMed]
8. Hekman, K.; Weir, S.; Fussman, C.; Lyon-Callo, S. *Health Risk Behaviors Among Arab Adults Within the State of Michigan: 2013 Arab Behavioral Risk Factor Survey*; Michigan Department of Health and Human Services, Lifecourse Epidemiology and Genomics Division and Health Disparities Reduction and Minority Health Section: Lansing, MI, USA, 2015.
9. Neumayer, H.; Weir, S.; Fussman, C.; McKane, P. *Health Risk Behaviors Among Arab Adults Within the State of Michigan: 2016 Arab Behavioral Risk Factor Survey*; Michigan Department of Health and Human Services, Lifecourse Epidemiology and Genomics Division and Health Disparities Reduction and Minority Health Section: Lansing, MI, USA, 2017.
10. Khoja, L.; Torres, H.G.; Hak, L.; Aljebori, M.; Phung, M.T.; Sakleh, A.; Stiffler, M.; Tariq, M.; Pearce, C.L. Cervical Cancer Screening and Vaccination Acceptability and Attitudes among Arab American Women in Southeastern Michigan: A Qualitative Study. *Sci. Rep.* **2024**, *14*, 13624. [CrossRef]
11. Dallo, F.J.; Ajrouch, K.J.; Al-Snih, S. The Ancestry Question and Ethnic Heterogeneity: The Case of Arab Americans. *Int. Migr. Rev.* **2008**, *42*, 505–517. [CrossRef]
12. Rahal, D.; Kurtz-Costes, B.; Volpe, V.V. Ethnic Identity in Arab Americans: Gender, Religious Upbringing, and Age Differences. *Soc. Identities* **2022**, *28*, 544–569. [CrossRef] [PubMed]
13. Ogale, Y.; Yeh, P.T.; Kennedy, C.E.; Toskin, I.; Narasimhan, M. Self-Collection of Samples as an Additional Approach to Deliver Testing Services for Sexually Transmitted Infections: A Systematic Review and Meta-Analysis. *BMJ Glob. Health* **2019**, *4*, e001349. [CrossRef]
14. Madzima, T.R.; Vahabi, M.; Lofters, A. Emerging Role of HPV Self-Sampling in Cervical Cancer Screening for Hard-to-Reach Women. *Can. Fam. Physician* **2017**, *63*, 597–601.
15. Serrano, B.; Ibáñez, R.; Robles, C.; Peremiquel-Trillas, P.; de Sanjosé, S.; Bruni, L. Worldwide Use of HPV Self-Sampling for Cervical Cancer Screening. *Prev. Med.* **2022**, *154*, 106900. [CrossRef]
16. Hennink, M.; Kaiser, B.N. Sample Sizes for Saturation in Qualitative Research: A Systematic Review of Empirical Tests. *Soc. Sci. Med.* **2022**, *292*, 114523. [CrossRef] [PubMed]
17. Brown, R. Best Practices for Conducting Virtual Focus Groups. *Publ. Scholarsh.* **2022**, *6*. Available online: https://source.sheridancollege.ca/ctl_publ/6 (accessed on 9 November 2024).
18. Dedoose. *Web Application for Managing, Analyzing, and Presenting Qualitative and Mixed Method Research Data*; SocioCultural Research Consultants, LLC: Los Angeles, CA, USA, 2021.
19. Chun Tie, Y.; Birks, M.; Francis, K. Grounded Theory Research: A Design Framework for Novice Researchers. *SAGE Open Med.* **2019**, *7*, 2050312118822927. [CrossRef] [PubMed]
20. Cofie, N.; Braund, H.; Dalgarno, N. Eight Ways to Get a Grip on Intercoder Reliability Using Qualitative-Based Measures. *Can. Med. Educ. J.* **2022**, *13*, 73–76. [CrossRef] [PubMed]

21. Gottschlich, A.; Rivera-Andrade, A.; Grajeda, E.; Alvarez, C.; Mendoza Montano, C.; Meza, R. Acceptability of Human Papillomavirus Self-Sampling for Cervical Cancer Screening in an Indigenous Community in Guatemala. *J. Glob. Oncol.* **2017**, *3*, 444–454. [CrossRef] [PubMed]
22. Gottschlich, A.; Nuntadusit, T.; Zarins, K.R.; Hada, M.; Chooson, N.; Bilheem, S.; Navakanitworakul, R.; Nittayaboon, K.; Virani, S.; Rozek, L.; et al. Barriers to Cervical Cancer Screening and Acceptability of HPV Self-Testing: A Cross-Sectional Comparison between Ethnic Groups in Southern Thailand. *BMJ Open* **2019**, *9*, e031957. [CrossRef] [PubMed]
23. Murchland, A.R.; Gottschlich, A.; Bevilacqua, K.; Pineda, A.; Sandoval-Ramírez, B.A.; Alvarez, C.S.; Ogilvie, G.S.; Carey, T.E.; Prince, M.; Dean, M.; et al. HPV Self-Sampling Acceptability in Rural and Indigenous Communities in Guatemala: A Cross-Sectional Study. *BMJ Open* **2019**, *9*, e029158. [CrossRef]
24. Microsoft Corporation. *Microsoft Excel 2018*; Microsoft Corporation: Redmond, WA, USA, 2018.
25. Posit Team. *RStudio: Integrated Development Environment for R*; Posit Team: Boston, MA, USA, 2024.
26. Dallo, F.J.; Kindratt, T.B. Disparities in Vaccinations and Cancer Screening among U.S.- and Foreign-Born Arab and European American Non-Hispanic White Women. *Womens Health Issues* **2015**, *25*, 56–62. [CrossRef] [PubMed]
27. Shah, S.M.; Ayash, C.; Pharaon, N.A.; Gany, F.M. Arab American Immigrants in New York: Health Care and Cancer Knowledge, Attitudes, and Beliefs. *J. Immigr. Minor. Health* **2008**, *10*, 429–436. [CrossRef]
28. Roman, L.; Meghea, C.; Ford, S.; Penner, L.; Hamade, H.; Estes, T.; Williams, K.P. Individual, Provider, and System Risk Factors for Breast and Cervical Cancer Screening Among Underserved Black, Latina, and Arab Women. *J. Women's Health* **2014**, *23*, 57–64. [CrossRef]
29. Williams, K.P.; Roman, L.; Meghea, C.I.; Penner, L.; Hammad, A.; Gardiner, J. Kin KeeperSM: Design and Baseline Characteristics of a Community-Based Randomized Controlled Trial Promoting Cancer Screening in Black, Latina, and Arab Women. *Contemp. Clin. Trials* **2013**, *34*, 312–319. [CrossRef]
30. Kusow, A.M.; Ajrouch, K.J.; Corra, M. Socioeconomic Achievement Among Arab Immigrants in the USA: The Influence of Region of Origin and Gender. *Int. Migr. Integr.* **2018**, *19*, 111–127. [CrossRef]
31. Tsegaye, A.T.; Lin, J.; Cole, A.M.; Szpiro, A.; Rao, D.W.; Walson, J.; Winer, R.L. Adherence and Correlates of Cervical Cancer Screening Among East African Immigrant Women in Washington State. *J. Racial Ethn. Health Disparities* **2024**. [CrossRef] [PubMed]
32. Romelus, J.; McLaughlin, C.; Ruggieri, D.; Morgan, S. A Narrative Review of Cervical Cancer Screening Utilization Among Haitian Immigrant Women in the U.S.: Health Beliefs, Perceptions, and Societal Barriers and Facilitators. *J. Immigr. Minor. Health* **2024**, *26*, 596–603. [CrossRef] [PubMed]
33. Adegboyega, A.; Kang, J.; Aroh, A.; Williams, L.B. Perceived Barriers to Pap Screening Influence Adherence to Screening Recommendations Among Black Women. *J. Women's Health* **2024**, *33*, 1482–1491. [CrossRef] [PubMed]
34. Chen, J.J.; Sarkar, I.N.; Hsu, E.; Dizon, D.S. An Intersectional Approach to Cervical Cancer Screening Disparities by Race/Ethnicity and Immigrant Status. *J. Women's Health* **2024**. [CrossRef] [PubMed]
35. Adebola, A.; Adaeze, A.; Adeyimika, D.; Lovoria, W.B.; Gia, M.-M. Experiences and Challenges of African American and Sub-Saharan African Immigrant Black Women in Completing Pap Screening: A Mixed Methods Study. *J. Racial Ethn. Health Disparities* **2024**, *11*, 1405–1417. [CrossRef] [PubMed]
36. Bishop, E.; Katz, M.L.; Reiter, P.L. Acceptability of Human Papillomavirus Self-Sampling Among a National Sample of Women in the United States. *Biores Open Access* **2019**, *8*, 65–73. [CrossRef] [PubMed]
37. Bansil, P.; Wittet, S.; Lim, J.L.; Winkler, J.L.; Paul, P.; Jeronimo, J. Acceptability of Self-Collection Sampling for HPV-DNA Testing in Low-Resource Settings: A Mixed Methods Approach. *BMC Public Health* **2014**, *14*, 596. [CrossRef]
38. Khoja, L.; Wang, Y.; Haque, S.E.; Ahsan, H.; Islam, T.; Munshi, S.U.; Hasan, A.K.M.R.; Islam, M.T.; Jharna, A.B.; Pearce, C.L. Understanding of Cervical Cancer, Acceptability of HPV Self-Collection, and Prevalence of HPV in a Semi-Urban Setting in Bangladesh. *PLoS Glob. Public Health* **2024**, *4*, e0003157. [CrossRef] [PubMed]
39. Vahabi, M.; Lofters, A. Muslim Immigrant Women's Views on Cervical Cancer Screening and HPV Self-Sampling in Ontario, Canada. *BMC Public Health* **2016**, *16*, 868. [CrossRef] [PubMed]
40. Abouhala, S.; Hamidaddin, A.; Taye, M.; Glass, D.J.; Zanial, N.; Hammood, F.; Allouch, F.; Abuelezam, N.N. A National Survey Assessing COVID-19 Vaccine Hesitancy Among Arab Americans. *J. Racial Ethn. Health Disparities* **2022**, *9*, 2188–2196. [CrossRef] [PubMed]
41. Kheil, M.H.; Jain, D.; Jomaa, J.; Askar, B.; Alcodray, Y.; Wahbi, S.; Brikho, S.; Kadouh, A.; Harajli, D.; Jawad, Z.N.; et al. COVID-19 Vaccine Hesitancy among Arab Americans. *Vaccines* **2022**, *10*, 610. [CrossRef]
42. Kaljee, L.M.; Dankerlui, D.; Pach, A.; Veer, S.; Alhawli, A.; Ayoub, R.; Toma, N.; Ahmed, R.; Chue, S.; Coral Arrua, C.; et al. COVID-19 Vaccine Hesitancy and Uptake among Caribbean, Latino/a, Native American, and Middle Eastern/North African Populations in Detroit and Surrounding Counties: A Qualitative Study. *J. Community Med. Public Health* **2023**, *7*, 374. [CrossRef]
43. Ayash, C.; Raad, N.; Finik, J.; Attia, N.; Nourredine, S.; Aragones, A.; Gany, F. Arab American Mothers' HPV Vaccination Knowledge and Beliefs. *J. Community Health* **2022**, *47*, 716–725. [CrossRef] [PubMed]

44. AlAbdulKader, A.M.; AlAsfour, A.; Golembiewski, M.; Gullett, H. Disparities in Cervical Cancer Screening among Arabic-Speaking Women Refugees. *Ethn. Health* **2023**, *28*, 1115–1127. [CrossRef]
45. Holmes, H.J.; Yamin, J.B.; Krohner, S.; Rapport, L.J.; Grekin, E.R.; Schubiner, H.; Lumley, M.A. Effects of a Sexual Health Interview among Arab American Women: An Experimental Disclosure Study. *Arch. Sex. Behav.* **2021**, *50*, 373–384. [CrossRef] [PubMed]
46. Santinele Martino, A.; Perrotta, A.; McGillion, B.J. Who Can You Trust These Days?: Dealing with Imposter Participants during Online Recruitment and Data Collection. *Qual. Res.* **2024**. [CrossRef]
47. Marrow, B.; Batalova, J. Middle Eastern and North African Immigrants in the United States. Available online: https://www.migrationpolicy.org/article/middle-eastern-and-north-african-immigrants-united-states (accessed on 2 October 2022).
48. Wardle, J. Socioeconomic Differences in Attitudes and Beliefs about Healthy Lifestyles. *J. Epidemiol. Community Health* **2003**, *57*, 440–443. [CrossRef] [PubMed]
49. Barbee, L.; Kobetz, E.; Menard, J.; Cook, N.; Blanco, J.; Barton, B.; Auguste, P.; McKenzie, N. Assessing the Acceptability of Self-Sampling for HPV among Haitian Immigrant Women: CBPR in Action. *Cancer Causes Control* **2010**, *21*, 421–431. [CrossRef] [PubMed]
50. Pretsch, P.K.; Spees, L.P.; Brewer, N.T.; Hudgens, M.G.; Sanusi, B.; Rohner, E.; Miller, E.; Jackson, S.L.; Barclay, L.; Carter, A.; et al. Effect of HPV Self-Collection Kits on Cervical Cancer Screening Uptake among under-Screened Women from Low-Income US Backgrounds (MBMT-3): A Phase 3, Open-Label, Randomised Controlled Trial. *Lancet Public Health* **2023**, *8*, e411–e421. [CrossRef]
51. Office of Management and Budget Revisions to OMB's Statistical Policy Directive No. 15: Standards for Maintaining, Collecting, and Presenting Federal Data on Race and Ethnicity. *Fed. Regist.* **2024**, *89*, 22182–22196.

cancers

MDPI

Article

Real-World Data on the Associations of Tricyclic Antidepressants and Selective Serotonin Reuptake Inhibitors with Gynecologic Cancer Risk

Ching-Huan Wang [1,†], Chih-Wei Huang [1,2,†], Nhi Thi Hong Nguyen [3,4], Ming-Chin Lin [1,5], Phung-Anh Nguyen [6,7,8,9], Md. Mohaimenul Islam [10], Shuo-Chen Chien [1] and Hsuan-Chia Yang [1,7,8,11,*]

1 Graduate Institute of Biomedical Informatics, College of Medical Science and Technology, Taipei Medical University, Taipei 110301, Taiwan; cwang226@alumni.jh.edu (C.-H.W.); gracehuang@tmu.edu.tw (C.-W.H.); arbiter@tmu.edu.tw (M.-C.L.); d610108001@tmu.edu.tw (S.-C.C.)
2 International Center for Health Information Technology (ICHIT), College of Medical Science and Technology, Taipei Medical University, Taipei 110301, Taiwan
3 Health Personnel Training Institute, University of Medicine and Pharmacy, Hue University, Hue City, Vietnam; nthnhi@hueuni.edu.vn
4 School of Nutrition and Health Sciences, College of Nutrition, Taipei Medical University, Taipei 110301, Taiwan
5 Department of Neurosurgery, Taipei Municipal Wanfang Hospital, Taipei Medical University, Taipei 110301, Taiwan
6 Graduate Institute of Data Science, College of Management, Taipei Medical University, Taipei 110301, Taiwan; alex0303@tmu.edu.tw
7 Clinical Big Data Research Center, Taipei Medical University Hospital, Taipei 110301, Taiwan
8 Research Center of Healthcare Industry Data Science, College of Management, Taipei Medical University, Taipei 110301, Taiwan
9 Clinical Data Center, Office of Data Science, Taipei Medical University, Taipei 110301, Taiwan
10 College of Pharmacy, University of Iowa, Iowa City, IA 52242, USA; mdmohaimenul-islam@uiowa.edu
11 Research Center of Big Data and Meta-Analysis, Taipei Municipal Wanfang Hospital, Taipei Medical University, Taipei 110301, Taiwan
* Correspondence: lovejog@tmu.edu.tw or itpharmacist@gmail.com
† These authors contributed equally to this work.

Simple Summary: Despite previous studies suggesting antidepressants' potential anti-cancer effects against gynecologic cancers, little is known about these effects across different age groups. This study aims to investigate the associations of individual tricyclic antidepressants (TCAs) and selective serotonin reuptake inhibitors (SSRIs) with the risk of gynecologic cancers. We found that TCA and SSRI use was significantly associated with lowered risks of gynecologic cancers particularly among women aged 40–64. In contrast, no significant protective effect against ovarian or uterine cancer was observed in elderly women. We concluded that middle-aged women may benefit the most from future research on drug repurposing antidepressants for gynecologic cancer prevention.

Abstract: Background: While the potential anti-cancer effects of antidepressants have been investigated, limited research has been conducted incorporating age-specific analyses for individual tricyclic antidepressants (TCAs) and selective serotonin reuptake inhibitors (SSRIs). This study aims to elucidate the associations between TCAs and SSRIs with gynecologic cancers and to identify specific age groups and cancer types that may benefit from the chemopreventive effects of these medications. **Methods**: A case-control study comprised 97,736 female patients diagnosed with gynecologic cancers between 2002 and 2016. Each newly diagnosed case of cervical, ovarian, or uterine cancer was matched with four controls. Both cases and controls were then stratified by age to perform subgroup analyses. Associations between antidepressant use and cancer risk were evaluated using multivariable conditional logistic regression models. **Results**: The TCA class was significantly associated with reduced risks of cervical, ovarian, and uterine cancers, displaying adjusted odds ratios (aORs) of 0.799, 0.775, and 0.813, respectively. The SSRI class also indicated reduced risks, with aORs of 0.736, 0.638, and 0.567 for the same cancer types. Particularly noteworthy were females aged 40–64, who demonstrated the most significant associations between gynecologic cancers and using TCAs or SSRIs. **Conclusions**: TCAs and SSRIs are associated with reduced risks of developing cervical, ovarian, and uterine cancers. The middle-aged population may have the most significant potential for future research on drug repurposing against gynecologic cancers, and both cervical and uterine cancers are potential targets for drug repurposing involving TCAs or SSRIs.

Keywords: tricyclic antidepressant; selective serotonin reuptake inhibitor; gynecologic cancer; cancer risk; real-world data

1. Introduction

Depression is a significant factor contributing to global disability, ranking third for females and fifth for males in 2017 [1]. It is estimated that depressive disorders affect approximately 264 million people worldwide, accounting for 3–4% of the global population [1]. The lifetime prevalence of depression is reported to be 10.8%, with a higher susceptibility observed in females compared to males [2]. Since 2000, the consumption of antidepressants has doubled [3]. Antidepressants are primarily formulated to target amine signaling pathways under the monoamine-deficiency hypothesis of depression [4]. Pharmacologically, these agents can be classified into tricyclic antidepressants (TCAs), selective serotonin reuptake inhibitors (SSRIs), and monoamine oxidase inhibitors (MAOIs), among others [5]. Alongside the growing demand for antidepressant medications, concerns have emerged regarding drug safety and the potential for repurposing existing drugs.

Research exploring the association between antidepressant use and cancer risk has been ongoing for several decades. The first substantial study on this topic can be traced back to 1995, when Steingart et al. reported ambiguous findings on antidepressants' dual potential for both promoting and inhibiting tumor growth, leaving the exact relationship inconclusive [6]. More recently, various hypotheses have been put forth to explain a potential association between antidepressants and gynecologic cancer risk, including increased release of gonadotropins and prolactin [7] as well as induced apoptosis of cancer cells [8]. However, empirical findings have been inconsistent. Some studies have suggested a potential risk reduction in ovarian [9] and endometrial [10] cancers with SSRI usage, whereas others have indicated an increased risk of ovarian cancer associated with TCAs [11]. Meanwhile, other research has reported that neither TCAs [7,10,12] nor SSRIs [7,12] show a significant association with ovarian [7], cervical [12], or endometrial cancers [10].

In Taiwan, the prevalence rates of uterine body, ovarian, and cervical cancers rank fifth, seventh, and ninth, respectively, among female cancers [13]. Previous observational studies have investigated the association between antidepressant use and the risk of gynecologic cancers; however, limited research has been conducted on age-specific effects related to individual TCAs or SSRIs. Therefore, our study aims to assess the chemopreventive potential of antidepressants with a focus on identifying age groups and types of gynecologic cancers that warrant further research. Utilizing real-world claims data, we examined the association between antidepressant use and the incidence of new cases of gynecologic cancers. Given that TCAs and SSRIs constitute the primary classes of antidepressants in our dataset and have the most extensive user base, we limited our focus to these two categories. We also stratified the subject population into three age subgroups to preliminarily observe variations in the impact of antidepressants on gynecologic cancers among users of different ages. The findings from this study may offer valuable insights for healthcare professionals and researchers interested in repurposing TCAs or SSRIs for the prevention of gynecologic cancers.

2. Methods

2.1. Study Design and Population

This study employed a case-control design. We recruited patients with a new cancer diagnosis between 2000 and 2016 (Figure 1). Individuals were excluded from the analysis if their age at cancer diagnosis was under 20 or if their data were missing, unidentifiable, or inconsistent. Additionally, subjects initially diagnosed with cancer in 2000 or 2001 were excluded, as the study required verification of antidepressant use for at least two years prior to the index date. The index date for each case was defined as the date on which the cancer diagnosis was confirmed. Each control was assigned the same index date as its corresponding case. The study examined the association between the risk of cervical, ovarian, and uterine cancers and the usage of two classes of antidepressants: TCAs and SSRIs. Cases and controls were stratified into four age groups: young adults (20–39 years), middle-aged adults (40–64 years), elderly adults ($\geq$ 65 years), and an overall age population (all subjects aged $\geq$ 20). Subsequent analyses were conducted to assess the gynecologic cancer risks associated with using individual TCAs or SSRIs.

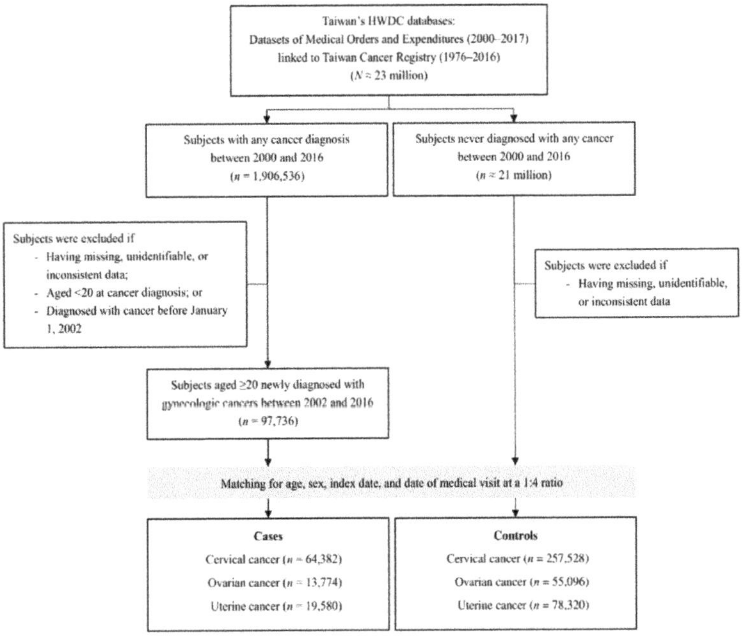

Figure 1. Flowchart of the study design.

2.2. Data Sources

We utilized databases from Taiwan's Health and Welfare Data Science Center (HWDC) that include medication and diagnosis information spanning 2000–2017 and cancer data from 1979 to 2016 (Figure 1). The HWDC operates as a centralized data repository overseen by Taiwan's Ministry of Health and Welfare (MOHW). This repository securely houses de-identified claims data from Taiwan's National Health Insurance (NHI) system [14]. The NHI system in Taiwan offers healthcare coverage to 99.9% of the population, encompassing over 23 million beneficiaries [15]. For this study, we accessed specific medical orders and expenditures datasets alongside the Taiwan Cancer Registry dataset containing pathologically confirmed cancer diagnoses. Disease categorization relied on the International Classification of Diseases, 9th Revision, Clinical Modification (ICD-9-CM) coding system [14,16].

2.3. Cases and Controls

Cases in this study were identified as patients who received an initial diagnosis of cervical, ovarian, or uterine cancer, as confirmed by the Taiwan Cancer Registry in 2002. These cancer diagnoses were validated through pathological assessment. We matched cases and controls in a 1:4 ratio, using criteria such as age, sex, index date, and date of medical visit [17]. Controls were defined as individuals who had never received a cancer diagnosis up to the end of 2016. Our analysis focused on the risk associated with specific gynecologic cancers, which are identified by their respective ICD-9-CM codes: cervical cancer (180.x), uterine cancer (182.x), and ovarian cancer (183.0).

2.4. Definitions of Exposure to Antidepressants

Exposure to antidepressants was defined as a prescription of these medications for at least 60 days within a two-year period preceding the index date. Based on this definition, exposure to TCAs and exposure to SSRIs were not mutually exclusive. Those who had used both TCAs and SSRIs were classified as exposed in both the TCA and SSRI groups, respectively. Antidepressants were classified according to the World Health Organization's Anatomical Therapeutic Chemical (ATC) Classification system. This study specifically examined TCAs (N06AAxx) and SSRIs (N06ABxx) [5] due to their prevalence among users. The TCA category included imipramine (N06AA02), clomipramine (N06AA04), amitriptyline (N06AA09), doxepin (A06AA12), dosulepin (N06AA16), and maprotiline (N06AA21). The SSRI category comprised fluoxetine (N06AB03), citalopram (N06AB04), paroxetine (N06AB05), sertraline (N06AB06), fluvoxamine (N06AB08), and escitalopram (N06AB10) [5].

2.5. Potential Confounding Factors

The potential confounders considered in this study were age, Charlson Comorbidity Index (CCI) score [18], and the usage of metformin (ATC code: A10BA02), aspirin (ATC code: B01AC06), or statins (ATC codes: C10AA). CCI comorbidities, excluding cancers (Table 1), were deemed confounders if a patient had been diagnosed with them on two or more occasions. Likewise, the use of metformin [19–21], aspirin [22–24], and statins (Table 1) [21,25] for a minimum of 60 days within the two-year period preceding the index date was considered confounding, owing to their potential association with cancer risks.

Table 1. Demographic characteristics.

Characteristics	Cervical Cancer				Ovarian Cancer				Uterine Cancer			
	Cases (n = 64,382)		**Controls (n = 257,528)**		**Cases (n = 13,774)**		**Controls (n = 55,096)**		**Cases (n = 19,580)**		**Controls (n = 78,320)**	
Age												
Mean ± SD	49.74	± 14.48	49.74	± 14.48	51.72	± 13.96	51.72	± 13.95	53.68	± 10.93	53.68	± 10.92
20–39 years, n (%)	17,861	(27.74)	71,444	(27.74)	2547	(18.49)	10,188	(18.49)	1905	(9.73)	7620	(9.73)
40–64 years, n (%)	34,581	(53.71)	138,324	(53.71)	8698	(63.15)	34,792	(63.15)	14,783	(75.50)	59,132	(75.50)
≥65 years, n (%)	11,940	(18.55)	47,760	(18.55)	2529	(18.36)	10,116	(18.36)	2892	(14.77)	11,568	(14.77)

Table 1. *Cont.*

Characteristics	Cervical Cancer				Ovarian Cancer				Uterine Cancer			
	Cases (n = 64,382)		**Controls (n = 257,528)**		**Cases (n = 13,774)**		**Controls (n = 55,096)**		**Cases (n = 19,580)**		**Controls (n = 78,320)**	
Comorbid conditions, n (%)												
Myocardial infarction	108	(0.17)	520	(0.20)	39	(0.28)	112	(0.20)	47	(0.24)	142	(0.18)
Congestive heart failure	896	(1.39)	4222	(1.64)	235	(1.71)	914	(1.66)	342	(1.76)	1195	(1.53)
Peripheral vascular disease	377	(0.59)	1884	(0.73)	97	(0.70)	405	(0.74)	128	(0.65)	643	(0.82)
Cerebrovascular disease	2347	(3.65)	11,388	(4.42)	545	(3.96)	2560	(4.65)	831	(4.24)	3518	(4.49)
Dementia	437	(0.68)	2409	(0.94)	102	(0.74)	497	(0.90)	88	(0.45)	521	(0.67)
Chronic pulmonary disease	1766	(2.74)	8427	(3.27)	370	(2.69)	1760	(3.19)	544	(2.78)	2478	(3.16)
Rheumatic disease	898	(1.39)	4141	(1.61)	219	(1.59)	935	(1.70)	246	(1.26)	1396	(1.78)
Peptic ulcer disease	6348	(9.86)	31,919	(12.39)	1814	(13.17)	7020	(12.74)	2083	(10.64)	10,536	(13.45)
Liver disease	2909	(4.52)	15,551	(6.04)	893	(6.48)	3420	(6.21)	1256	(6.41)	5337	(6.81)
Diabetes	6452	(10.02)	34,379	(13.35)	1664	(12.08)	8312	(15.09)	3506	(17.91)	12,780	(16.32)
Hemiplegia or paraplegia	103	(0.16)	481	(0.19)	30	(0.22)	140	(0.25)	24	(0.12)	149	(0.19)
Renal disease	1272	(1.98)	5543	(2.15)	281	(2.04)	1322	(2.40)	466	(2.38)	2015	(2.57)
AIDS/HIV infection	38	(0.06)	42	(0.02)	0	(0.00)	9	(0.02)	0	(0.00)	13	(0.02)
CCI scores												
Mean (SD) *	0.40	(0.87)	0.50	(0.92)	0.49	(0.94)	0.53	(0.94)	0.53	(0.94)	0.56	(0.94)
Other drugs, n (%)												
Metformin	3990	(6.20)	23,451	(9.11)	1192	(8.65)	6186	(11.23)	2408	(12.30)	10,291	(13.14)
Aspirin	3775	(5.86)	19,109	(7.42)	906	(6.58)	4436	(8.05)	1545	(7.89)	6401	(8.17)
Statin	3708	(5.76)	20,402	(7.92)	943	(6.85)	5125	(9.30)	2162	(11.04)	7891	(10.08)

AIDS, acquired immunodeficiency syndrome; HIV, human immunodeficiency virus; n, number; N, total number; SD, standard deviation. * All CCI scores are ≥0. Due to the right-skewed distribution of comorbidities, mean ± SD may appear to allow negative values, but no negative values are present in the dataset.

2.6. Statistical Analysis

Descriptive statistics were used to describe the participants' demographic characteristics. Continuous variables were presented as means and standard deviations; categorical variables were presented as numbers and percentages. The cancer risk associated with each medication or class was evaluated using the adjusted odds ratio (aOR) derived from multivariable conditional logistic regression. The significance level was set at $\alpha = 0.05$ (two-sided). A two-sided p-value less than 0.05 was considered statistically significant. Data were analyzed using SAS 9.4 (SAS Institute Inc., Cary, NC, USA).

2.7. Ethics Approval Statement

The study protocol was approved by the Joint Institutional Review Board of Taipei Medical University, Taipei, Taiwan (TMU-JIRB N202003609).

3. Results

3.1. Participant Characteristics

Between 2000 and 2016, we identified 1,906,536 new cases of cancer. Of these, 97,736 individuals were newly diagnosed with gynecologic cancer at the age of 20 or older between 2002 and 2016. This cohort included 64,382 cases of cervical cancer, 13,774 of ovarian cancer, and 19,580 of uterine cancer (Figure 1). The mean ages at first diagnosis were 49.74 years for cervical cancer, 51.75 years for ovarian cancer, and 53.68 years for uterine cancer (Table 1). The middle-aged demographic was the most prevalent across all three cancer types, constituting 53.71%, 63.15%, and 75.50% of cervical, ovarian, and uterine cancer cases, respectively. Relative to the elderly group, the younger cohort displayed a higher proportion of cervical cancer cases (27.74% compared to 18.55%), a similar proportion of ovarian cancer cases (18.49% compared to 18.36%), and a lower proportion of uterine cancer cases (9.73% compared to 14.77%).

3.2. Associations of Gynecologic Cancers with the TCA and SSRI Classes

Table 2 presents the associations between gynecologic cancer risks and the use of TCAs and SSRIs. TCA and SSRI classes were associated with statistically significant reductions in the risk of gynecologic cancers. Specifically, TCAs demonstrated a decreased risk for cervical cancer (aOR = 0.799; 95% CI: 0.751, 0.850; $p < 0.0001$), ovarian cancer (aOR = 0.775; 95% CI: 0.681, 0.882; $p = 0.0001$), and uterine cancer (aOR = 0.813; 95% CI: 0.732, 0.903; $p = 0.0001$). Similarly, SSRIs were associated with reduced risks for cervical cancer (aOR = 0.736; 95% CI: 0.693, 0.782; $p < 0.0001$), ovarian cancer (aOR = 0.638; 95% CI: 0.559, 0.728; $p < 0.0001$), and uterine cancer (aOR = 0.567; 95% CI: 0.505, 0.636; $p < 0.0001$). In the age-stratified analysis, TCA use demonstrated significant reductions in cancer risks for cervical cancer in individuals aged $\geq$ 40, for ovarian cancer in those aged 40–64, and for uterine cancer in those aged < 65. However, there were no significant associations between TCA use and ovarian or uterine cancers in the older population. On the contrary, SSRIs consistently exhibited a reduced risk of cervical, ovarian, and uterine cancers across all age groups, except for uterine cancer in individuals aged $\geq$ 65. Table 3 summarizes the statistical significance of aORs for gynecologic cancers associated with each TCA and SSRI, stratified by age.

Table 2. Risk of gynecologic cancers associated with the TCA and the SSRI classes, stratified by age.

	Cervical Cancer (64,382 Cases; 257,528 Controls)			Ovarian Cancer (13,774 Cases; 55,096 Controls)			Uterine Cancer (19,580 Cases; 78,320 Controls)		
Antidepressants	***n* (Case/Ctrl)**	**Crude OR (95% CI)**	**Adjusted OR [a] (95% CI)**	***n* (Case/Ctrl)**	**Crude OR (95% CI)**	**Adjusted OR [a] (95% CI)**	***n* (Case/Ctrl)**	**Crude OR (95% CI)**	**Adjusted OR [a] (95% CI)**
TCAs									
Overall	1249/6800	0.729 (0.686, 0.775)	0.799 (0.751, 0.850)	282/1521	0.736 (0.647, 0.837)	0.775 (0.681, 0.882)	431/2177	0.787 (0.709, 0.874)	0.813 (0.732, 0.903)
Aged 20–39	143/681	0.839 (0.700, 1.005)	0.876 (0.730, 1.050)	10/66	0.605 (0.310, 1.177)	0.605 (0.310, 1.179)	6/66	0.362 (0.157, 0.835)	0.327 (0.140, 0.765)
Aged 40–64	618/3731	0.656 (0.602, 0.715)	0.740 (0.678, 0.807)	146/961	0.601 (0.504, 0.716)	0.655 (0.549, 0.782)	274/1576	0.690 (0.606, 0.785)	0.724 (0.636, 0.825)
Aged ≥ 65	488/2388	0.810 (0.733, 0.894)	0.867 (0.784, 0.959)	126/494	1.021 (0.836, 1.248)	1.015 (0.830, 1.242)	151/535	1.136 (0.944, 1.367)	1.132 (0.940, 1.363)
SSRIs									
Overall	1262/7154	0.700 (0.659, 0.743)	0.736 (0.693, 0.782)	262/1656	0.626 (0.549, 0.714)	0.638 (0.559, 0.728)	339/2412	0.554 (0.494, 0.622)	0.567 (0.505, 0.636)
Aged 20–39	355/1650	0.858 (0.764, 0.963)	0.878 (0.782, 0.987)	28/202	0.550 (0.369, 0.818)	0.553 (0.371, 0.825)	15/194	0.304 (0.179, 0.515)	0.313 (0.184, 0.530)
Aged 40–64	627/4068	0.609 (0.560, 0.663)	0.649 (0.596, 0.707)	176/1136	0.612 (0.521, 0.718)	0.641 (0.545, 0.753)	244/1863	0.516 (0.451, 0.590)	0.531 (0.464, 0.608)
Aged ≥ 65	280/1436	0.775 (0.680, 0.882)	0.816 (0.716, 0.930)	58/318	0.723 (0.545, 0.960)	0.699 (0.525, 0.930)	80/355	0.899 (0.703, 1.149)	0.897 (0.700, 1.149)

[a] Adjusted for age, the Charlson Comorbidity Index, and use of medications (metformin, aspirin, and statins). CI, confidence interval; ctrl, control; n, number of both cases and controls exposed to TCAs or SSRIs; OR, odds ratio; SSRIs, selective serotonin reuptake inhibitors; TCAs, tricyclic antidepressants.

Table 3. Summary of significantly lowered gynecologic cancer risks associated with TCAs or SSRIs.

	Cervical Cancer				Ovarian Cancer				Uterine Cancer			
Age Groups	**Overall**	**20–39**	**40–64**	**≥65**	**Overall**	**20–39**	**40–64**	**≥65**	**Overall**	**20–39**	**40–64**	**≥65**
TCAs	****		****	**	***		****		***	**	****	
Imipramine	**		**				**					
Amitriptyline			**				*		**		**	
Doxepin	*		*		*				**		*	
SSRIs	****	*	****	**	****	**	****	*	****	****	****	
Fluoxetine	****		****		***	**	*		****		****	
Citalopram	**		**		*				****		****	
Paroxetine	****		****	**					***		***	
Sertraline	****		****		***		**		****	*	****	
Fluvoxamine	*		*						*		*	
Escitalopram	****		****		**		***		****	*	****	

p-values of adjusted odds ratios: * $p < 0.05$; ** $p < 0.01$; *** $p < 0.001$; **** $p < 0.0001$. SSRIs, selective serotonin reuptake inhibitors; TCAs, tricyclic antidepressants.

3.3. Associations of Gynecologic Cancers with Individual TCAs and SSRIs

For cervical cancer, all individual TCAs and SSRIs, except for amitriptyline, were associated with a significant reduction in cancer risk (Figure 2). Age-specific analysis revealed that females between the ages of 20–39 exhibited no statistically significant association between cervical cancer risk and the use of TCAs or SSRIs. Conversely, females aged 40–64 who were treated with any TCA or SSRI showed a significantly decreased risk of developing cervical cancer. Among the elderly, paroxetine was the sole agent associated with a significant reduction in cervical cancer risk (aOR = 0.608; $p = 0.0043$).

Antidepressants	*n* (case/ctrl)	Adjusted odds ratio (95% CI) Overall	Aged 20–39	Aged 40–64	Aged ≥ 65
TCAs					
Imipramine	487/2534	0.872 (0.790, 0.962)	0.783 (0.522, 1.175)	0.812 (0.702, 0.939)	0.939 (0.814, 1.082)
Amitriptyline	156/819	0.846 (0.712, 1.005)	0.901 (0.548, 1.482)	0.715 (0.561, 0.913)	1.057 (0.797, 1.402)
Doxepin	62/350	0.738 (0.563, 0.968)	0.713 (0.364, 1.396)	0.672 (0.462, 0.977)	0.920 (0.565, 1.497)
SSRIs					
Fluoxetine	380/2043	0.782 (0.700, 0.873)	0.840 (0.681, 1.037)	0.703 (0.604, 0.819)	0.973 (0.760, 1.246)
Citalopram	133/737	0.757 (0.629, 0.912)	0.909 (0.636, 1.299)	0.709 (0.549, 0.916)	0.719 (0.473, 1.091)
Paroxetine	222/1408	0.663 (0.575, 0.764)	0.899 (0.689, 1.172)	0.594 (0.489, 0.722)	0.608 (0.432, 0.855)
Sertraline	341/1876	0.771 (0.687, 0.866)	0.883 (0.695, 1.121)	0.698 (0.593, 0.823)	0.820 (0.653, 1.029)
Fluvoxamine	73/403	0.748 (0.581, 0.962)	1.175 (0.745, 1.851)	0.641 (0.455, 0.902)	0.654 (0.335, 1.277)
Escitalopram	217/1297	0.700 (0.605, 0.809)	0.867 (0.669, 1.123)	0.530 (0.425, 0.661)	0.963 (0.718, 1.291)

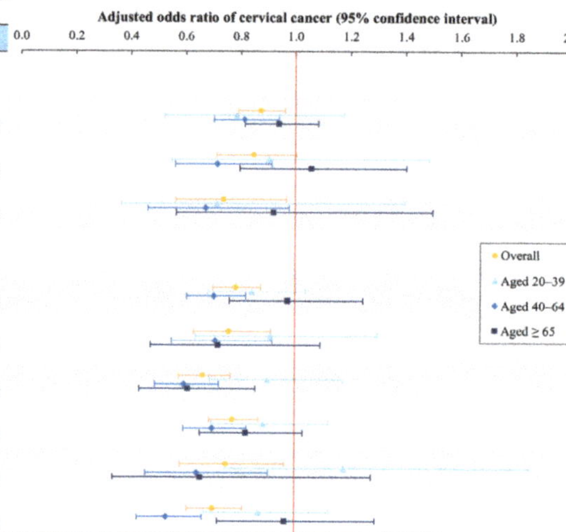

CI, confidence interval; ctrl, control; *n*, numbers of cases and controls exposed to TCAs or SSRIs; N/A, not applicable; SSRI, selective serotonin reuptake inhibitors; TCA, tricyclic antidepressants.

Figure 2. Cervical cancer risks associated with a single TCA or SSRI, stratified by age.

Although the TCAs and SSRIs as classes were associated with significant reductions in ovarian cancer risk, the use of individual TCAs or SSRIs did not indicate a lowered risk of ovarian cancer (Figure 3). Across the general population, doxepin (aOR = 0.492; p = 0.0449), fluoxetine (aOR = 0.619; p = 0.0003), citalopram (aOR = 0.622; p = 0.0387), sertraline (aOR = 0.623; p = 0.0005), and escitalopram (aOR = 0.654; p = 0.0024) exhibited associations with decreased ovarian cancer risk. Notably, significant reductions in ovarian cancer risk were observed in 20-to-39-year-old users of fluoxetine (aOR = 0.287; p = 0.0073) and 40-to-64-year-old users of imipramine (aOR = 0.616; p = 0.0028), amitriptyline (aOR = 0.509; p = 0.0140), fluoxetine (aOR = 0.710; p = 0.0231), sertraline (aOR = 0.630; p = 0.0056), and escitalopram (aOR = 0.515; p = 0.0004). However, no significant reductions were observed in females aged 65 or older.

Antidepressants	*n* (case/ctrl)	Adjusted odds ratio (95% CI) Overall	Aged 20–39	Aged 40–64	Aged ≥ 65
TCAs					
Imipramine	119/560	0.911 (0.746, 1.112)	1.072 (0.354, 3.243)	0.616 (0.449, 0.847)	1.263 (0.962, 1.656)
Amitriptyline	32/177	0.750 (0.514, 1.094)	N/A	0.509 (0.297, 0.872)	1.432 (0.821, 2.499)
Doxepin	9/76	0.492 (0.246, 0.984)	N/A	0.669 (0.302, 1.485)	0.359 (0.084, 1.539)
SSRIs					
Fluoxetine	67/440	0.619 (0.478, 0.802)	0.287 (0.115, 0.714)	0.710 (0.528, 0.954)	0.608 (0.312, 1.187)
Citalopram	22/142	0.622 (0.396, 0.976)	N/A	0.678 (0.405, 1.136)	0.804 (0.305, 2.120)
Paroxetine	65/337	0.784 (0.600, 1.024)	0.618 (0.262, 1.459)	0.745 (0.536, 1.036)	1.089 (0.631, 1.880)
Sertraline	63/412	0.623 (0.477, 0.813)	0.535 (0.228, 1.259)	0.630 (0.454, 0.873)	0.647 (0.374, 1.119)
Fluvoxamine	13/89	0.617 (0.344, 1.108)	0.368 (0.047, 2.851)	0.633 (0.324, 1.236)	0.790 (0.173, 3.612)
Escitalopram	60/374	0.654 (0.497, 0.860)	1.249 (0.666, 2.345)	0.515 (0.356, 0.745)	0.786 (0.449, 1.375)

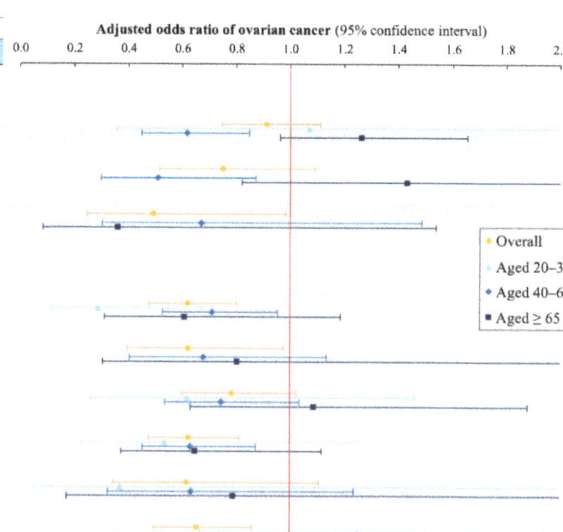

CI, confidence interval; ctrl, control; *n*, numbers of cases and controls exposed to TCAs or SSRIs; N/A, not applicable; SSRI, selective serotonin reuptake inhibitors; TCA, tricyclic antidepressants.

Figure 3. Ovarian cancer risks associated with a single TCA or SSRI, stratified by age.

Regarding uterine cancer, the risk was generally reduced in all females treated with TCAs or SSRIs, as well as those aged 40–64, except imipramine users (Figure 4). In the younger age group, sertraline (aOR = 0.229; p = 0.0130) and escitalopram (aOR = 0.173; p = 0.0152) significantly reduced uterine cancer risk. In contrast, no individual TCA or SSRI exhibited a statistically significant association with uterine cancer risk in the elderly population.

Antidepressants	*n* (case/ctrl)	Adjusted odds ratio (95% CI) Overall	Aged 20–39	Aged 40–64	Aged ≥ 65
TCAs					
Imipramine	186/ 793	0.969 (0.825, 1.138)	0.200 (0.026, 1.537)	0.899 (0.733, 1.103)	1.177 (0.902, 1.536)
Amitriptyline	34/ 227	0.614 (0.427, 0.881)	0.300 (0.035, 2.554)	0.527 (0.331, 0.841)	0.868 (0.473, 1.591)
Doxepin	12/ 115	0.425 (0.234, 0.772)	N/A	0.473 (0.245, 0.913)	0.358 (0.084, 1.525)
SSRIs					
Fluoxetine	96/ 635	0.617 (0.497, 0.765)	0.495 (0.225, 1.086)	0.599 (0.467, 0.768)	0.831 (0.492, 1.403)
Citalopram	25/ 239	0.424 (0.281, 0.640)	N/A	0.357 (0.218, 0.587)	0.915 (0.423, 1.979)
Paroxetine	68/ 444	0.626 (0.485, 0.809)	0.328 (0.101, 1.063)	0.549 (0.406, 0.743)	1.426 (0.817, 2.488)
Sertraline	90/ 617	0.598 (0.479, 0.747)	0.229 (0.071, 0.732)	0.574 (0.440, 0.748)	0.854 (0.548, 1.330)
Fluvoxamine	16/ 127	0.513 (0.305, 0.864)	0.572 (0.070, 4.647)	0.491 (0.270, 0.893)	0.637 (0.188, 2.154)
Escitalopram	71/ 600	0.485 (0.379, 0.621)	0.173 (0.042, 0.714)	0.471 (0.353, 0.628)	0.726 (0.433, 1.220)

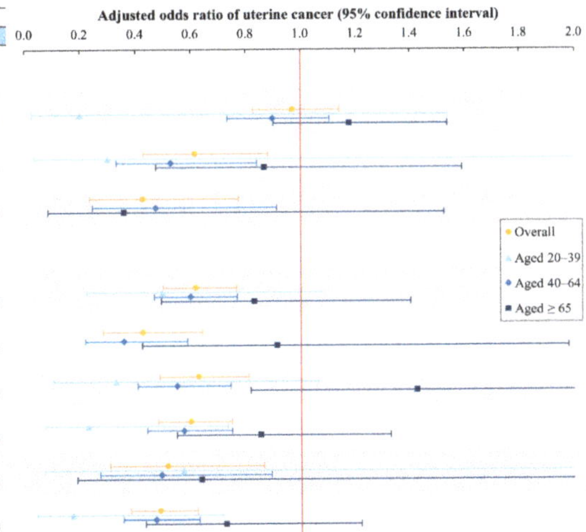

CI, confidence interval; ctrl, control; *n*, numbers of cases and controls exposed to TCAs or SSRIs; N/A, not applicable; SSRI, selective serotonin reuptake inhibitors; TCA, tricyclic antidepressants.

Figure 4. Uterine cancer risks associated with a single TCA or SSRI, stratified by age.

4. Discussion

Our findings indicate a potential protective effect of TCAs and SSRIs against gynecologic malignancies, specifically in women aged 40–64 (Table 3). Both TCAs and SSRIs significantly reduced the risk of cervical cancer among users in this age group. In contrast, in the elderly population, only paroxetine was associated with a significant reduction in the risk of cervical cancer. Concerning ovarian cancer, specific agents—imipramine, amitriptyline, fluoxetine, sertraline, and escitalopram—were associated with a statistically significant reduction in cancer risk for women aged 40–64, while fluoxetine was effective in those aged 20–39. In the context of uterine cancer, all individual TCAs and SSRIs, except for imipramine, were associated with a significant reduction in risk for women aged 40–64. In younger individuals, however, this risk reduction was observed only in users of sertraline and escitalopram. Notably, no single TCA or SSRI was found to significantly mitigate the risks of ovarian and uterine cancers in elderly women.

In addition, our study revealed a clear age-dependent trend in antidepressant use. Among women aged 20–39 years, SSRIs were more frequently prescribed than TCAs, likely due to their more favorable safety and side-effect profiles. In the middle-aged group, the usage of TCAs and SSRIs was approximately balanced. Interestingly, in the elderly population, TCAs were used more frequently than SSRIs. This pattern reflects real-world prescribing practices in Taiwan, where SSRIs are preferred for younger patients, while TCAs remain common in older adults not only for long-standing depression but also for non-psychiatric indications such as neuropathic pain, including diabetic neuropathy and postherpetic neuralgia.

4.1. Comparison with Relevant Studies and Possible Mechanisms

4.1.1. Cervical Cancer

To date, few studies have investigated the effects of antidepressants on the risk of cervical cancer. Our findings suggest that both TCA and SSRI classes are associated with a significantly decreased risk of cervical cancer. In contrast, a study conducted in Taiwan using the NHI Databases found no significant link between any major class of antidepressants and invasive cervical cancer [12]. The disparities in findings may be attributed to variations in study design, including the study's duration, the study participants' characteristics, and adjustments made for confounding variables. An in vitro study demonstrated that amitriptyline, a TCA, exerts cytotoxic effects on cervical cancer cells by inducing apoptosis, elevating reactive oxygen species levels, and inflicting irreversible mitochondrial damage [26]. Moreover, fluoxetine, an SSRI, has been shown to induce G0/G1 cell cycle arrest in cervical cancer cell lines and trigger apoptotic cell death by downregulating cyclin-dependent kinases regulatory subunit 1 (CKS1), a key protein in cell cycle regulation [27]. In our cohort, individuals aged 40–64 years who were using amitriptyline or fluoxetine exhibited a lower risk of cervical malignancy.

4.1.2. Ovarian Cancer

The relationship between ovarian cancer risk and the use of TCA and SSRI has yielded inconsistent results in previous research. A Danish case-control study that utilized nationwide registries suggested a decreased risk of epithelial ovarian cancer in individuals taking SSRIs, particularly citalopram; however, no such association was observed in TCA users [9]. In alignment with these findings, our study also showed similar results for SSRIs and citalopram. Conversely, several other studies have failed to identify any correlation between antidepressant use and ovarian cancer risk [28–32]. Two meta-analyses that pooled data from multiple studies concluded that neither TCAs nor SSRIs significantly impact ovarian cancer risk [7,11]. One possible explanation for these conflicting findings may lie in the biphasic effect of antidepressants on cancer risk [7,11]. Low-dose antidepressant administration has been shown to stimulate malignant cell proliferation, while high-dose administration exhibits an inhibitory effect [7,11]. Our study's findings of reduced risk might be partially attributed to our selection criteria, which required study participants to have used antidepressants for a minimum of 60 days within the two years preceding the index date. In vitro studies have supported these observations and shed light on the underlying mechanisms. For instance, imipramine, a TCA, was found to inhibit growth in the SK-OV-3 ovarian cancer cell line by blocking Ether-à-go-go (Eag) channels, voltage-gated potassium channels that suppress apoptosis in ovarian cancer cells [33]. Similarly, fluoxetine, an SSRI, was shown to induce cell death in SK-OV-3 and OVCAR-3 epithelial ovarian cancer cell lines by altering mitochondrial membrane permeability, triggering the release of cytochrome c, and activating caspase-3 [34]. At higher concentrations, fluoxetine also reduced the viability of ovarian granulosa tumor COV434 cell lines by affecting cytoplasmic Ca^{2+} accumulation and reducing intracellular cyclic AMP responses and ATP levels [35]. These in vitro observations may help elucidate our study results, which found a lower risk of ovarian cancer in imipramine users aged 40–64 and in fluoxetine users across all age groups except the elderly.

4.1.3. Uterine Cancer

Limited research exists concerning the association between antidepressant use and uterine cancer risk. A case-control study focusing on Taiwanese cohorts categorized endometrial cancer patients into various subgroups according to their cumulative exposure to antidepressants [36]. This study found no significant association across subgroups with exposure to either SSRIs or TCAs. In contrast, our study revealed that both TCAs and SSRIs were associated with a significantly reduced risk of uterine cancer across all age categories, including younger and middle-aged groups. An in vitro investigation has indicated that amitriptyline significantly diminishes the viability of HTB114 uterine leiomyosarcoma cells [37]. This is accomplished by upregulating the p75 neurotrophin receptor (p75NTR), a tumor necrosis factor (TNF) receptor superfamily member. This upregulation leads to the downregulation of neurotrophin-selective tropomyosin receptor kinase ($TrKA^+$)-prosurvival signaling, ultimately inducing $p75^{NTR}$-dependent apoptosis via caspase-3 [37]. Our study also identified a considerable reduction in uterine cancer risk among both the overall age group and those aged 40 to 64 years. The disparities between our findings and previous research could be ascribed to differences in the study populations and the adjusted confounders.

4.1.4. Elderly Population

Our results showed that most individual TCAs and SSRIs were associated with a significantly reduced risk of the three gynecologic cancers in women aged 40–64, but not among the elderly population, except for paroxetine in cervical cancer (Table 3). This might partly be explained by hormonal changes and immunosenescence in elderly women.

Pharmacologically, both TCAs and SSRIs increase serotonin activity by inhibiting serotonin reuptake [38,39]. An increased serotonin level can induce the secretion of prolactin [40] and estrogen [41,42]. Although previous research has demonstrated that prolactin and sex hormones are associated with an increased risk of gynecologic cancers [43–46], TCAs and SSRIs may exert additional hormonal regulatory mechanisms that could mitigate the pro-tumor influences of prolactin and estrogen, thereby contributing to a net antitumor effect. For instance, fluoxetine and sertraline can exhibit weak estrogenic or antiestrogenic effects depending on the concentration [47]. Fluoxetine has also been reported to downregulate estrogen receptor expression in vitro [48,49], while its effect on estradiol levels varies across animal models [49]. Sertraline and paroxetine enhance estradiol-mediated transcription at low concentrations [47], whereas paroxetine reduces estradiol levels dose-dependently [49]. Amitriptyline may improve ovarian morphology and normalize estradiol levels in rats with polycystic ovarian syndrome [49]. Thus, if the potential chemoprotective effects of TCAs and SSRIs against gynecologic cancers are mediated through hormonal modulation in middle-aged women, then these effects can be diminished in postmenopausal women due to the typically low baseline levels of these hormones.

Beyond hormonal factors, immunosenescence may outweigh the potential chemoprotective effect of TCAs and SSRIs in the elderly population. Age-related impairment of immune surveillance in elderly women weakens the ability to detect and eliminate precancerous cells. Interestingly, some antidepressants may have immunomodulatory effects. Fluoxetine has been shown to enhance cytotoxic T cell responses by reducing programmed cell death-ligand 1 (PD-L1) expression on tumor cells through the depletion of peripheral serotonin [50]. TCAs and SSRIs might lower the risks of gynecologic cancers, but this effect may be attenuated by immunosenescence in elderly women.

4.2. Interpretation of Our Findings for Future Research

The primary objective of this study is to identify associations, rather than establish causality, between antidepressant use and gynecologic cancers. We aim to provide comprehensive data findings that could enrich the existing evidence base and provide a robust foundation for subsequent research. Our data indicate associations of reduced cancer risk with both TCA and SSRI classes, thereby suggesting these medications as potential candidates for further scrutiny in cancer prevention efforts. Previous studies have highlighted multiple beneficial activities of antidepressants that could be leveraged for cancer treatment [51,52]. Considering that existing literature predominantly employs either in vitro or observational study designs [7,9,11,26,27], more comprehensive studies grounded in large-scale, real-world data are essential for gaining nuanced insights into the causal mechanisms between gynecologic cancers and antidepressant use. Our findings reveal that the middle-aged demographic exhibits the most promising potential for drug repurposing. When evaluating individual TCAs or SSRIs, it is noteworthy that the elderly group did not show a statistically significant association with gynecologic cancer risks for any single medication, with the exception of reduced cervical cancer risk related to paroxetine. This absence of significance could be attributed either to the ineffectiveness of TCAs and SSRIs in chemoprevention due to hormonal changes or immunosenescence among older females or to the limited sample size compared to the middle-aged group. Likewise, the limited number of participants in certain subgroups, particularly among young women, may reduce the statistical power for subgroup analyses and lead to statistically non-significant findings even if true associations exist. Furthermore, in middle-aged women, almost all individual medications were associated with significant reductions in both cervical and uterine cancer risks. As such, these two types of cancer may serve as prospective targets in future research focused on the repurposing of TCAs and SSRIs for chemoprevention. In addition, exposures were treated independently for each antidepressant, irrespective of concurrent or sequential use of multiple agents in an individual participant. Future research could investigate differences in chemoprotective effects between users of a single antidepressant and those using multiple agents, as well as the influence of concurrent versus sequential use of multiple antidepressants.

4.3. Limitations

This study has several limitations that warrant discussion. First, the research design allows us to identify associations, not causality, between individual antidepressants and cancer risks, calling for additional studies to elucidate underlying mechanisms [53,54]. Secondly, information about patients' lifestyles, medication adherence, and laboratory data may not be available from the HWDC databases. However, the impact of this missing data is mitigated by verifying all cancer diagnoses through pathology reports and the substantial sample size.

Despite these limitations, our research has notable strengths rooted in high-quality registry data. Medical prescriptions are required for antidepressant medications in Taiwan, ensuring their comprehensive inclusion in the HWDC databases. Moreover, cancer diagnoses were rigorously verified using the Taiwan Cancer Registry database, confirming each diagnosis through pathology reports. Lastly, our extensive database, encompassing records from approximately 23 million patients, enabled age-based subject stratification and provided for a thorough analysis of the effects of individual TCAs and SSRIs.

5. Conclusions

Both TCAs and SSRIs are associated with significantly reduced risks of cervical, ovarian, and uterine cancers. The middle-aged demographic presents the most promising avenue for future research on drug repurposing against gynecologic cancers. Specifically, cervical and uterine cancers emerge as potential targets for repurposing individual TCAs or SSRIs.

To our knowledge, this is the first study utilizing extensive real-world data that includes age subgroup analysis for individual antidepressant agents against various types of cancers. Our findings underscore the promise of both TCA and SSRI classes in repurposing medications to prevent gynecologic cancers. They also advocate for future interventional research, such as randomized controlled trials, to clarify TCAs and SSRIs' precise antineoplastic mechanisms and efficacy against these cancers. This study serves as the empirical foundation for generating hypotheses for future animal studies.

Author Contributions: C.-H.W.: Conceptualization, methodology, writing-original draft, and writing—review and editing. C.-W.H.: Writing—original draft, and writing—review and editing. N.T.H.N.: Methodology, visualization, writing—original draft, and writing—review and editing. M.-C.L.: Conceptualization, methodology, writing—original draft, and writing—review and editing. P.-A.N.: Writing—original draft, and writing—review and editing. M.M.I.: Writing—original draft preparation, and writing—review and editing. S.-C.C.: Writing-original draft preparation, and writing—review and editing. H.-C.Y.: Conceptualization, data curation, formal analysis, methodology, visualization, writing-original draft, and writing—review and editing. All authors have read and agreed to the published version of the manuscript.

Funding: This study received grants from the National Science and Technology Council (NSTC) under NSTC 110-2320-B-038-029-MY3 and NSTC 113-2628-B-038-007-MY3.

Institutional Review Board Statement: The study protocol was approved by the Joint Institutional Review Board of Taipei Medical University, Taipei, Taiwan (TMU-JIRB N202003609).

Informed Consent Statement: Since this study utilizes de-identified data, patient information is not identifiable; therefore, patient informed consent is not required. However, the datasets developed and used in this study are not publicly accessible. They can be obtained from Taiwan's Ministry of Health and Welfare upon reasonable request.

Data Availability Statement: The datasets developed and used in the present study are not publicly accessible, but they are available the Taiwan's Ministry of Health and Welfare upon reasonable request.

Conflicts of Interest: The authors declared no conflicts of interest.

References

1. GBD 2017 Disease and Injury Incidence and Prevalence Collaborators. Global, regional, and national incidence, prevalence, and years lived with disability for 354 diseases and injuries for 195 countries and territories, 1990–2017: A systematic analysis for the Global Burden of Disease Study 2017. *Lancet* **2018**, *392*, 1789–1858. [CrossRef] [PubMed]
2. Lim, G.Y.; Tam, W.W.; Lu, Y.; Ho, C.S.; Zhang, M.W.; Ho, R.C. Prevalence of Depression in the Community from 30 Countries between 1994 and 2014. *Sci. Rep.* **2018**, *8*, 2861. [CrossRef]
3. Health at a Glance 2019: OECD Indicators. Pharmaceutical Consumption. 7 November 2019. Available online: https://www.oecd-ilibrary.org/sites/43146d4b-en/index.html?itemId=/content/component/43146d4b-en (accessed on 13 April 2021).
4. Belmaker, R.H.; Agam, G. Major depressive disorder. *N. Engl. J. Med.* **2008**, *358*, 55–68. [CrossRef]
5. ATC/DDD Index 2021. 2021. Available online: https://www.whocc.no/atc_ddd_index/ (accessed on 18 January 2021).
6. Steingart, A.B.; Cotterchio, M. Do antidepressants cause, promote, or inhibit cancers? *J. Clin. Epidemiol.* **1995**, *48*, 1407–1412. [CrossRef] [PubMed]
7. Huo, Y.L.; Qiao, J.M.; Gao, S. Association between antidepressant medication use and epithelial ovarian cancer risk: A systematic review and meta-analysis of observational studies. *Br. J. Clin. Pharmacol.* **2018**, *84*, 649–658. [CrossRef]
8. Radin, D.P.; Patel, P. A current perspective on the oncopreventive and oncolytic properties of selective serotonin reuptake inhibitors. *Biomed. Pharmacother.* **2017**, *87*, 636–639. [CrossRef] [PubMed]
9. Mørch, L.S.; Dehlendorff, C.; Baandrup, L.; Friis, S.; Kjaer, S.K. Use of antidepressants and risk of epithelial ovarian cancer. *Int. J. Cancer* **2017**, *141*, 2197–2203. [CrossRef]
10. Sperling, C.D.; Aalborg, G.L.; Dehlendorff, C.; Friis, S.; Mørch, L.S.; Kjaer, S.K. Use of antidepressants and endometrial-cancer risk: A nationwide nested case-control study. *Int. J. Epidemiol.* **2021**, *51*, 799–806. [CrossRef]
11. Cosgrove, L.; Shi, L.; Creasey, D.E.; Anaya-McKivergan, M.; Myers, J.A.; Huybrechts, K.F. Antidepressants and breast and ovarian cancer risk: A review of the literature and researchers' financial associations with industry. *PLoS ONE* **2011**, *6*, e18210. [CrossRef]
12. Chan, H.L.; Hsieh, Y.H.; Lin, C.F.; Liang, H.Y.; Huang, K.Y.; Chiu, W.C.; Lee, Y.; McIntyre, R.S.; Chen, V.C. Invasive Cervical Cancer and Antidepressants: A Nationwide Population-Based Study. *Medicine* **2015**, *94*, e1866. [CrossRef]
13. Cancer Registry Annual Report, 2019, Taiwan. 13 January 2022. Available online: https://www.hpa.gov.tw/Pages/List.aspx?nodeid=269 (accessed on 16 February 2022).

14. Hsieh, C.Y.; Su, C.C.; Shao, S.C.; Sung, S.F.; Lin, S.J.; Kao Yang, Y.H.; Lai, E.C. Taiwan's National Health Insurance Research Database: Past and future. *Clin. Epidemiol.* **2019**, *11*, 349–358. [CrossRef] [PubMed]
15. Universal Health Coverage in Taiwan. 2018. Available online: https://www.nhi.gov.tw/English/Content_List.aspx?n=4D70518 40BF42F52&topn=ED4A30E51A609E49 (accessed on 20 April 2021).
16. ICD-9-CM and ICD-10-CM/PCS Mapping Table. 1 May 2020. Available online: https://www.nhi.gov.tw/Content_List.aspx?n= C1C92AB9ED30A9FD&topn=23C660CAACAA159D (accessed on 18 January 2021).
17. Grimes, D.A.; Schulz, K.F. Compared to what? Finding controls for case-control studies. *Lancet* **2005**, *365*, 1429–1433. [CrossRef] [PubMed]
18. Deyo, R.A.; Cherkin, D.C.; Ciol, M.A. Adapting a clinical comorbidity index for use with ICD-9-CM administrative databases. *J. Clin. Epidemiol.* **1992**, *45*, 613–619. [CrossRef]
19. Chen, Y.H.; Wang, P.H.; Chen, P.N.; Yang, S.F.; Hsiao, Y.H. Molecular and Cellular Mechanisms of Metformin in Cervical Cancer. *Cancers* **2021**, *13*, 2545. [CrossRef] [PubMed]
20. Gadducci, A.; Biglia, N.; Tana, R.; Cosio, S.; Gallo, M. Metformin use and gynecological cancers: A novel treatment option emerging from drug repositioning. *Crit. Rev. Oncol. Hematol.* **2016**, *105*, 73–83. [CrossRef]
21. Lee, D.Y.; Lee, T.S. Associations between metabolic syndrome and gynecologic cancer. *Obstet. Gynecol. Sci.* **2020**, *63*, 215–224. [CrossRef]
22. Wang, Y.; Zhao, J.; Chen, X.; Zhang, F.; Li, X. Aspirin use and endometrial cancer risk: A meta-analysis and systematic review. *Ann. Transl. Med.* **2020**, *8*, 461. [CrossRef]
23. Zhang, D.; Bai, B.; Xi, Y.; Wang, T.; Zhao, Y. Is aspirin use associated with a decreased risk of ovarian cancer? A systematic review and meta-analysis of observational studies with dose-response analysis. *Gynecol. Oncol.* **2016**, *142*, 368–377. [CrossRef]
24. Joharatnam-Hogan, N.; Cafferty, F.H.; Macnair, A.; Ring, A.; Langley, R.E. The role of aspirin in the prevention of ovarian, endometrial and cervical cancers. *Womens Health* **2020**, *16*, 1745506520961710. [CrossRef]
25. Markowska, A.; Antoszczak, M.; Markowska, J.; Huczyński, A. Statins: HMG-CoA Reductase Inhibitors as Potential Anticancer Agents against Malignant Neoplasms in Women. *Pharmaceuticals* **2020**, *13*, 422. [CrossRef]
26. Cordero, M.D.; Sánchez-Alcázar, J.A.; Bautista-Ferrufino, M.R.; Carmona-López, M.I.; Illanes, M.; Ríos, M.J.; Garrido-Maraver, J.; Alcudia, A.; Navas, P.; de Miguel, M. Acute oxidant damage promoted on cancer cells by amitriptyline in comparison with some common chemotherapeutic drugs. *Anticancer Drugs* **2010**, *21*, 932–944. [CrossRef] [PubMed]
27. Krishnan, A.; Hariharan, R.; Nair, S.A.; Pillai, M.R. Fluoxetine mediates G0/G1 arrest by inducing functional inhibition of cyclin dependent kinase subunit (CKS)1. *Biochem. Pharmacol.* **2008**, *75*, 1924–1934. [CrossRef] [PubMed]
28. Wu, C.S.; Lu, M.L.; Liao, Y.T.; Lee, C.T.; Chen, V.C. Ovarian cancer and antidepressants. *Psychooncology* **2015**, *24*, 579–584. [CrossRef] [PubMed]
29. Moorman, P.G.; Berchuck, A.; Calingaert, B.; Halabi, S.; Schildkraut, J.M. Antidepressant medication use [corrected] and risk of ovarian cancer. *Obstet. Gynecol.* **2005**, *105*, 725–730. [CrossRef]
30. Dublin, S.; Rossing, M.A.; Heckbert, S.R.; Goff, B.A.; Weiss, N.S. Risk of epithelial ovarian cancer in relation to use of antidepressants, benzodiazepines, and other centrally acting medications. *Cancer Causes Control* **2002**, *13*, 35–45. [CrossRef]
31. Harlow, B.L.; Cramer, D.W.; Baron, J.A.; Titus-Ernstoff, L.; Greenberg, E.R. Psychotropic medication use and risk of epithelial ovarian cancer. *Cancer Epidemiol. Biomarkers Prev.* **1998**, *7*, 697–702.
32. Haukka, J.; Sankila, R.; Klaukka, T.; Lonnqvist, J.; Niskanen, L.; Tanskanen, A.; Wahlbeck, K.; Tiihonen, J. Incidence of cancer and antidepressant medication: Record linkage study. *Int. J. Cancer* **2010**, *126*, 285–296. [CrossRef]
33. Asher, V.; Warren, A.; Shaw, R.; Sowter, H.; Bali, A.; Khan, R. The role of Eag and HERG channels in cell proliferation and apoptotic cell death in SK-OV-3 ovarian cancer cell line. *Cancer Cell Int.* **2011**, *11*, 6. [CrossRef]
34. Lee, C.S.; Kim, Y.J.; Jang, E.R.; Kim, W.; Myung, S.C. Fluoxetine induces apoptosis in ovarian carcinoma cell line OVCAR-3 through reactive oxygen species-dependent activation of nuclear factor-kappaB. *Basic. Clin. Pharmacol. Toxicol.* **2010**, *106*, 446–453. [CrossRef]
35. Nguyen, T.M.D.; Klett, D.; Combarnous, Y. Fluoxetine affects cytosolic cAMP, ATP, Ca(2+) responses to forskolin, and survival of human ovarian granulosa tumor COV434 cells. *Korean J. Physiol. Pharmacol.* **2021**, *25*, 189–195. [CrossRef]
36. Lin, C.F.; Chan, H.L.; Hsieh, Y.H.; Liang, H.Y.; Chiu, W.C.; Huang, K.Y.; Lee, Y.; McIntyre, R.S.; Chen, V.C. Endometrial cancer and antidepressants: A nationwide population-based study. *Medicine* **2016**, *95*, e4178. [CrossRef] [PubMed]
37. Pula, G.; Pistilli, A.; Montagnoli, C.; Stabile, A.M.; Rambotti, M.G.; Rende, M. The tricyclic antidepressant amitriptyline is cytotoxic to HTB114 human leiomyosarcoma and induces p75(NTR)-dependent apoptosis. *Anticancer Drugs* **2013**, *24*, 899–910. [CrossRef] [PubMed]
38. Hirsch, M.; Birnbaum, R.J. Tricyclic and Tetracyclic Drugs: Pharmacology, Administration, and Side Effects. 11 December 2023. Available online: https://www.uptodate.com/contents/tricyclic-and-tetracyclic-drugs-pharmacology-administration-and-side-effects (accessed on 25 April 2025).

39. Hirsch, M.; Birnbaum, R.J. Selective Serotonin Reuptake Inhibitors: Pharmacology, Administration, and Side Effects. 31 July 2023. Available online: https://www.uptodate.com/contents/selective-serotonin-reuptake-inhibitors-pharmacology-administration-and-side-effects (accessed on 25 April 2025).
40. Emiliano, A.B.; Fudge, J.L. From galactorrhea to osteopenia: Rethinking serotonin-prolactin interactions. *Neuropsychopharmacology* **2004**, *29*, 833–846. [CrossRef]
41. Monroy, J.; Cortés, O.D.; Domínguez, R.; Mendoza-Garrido, M.E.; Gallegos, E.; Cárdenas, M.; Aragón, A.; Ayala, M.E. The differential sensitivity of the hypothalamic-hypophysial-ovarian axis to 5-hydroxytryptophan alters the secretion of estradiol. *Exp. Physiol.* **2024**, *109*, 365–379. [CrossRef] [PubMed]
42. Bendis, P.C.; Zimmerman, S.; Onisiforou, A.; Zanos, P.; Georgiou, P. The impact of estradiol on serotonin, glutamate, and dopamine systems. *Front. Neurosci.* **2024**, *18*, 1348551. [CrossRef]
43. Clendenen, T.V.; Arslan, A.A.; Lokshin, A.E.; Liu, M.; Lundin, E.; Koenig, K.L.; Berrino, F.; Hallmans, G.; Idahl, A.; Krogh, V.; et al. Circulating prolactin levels and risk of epithelial ovarian cancer. *Cancer Causes Control* **2013**, *24*, 741–748. [CrossRef]
44. Levina, V.V.; Nolen, B.; Su, Y.; Godwin, A.K.; Fishman, D.; Liu, J.; Mor, G.; Maxwell, L.G.; Herberman, R.B.; Szczepanski, M.J.; et al. Biological significance of prolactin in gynecologic cancers. *Cancer Res.* **2009**, *69*, 5226–5233. [CrossRef]
45. Rodriguez, A.C.; Blanchard, Z.; Maurer, K.A.; Gertz, J. Estrogen Signaling in Endometrial Cancer: A Key Oncogenic Pathway with Several Open Questions. *Horm. Cancer* **2019**, *10*, 51–63. [CrossRef]
46. Ramírez-de-Arellano, A.; Villegas-Pineda, J.C.; Hernández-Silva, C.D.; Pereira-Suárez, A.L. The Relevant Participation of Prolactin in the Genesis and Progression of Gynecological Cancers. *Front. Endocrinol.* **2021**, *12*, 747810. [CrossRef]
47. Pop, A.; Lupu, D.I.; Cherfan, J.; Kiss, B.; Loghin, F. Estrogenic/antiestrogenic activity of selected selective serotonin reuptake inhibitors. *Clujul Med.* **2015**, *88*, 381–385. [CrossRef]
48. Lupu, D.; Varshney, M.K.; Mucs, D.; Inzunza, J.; Norinder, U.; Loghin, F.; Nalvarte, I.; Rüegg, J. Fluoxetine Affects Differentiation of Midbrain Dopaminergic Neurons In Vitro. *Mol. Pharmacol.* **2018**, *94*, 1220–1231. [CrossRef] [PubMed]
49. Pavlidi, P.; Kokras, N.; Dalla, C. Antidepressants' effects on testosterone and estrogens: What do we know? *Eur. J. Pharmacol.* **2021**, *899*, 173998. [CrossRef] [PubMed]
50. Schneider, M.A.; Heeb, L.; Beffinger, M.M.; Pantelyushin, S.; Linecker, M.; Roth, L.; Lehmann, K.; Ungethüm, U.; Kobold, S.; Graf, R.; et al. Attenuation of peripheral serotonin inhibits tumor growth and enhances immune checkpoint blockade therapy in murine tumor models. *Sci. Transl. Med.* **2021**, *13*, eabc8188. [CrossRef]
51. Rácz, B.; Spengler, G. Repurposing Antidepressants and Phenothiazine Antipsychotics as Efflux Pump Inhibitors in Cancer and Infectious Diseases. *Antibiotics* **2023**, *12*, 137. [CrossRef] [PubMed]
52. Song, Y.; Yang, X.; Yu, B. Repurposing antidepressants for anticancer drug discovery. *Drug Discov. Today* **2022**, *27*, 1924–1935. [CrossRef]
53. Yang, H.C.; Islam, M.M.; Nguyen, P.A.A.; Wang, C.H.; Poly, T.N.; Huang, C.W.; Li, Y.J. Development of a Web-Based System for Exploring Cancer Risk With Long-term Use of Drugs: Logistic Regression Approach. *JMIR Public. Health Surveill.* **2021**, *7*, e21401. [CrossRef]
54. Wu, C.C.; Islam, M.M.; Nguyen, P.A.; Poly, T.N.; Wang, C.H.; Iqbal, U.; Li, Y.J.; Yang, H.C. Risk of cancer in long-term levothyroxine users: Retrospective population-based study. *Cancer Sci.* **2021**, *112*, 2533–2541. [CrossRef]

MDPI AG
Grosspeteranlage 5
4052 Basel
Switzerland
Tel.: +41 61 683 77 34

Cancers Editorial Office
E-mail: cancers@mdpi.com
www.mdpi.com/journal/cancers

www.ingramcontent.com/pod-product-compliance
Lightning Source LLC
Chambersburg PA
CBHW061548210926
51850CB00009B/2034
* 9 7 8 3 7 2 5 8 7 7 9 3 5 *